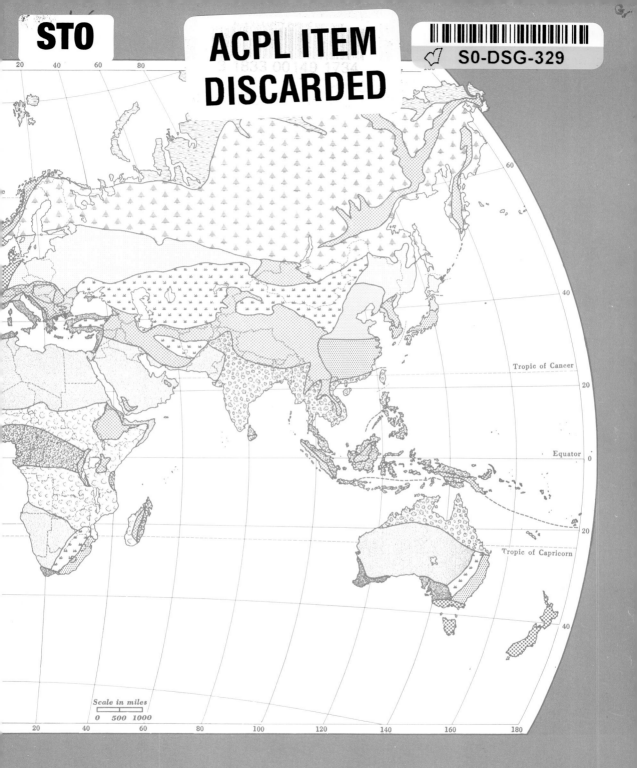

key to the detailed maps found in each of the regional chapters.

WORLD REGIONAL GEOGRAPHY

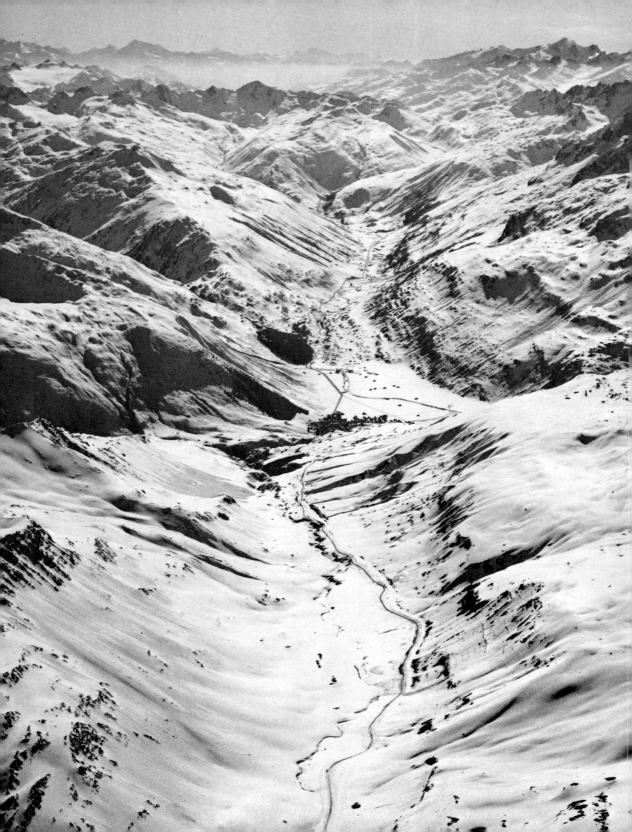

fourth edition

Oliver H. Heintzelman
Department of Geography, Oregon State University

Richard M. Highsmith, Jr.
Department of Geography, Oregon State University

WORLD REGIONAL GEOGRAPHY

PRENTICE-HALL, INC., ENGLEWOOD CLIFFS, NEW JERSEY

Library of Congress Cataloging in Publication Data

HEINTZELMAN, OLIVER HARRY
    World regional geography.

    Includes bibliographical references.
    1. Physical geography. I. Highsmith, Richard Morgan II. Title.
GB56.H4 1973   910′.02   72-6729
ISBN 0-13-969006-9

© 1973, 1967, 1963, 1955 by Prentice-Hall, Inc., Englewood Cliffs, New Jersey

All rights reserved. No part of this book may be reproduced in any form or by any means without permission in writing from the publisher.

PRINTED IN THE UNITED STATES OF AMERICA

10 9 8 7 6 5 4 3 2 1

Prentice-Hall International, Inc., London
Prentice-Hall of Australia, Pty. Ltd., Sydney
Prentice-Hall of Canada, Ltd., Toronto
Prentice-Hall of India Private Limited, New Delhi
Prentice-Hall of Japan, Inc., Tokyo

Credits for part opening photographs: *Part One,* Oliver H. Heintzelman
Part Two, Brazilian Government Trade Bureau
Part Three, Wide World Photos
Part Four, J. B. Guss from Black Star

1719787

To **J. Granville Jensen,** *friend and colleague,
this book is sincerely dedicated*

CONTENTS

PREFACE xi

PART ONE

THE STUDY
OF
THE MAN-EARTH
SYSTEM

one

GEOGRAPHY, ITS SCOPE AND METHOD 3

GEOGRAPHY IN THE HISTORICAL VIEW 5
CURRENT THEMES 6
METHODS OF STUDY 8
TERMS AND DEFINITIONS HELPFUL IN THE UNDERSTANDING OF MAPS AND PROJECTIONS 14
MAP PROJECTIONS 15
SELECTED PERIODICALS IN ENGLISH 21

two

MAN, CULTURE, AND THE EARTH 22

POPULATION 23
SOCIAL ORGANIZATION 32
DEVELOPMENT OF TECHNOLOGY 35
SUMMARY 36

three

THE PHYSICAL EARTH AND MAN 37

THE FRAME OF REFERENCE 37
NATURAL RESOURCES, A CULTURAL EVALUATION OF ENVIRONMENTS 68

four
INTRODUCTION TO WORLD REGIONS 72

WORLD REGIONAL APPROACH 73
THE PRESENTATION FORMAT 75

PART TWO

THE TROPICAL REALM

five
RAINY TROPICS 83

LOCATION 84
PHYSICAL ENVIRONMENT 86
OCCUPANCE IN THE RAINY TROPICS 95

six
WET-DRY TROPICS 110

LOCATION 110
PHYSICAL ENVIRONMENT 111
OCCUPANCE IN THE WET-DRY TROPICS 121

seven
MONSOON TROPICS 134

LOCATION 134
PHYSICAL ENVIRONMENT 136
OCCUPANCE IN THE MONSOON TROPICS 141

eight
TROPICAL DESERTS 152

DESERT CAUSES 152
GENERAL LOCATION 153
PHYSICAL ENVIRONMENT 157
OCCUPANCE IN THE TROPICAL DESERT 164

nine
TROPICAL HIGHLANDS 178

LOCATION 178
PHYSICAL ENVIRONMENT 180
OCCUPANCE IN THE TROPICAL HIGHLANDS 185

PART THREE

THE MIDDLE LATITUDE REALM

ten
DRY SUMMER SUBTROPICS 195

LOCATION 196
PHYSICAL ENVIRONMENT 199
OCCUPANCE IN THE DRY SUMMER SUBTROPICS 205

HUMID SUBTROPICS 220

LOCATION 220
PHYSICAL ENVIRONMENT 225
OCCUPANCE IN THE HUMID SUBTROPICS 235

twelve
LONG SUMMER HUMID CONTINENTALS 256

LOCATION 256
PHYSICAL ENVIRONMENT 259
OCCUPANCE IN THE LONG SUMMER HUMID CONTINENTALS 265

thirteen
SHORT SUMMER HUMID CONTINENTALS 282

LOCATION 282
PHYSICAL ENVIRONMENT 284
OCCUPANCE IN THE SHORT SUMMER HUMID CONTINENTALS 292

fourteen
MARINE WEST COASTS 306

LOCATION 306
PHYSICAL ENVIRONMENT 307
OCCUPANCE IN THE MARINE WEST COASTS 319

fifteen
DRY CONTINENTALS 337

LOCATION 337
PHYSICAL ENVIRONMENT 341
OCCUPANCE IN THE DRY CONTINENTALS 351

sixteen
MIDDLE LATITUDE HIGHLANDS 368

MOUNTAIN INFLUENCES 368
DISTRIBUTION 370
PHYSICAL ENVIRONMENT 371
OCCUPANCE IN THE MIDDLE LATITUDE HIGHLANDS 379

seventeen
SUBARCTICS 390

LOCATION 390
PHYSICAL ENVIRONMENT 393
OCCUPANCE IN THE SUBARCTICS 398

PART FOUR

THE POLAR REALM

eighteen
POLAR LANDS 411

LOCATION 411
PHYSICAL ENVIRONMENT 414
OCCUPANCE IN THE POLAR LANDS 419

INDEX 423

PREFACE

During the two decades that have elapsed since the first edition of this book was formulated, geography has advanced significantly as a major field of study leading to professional careers, and as a supporting field of study for liberal education. The increased number of geographers, working at college and university posts, and in business, industry, and government positions, has permitted considerable scholarly attention to be turned to the explication of the conceptual structure and overriding problems of the discipline, to clarification and codification of its dominant approaches and themes, and to improvement of its methodology. Concomitantly, changes have occurred in the preparation of entering college students. As a result of exposure to mass communication media, travel experience, and improving geography instruction in public schools, students have greater awareness of the world. They are also more familiar with the formulation and analysis of problems and with the methods of science.

As a result of these developments, college geography, even in survey courses, can be taught at a higher level of sophistication than was possible when the first edition of this book appeared. Thus, to the original purpose of presenting a regional survey of the earth as the home of man, the third edition added another major purpose: that of central questions, conceptual approaches, and methods. These are analyzed in chapters 1, 2, and 3 which focus upon the elements of the man-earth system (the natural environment and human culture) in terms of geographic interests—the intersects of place, space, and time. Chapter 4 discusses regionalization and sets the stage for the fourteen regional chapters, which were brought formally into the new conceptual orientation in the fourth edition. The regional chapters, employing this conceptual background, systematically analyze the earth as the human habitat: its physical character, human modification, organization, and use. The revisions for the fourth edition have primarily been directed toward clearer expression of concepts, updating of subject matter, and improvement of pictures.

The authors are indebted to many individuals: users who offered suggestions; unknown (to us) critics who reviewed our manuscript; our colleagues at Oregon State

University who offered numerous useful suggestions; and many who have contributed to the literature of geography. Finally, however, we take full responsibility for interpretations herein presented.

Oliver H. Heintzelman
Richard M. Highsmith, Jr.

WORLD REGIONAL GEOGRAPHY

PART ONE

THE STUDY OF
THE MAN-EARTH SYSTEM

A major service geography offers to general education is the development of a perspective that leads to understanding the earth as the home of man. Many other disciplines deal in global terms with either aspects of man and culture or aspects of the physical-biotic world. Geography alone, however, views both man and the earth, and makes its unique contribution by synthesizing otherwise unrelated knowledge for the distinctive purpose of providing and understanding of the man-earth system.

Geography is, therefore, concerned with two great sets of variables, those stemming from man and his culture and those stemming from the earth and nature. Before we turn attention to an examination of the earth as the home of man within the framework of regions, it will be of value to establish more clearly the point of view and methods of geography and to obtain an insight into the major elements of the man-earth system. These features will occupy our attention through the first four chapters.

one

GEOGRAPHY, ITS SCOPE AND METHOD

Man is a creature of insatiable curiosity. His inquisitiveness has led him to ask questions and seek answers about himself, his neighbors, the earth on which he lives, the composition of the matter surrounding him, other terrestrial life forms, and the sun and other elements of the cosmos. Throughout time he has systematized his learning into branches with core interests and methods of study. These are commonly called academic disciplines and are grouped into broad order categories of the arts, humanities, social sciences, physical sciences, biological sciences, and earth sciences. With the advance of education and technology, learning at the academic level has been particularized into an increasing number of subdisciplines, cross-disciplines, and professional fields. Such a development is not surprising in view of the advancement of knowledge, the need for even more knowledge, and the growing requirements for scientific and technical services. Today most of our larger universities offer graduate work in more than one hundred different fields.

Because of the interrelationships existing among the components of physical reality and human experience, an absolutely definitive classification of learning into disciplines is impossible. Nevertheless, it is upon the basis of elements of disunity in the universal that the highly specialized academic fields have been developed. Geography, along with history and anthropology, stands in contrast to this narrowing of focus. All three are broad-based and contain some features of commonality with a number of other disciplines.

In its major mission of providing an understanding of the earth as the home of man, geography draws substance from the realities of unity in the associations of man and his earth habitat. Thus the nature of the total scope of geographical inquiry is such that as an academic discipline it does not fit clearly into one of the standard groups. Although many American geographers consider themselves primarily social scientists because their interests are anthropocentric, most recognize and respect the ties with the earth sciences and biology. In fact, some wish to class themselves in these categories. Moreover, a substantial number stress their common grounds with the humanities. In a sense, geography forms a "bridge" between the disciplines concerned with man and those concerned with the earth. More important than classification as such, however, is the basic fact that geography does stand the test of discipline identification: It focuses upon an overriding problem, it has an organizing concept, and it has methods of study.

It has been stated in a kind of verbal shorthand that the overriding problem, or transcendent issue, of geography "is that of a full understanding of the vast system on the earth's surface comprising man and the natural environment" and that the organizing concept is the "spatial distribution and spatial relations" of the man-earth system and subsystems.[1] In this context, the term "system" is taken to mean "a functional entity composed of interacting, interdependent parts,"[2] and the term "spatial" denotes surficial area including the atmospheric envelope. Thus geography "derives its substance from man's sense of place (or particular area of the earth's surface) and from curiosity about the surface (including man and his works) and the atmospheric envelope of this planet."[3]

In commenting on the core problem of geography and its values, the authors cited in footnote 1 say: "The three great parameters for any scientific problem, albeit in varying dimensions and attributes, are time, space, and composition of matter. For the problem it treats, that of the man-environment system, geography is concerned primarily with space in time. It seeks to explain how the subsystems of the physical environment are organized on the earth's surface and how man distributes himself over the earth in his space relation to physical features and to other men. Space and space relations indeed impose one of the great mediators of the characteristics of any part of the system at any point on the earth's surface, and as the only one traditionally concerned with system interrelations within the space of the earth's surface, geography has a significant place in satisfying man's scientific curiosity."

The same statement also points out the following: "Geographers believe that correlations of spatial distributions, considered both statically and dynamically, may be the most ready keys to understanding existing or developing life systems, social systems, or environmental changes." They further believe that geography has significantly contributed in the past to the foundations of knowledge necessary for an understanding of the subsystems of the man-environment system. "Progress was gradual, however, because geographers were few, rigorous methods for analyzing multivariate problems and systems concepts were developed only recently, and few branches of science were committed to study of the man-environment system."[4]

In recent years geography has profited from the growing interest of other sciences in systems theory and study techniques as well as from concerted efforts along these lines by the increasing number of practicing geographers. The results in improved methodology and improved communication have been generally rewarding, and the future of geography appears brighter than ever. The growth in world population and the attendant demands for more food, raw material, and living space, as well as for improvements in all kinds of linkages between peoples, will continue to require a greater

---

[1] See "The Science of Geography," *Publication 1277* (Washington, D.C.: 1965), National Academy of Sciences–National Research Council, especially pp. 1–11.

[2] *Ibid.,* p. 9.

[3] See also Jan O. M. Broek, *Geography, Its Scope and Spirit* (Columbus, Ohio: Charles E. Merrill Books, Inc., 1965), pp. 5–6. This reference considers geography among the social sciences. This treatise and the reference cited in footnote 1 together present a good insight into the nature and interests of modern geography.

[4] "The Science of Geography," p. 9.

understanding of the man-earth system and the relation of its parts in a spatial perspective. It appears axiomatic that, as more and more people press increasingly harder into the already limited space of the earth, it will be necessary for environmental deterioration to be halted and for human occupance to become more ordered.

Geography in the Historical View

Although the roots of geography reach deeper into the past, the record begins with the ancient Greeks. They placed their knowledge of the earth in "geographic settings" and even dealt to some extent with the relationships among phenomena occupying mutual space. The Greeks, moreover, recognized the earth as a sphere, calculated its circumference, laid out temperature zones (torrid, temperate, and frigid), and developed a system of lines of latitude and longitude, making possible accurate location of earth features on a "real map."

Although during the Dark Ages in medieval Europe there was essentially a halt, even a regression, in the advance of scientific inquiry, geographical knowledge was extended to a considerable extent as the map of the known world was filled in. Moreover, several Moslem scholars made significant contributions between the eighth and fifteenth centuries, especially to the knowledge of the East and to the refinement of climatic classification. The Renaissance brought a reawakening of scientific inquiry. In geography there was a revival of classical thought and, of course, a vast increase in knowledge of the earth's surface through the lively activity in exploration.

The forerunner of modern geographers appeared in the seventeenth century. Bernhardus Varenius, spurred by a recognition of the need for an improved organization of geographic knowledge, published his *Geographia Generalis* in Amsterdam in 1650. He pointed to dualism in geographic interests and noted that on the one hand the field deals with phenomena and processes of a purely physical nature that are subject to the methods and exact measurements of science and on the other hand with phenomena that are social-cultural and not subject to the same kind of verification. He proposed a division of geography into general and special categories, the former to deal with physical aspects and the latter with areas of the earth deriving their personality from the interaction or interrelations of both physical and human processes. Varenius died at the age of twenty-eight, however, before he had had time to amplify his ideas on social-cultural geography in writing.

As Broek suggests, in tracing the development of geographic thought, "Varenius had presented the structure of geography as a scientific discipline."[5] But it was Immanuel Kant (1724–1804) who placed geography within the then contemporary philosophy of science. Working from a post at the University of Königsberg, he reasoned that all knowledge can be organized and studied from three points of view: (1) the studies grouped according to facts or objects—the "systematic sciences" such as botany (plants), pedology (soils), and sociology (social groups); (2) the study of facts in their relationship through time—the "historical sciences"; and (3) "the study of things as they are associated in space"—the "geographical sciences."

During the nineteenth century other European scholars steadily added to and

---

[5] Broek, *Geography,* pp. 10–20.

molded the substance of geography into scientific pattern. Alexander Von Humboldt (1769–1859) added to the stature of the discipline as a field study and to the depth especially respecting vegetation, temperature, and altitudinal relationships. Carl Ritter (1779–1859), as Broek points out, complemented Von Humboldt by focusing upon the earth as the home of man, or, more particularly, upon human activity in regional context.

The number of practitioners grew during the latter part of the nineteenth century, and scientific concepts in general were stimulated by the work of Charles Darwin. In the United States John Wesley Powell (1834–1902), first chief of the Geological Survey, did pioneer work in landform description and explanation based upon experience in the western lands. George Perkins Marsh (1801–1882), in his *Man and Nature; or Physical Geography as Modified by Human Action* (1864), sought to "indicate the character and, approximately, the extent of the changes produced by human action in the physical conditions of the globe we inhabit...."[6] This book, aimed at a conservation cause, introduced a theme that has remained of prime interest to geographers.

Friedrich Ratzel (1844–1904), a German with particular interest in human geography, did much to bring order to geographic study. Trained in the methods of the natural sciences and a student of anthropology, he organized the subject matter on the basis of physical features and attempted to link human and cultural features with them. Although his second volume of *Anthropogeographic* (1891) clarified his appreciation of cultural and historical factors, he was given credit for the idea of human geography as the study of the influence of environment on man. Ellen Churchill Semple, a student under Ratzel, through her writings and her teaching at Clark University and the University of Chicago, articulated this interpretation in the United States.[7] In the early years of this century Semple's views were supported by a number of other American geographers.

In a fuller treatment of the evolution of geographic thought, many other early contributors would be added to the list of names. On this list would appear Germans, Frenchmen, and Englishmen, as well as Americans. For our purposes, however, it is sufficient to note that with the greater number of geographers, as well as with the increasing numbers of workers in the systematic disciplines, have come major advancement in the knowledge of the man-earth system and the recognition that man, through culture, is a dynamic agent in his environmental relationships—both in the selection as well as in the modification of those earth features which may affect his life.

Current Themes

Although spatial distributions and spatial relations of the man-earth system have been a central theme of geography for many years, geographers have increasingly specialized their interests in particular aspects of system characteristics, patterns, processes, and/or

---

[6] George Perkins Marsh, *Man and Nature; or Physical Geography as Modified by Human Action*, ed. David Lowenthal (Cambridge: The Belknap Press of Harvard University Press, 1965). Quote is from the preface.

[7] See Ellen Churchill Semple, *Influences of Geographic Environment on the Basis of Ratzel's System of Anthropo-Geography* (New York: Holt, Rinehart & Winston, Inc., 1911).

analysis. Although, in total, these interests do cover the wider spectrum of the discipline, the individual specialities can be grouped into "clusters," or related problem areas. These have been commonly identified by the addition of an adjective prefix; hence have evolved such designations as cultural, physical, economic, settlement, and political geography. The indication is, then, that study has its focus upon the space relations of a particular aspect of the man-earth system or upon a particular group of features resulting from one kind of process.[8] Included in each speciality are several different study or interest orientations. A physical geographer may have prime interest in climate, landforms, vegetation, and so on, or problems in the interfaces of these; a cultural geographer may focus upon any one of the features on the earth's surface that has been produced or modified by human action or upon the spatial attributes, relations, or processes of diffusion of a particular cultural trait; and an economic geographer may specialize in any of the activities by which man wins a living or produces wealth, in systems of linkages and exchanges between economies, or even in theoretical problems of location and development.

Actually, within the general topical clusters specific interests of research geographers vary markedly. For example, some physical geographers specialize in physical elements and their spatial distributions and relations in the physical system per se, whereas others may approach the same elements with a view toward examining their possible influence upon man. The topical dimensions of agricultural geography interest some from the point of view of cultural geography; and even physical geographers may have an interest in the role of agriculture in the alteration of the physical system. Some geographers are interested in the character and cause of present physical or human conditions, some are more interested in processes, others in types, and still others in the regional differences of objects of their concern. Increasing numbers are concentrating upon the application of geography and its methodology to the rationalization of economic development and the ordering of human occupance.

In summation the topical concerns of research geographers at the present time can be categorized into three thematic approaches: (1) ecological studies concerned with man-earth interrelations (the stress of this book); (2) studies concerned with the determination of uniqueness of place or area; and (3) studies concerned with the spatial organization of society, especially the search for order and pattern.[9]

Regional geography, focusing upon a particular area, has concern for the total man-earth system. World courses taught in colleges at the introductory level, such as the course for which this book is designed, commonly employ the regional approach, as do courses concerned with continents or with broad areas of the earth. But many kinds of regions, based upon the areal distribution and relations of some feature or features of homogeneity, can be recognized; thus regions are also used in many topical studies. Moreover, the regional approach has applied value in solving area problems and in development planning. Regional geography will be discussed at greater length in chapter 4, which establishes the background for the body of this book.

---

[8] See P. E. James, "American Geography at Mid-Century," in P. E. James, ed., *New Viewpoints in Geography,* Twenty-ninth Yearbook of the National Council for the Social Studies (Washington, D.C., 1959), p. 10.

[9] For additional information on the last theme see Edward J. Taaffe, ed., *Geography* (Englewood Cliffs, N.J.: Prentice-Hall, Inc., 1970) and Richard L. Morrill, *The Spatial Organization of Society* (Belmont, Calif.: Wadsworth Publishing Company, 1970).

Methods of Study

Geography is a way of looking at the earth as the human habitat. As a field of inquiry, its concerns center upon: the physical-biotic system of the earth, the earth as a stage and support for human occupance (or the earth as the human habitat), how and why man has distributed himself over the earth, how he has organized himself and the earth to meet his needs, and what changes he has brought about in the physical earth. These concerns imply much more than a simple inventory of the content or characteristics of places; more importantly, they have to do with understanding the processes underlying the spatial associations of phenomena. The real overall mission is to provide a synthesis of spatial systems and the relationships between systems.

Although the content of the total scope of geography is diverse and complex and calls for various methods of study, there are overriding guidelines that lay out the general frames of reference or the themes and patterns of problems of inquiry. (1) Inherent in the base problem of the discipline—understanding the spatial aspects of the man-earth system—is the recognition that the phenomena that occupy mutual space are not heaped together without order; they do exist in associations that are subject to study and comprehension. (2) Geographers recognize man as being capable of a marked degree of ecological dominance. Although the earth provides the stage and the basic features for human support and the nature of this physical stage varies from place to place, each society in each time interprets its particular surroundings as to values and limitations; and in organizing and using natural features man has capacity to bring about many changes in the physical-biotic system—through clearing natural vegetation, planting crops, shaping surfaces, constructing buildings, and so on. (3) The regional concept, together with well-founded ecological principles, provides a framework for study of likeness, differences, and relations. It serves to bring into focus the aerial dimensions of the topic under examination. (4) The organizing concept of spatial distributions and spatial relations delimits the scope of the topic under examination.

During the years of the discipline's evolution, geographers have developed a considerable background of spatial knowledge. They have also developed models for analytic procedures. Empirical-inductive methods have been traditional in regional research. In determining the content of space on the earth's surface and in delineating the characteristics and relations of natural and cultural features occupying that space, six steps have commonly been followed: (1) determination of the physical matrix; (2) identification of specific space content; (3) identification of generic relations; (4) identification of genetic relations; (5) determination of covariance; and (6) integration of data on site, form, and process in order to show the full pattern of earth space relations in any given area or areas.[10] Although these procedures continue to be of great value in many facets of geographic research, the advancement of topical, theoretical, and applied interests has called for additional procedures.

Much geographic research today seeks solutions to specific problems. In addition to traditional questions—such as where are the

[10] Edward A. Ackerman, "Geography As a Fundamental Research Discipline," *Research Paper No. 53* (Department of Geography, The University of Chicago, June, 1958), p. 27. This publication is recommended for a fuller background on geographic research and methodology.

features of interest located, why are they there, how did they come about, and what are their implications?—a good deal of attention is focused upon answers to questions like: Is there order in the distributional pattern of the feature or form of interest? What are potential locations for particular developments? What is the most rational occupance or development form for a given area? What are the potential results of a given physical or cultural process, such as soil erosion, associated with a particular type of land use or an evolving occupance pattern resulting from a particular space-adjusting technique? With recent advancements, considerable attention is also being devoted to developing the theoretical framework of the discipline as well as to techniques for more exacting quantification. This work has brought a significant number of geographers into research aimed at improving the theoretical-deductive methodology of the discipline. For the most part, their research has centered upon abstract problems of distribution, location, processes, covariation of processes reflected in space relations, and the integration of process and site data. As a result, techniques of observation, classification, and interpretation are steadily improving, and quantifiable analytical models have been developed for some types of problem studies.

Fundamentally, geographic research is rooted in field study. The practicing geographer must be a facile observer and an able interviewer; he must be able to produce qualitative descriptions and quantitative measurements. Thus he must be able to identify and evaluate the parts of the man-earth system, to recognize relationships, to classify for recording, and to employ appropriate tools, such as the plane table, compass, clinometer, map, and so on. For many areas and many problems, however, part of the field work may be available through the work of such other specialists as the soil scientist, the land-use or agricultural specialist, the demographer, the planner, the census-taker, and so on. In these instances the geographer must be able to search out such materials, judge their validity, and employ them properly. Moreover, the increasing availability of air-photograph and other forms of remote sensor coverage and of land-use and other types of maps has greatly aided basic research; and the present-day geographer is required to be competent in the interpretation and use of these sources. In studies dealing with change, geographers have adapted methods common to the study of history and other sciences. These methods have been fruitful in bringing to light the genesis and relations of existing distributional patterns as well as in reconstruction of geographies of the past. It has already been implied that many of the current research problems require competence in statistical techniques.

Although it is considered valuable that the student utilizing this book as a text have at least a glimpse into the research methodology of the discipline, it is recognized that he will have little additional exposure unless he takes more advanced courses. Unfortunately, perhaps, books of this level are intended primarily as survey introductions and are based upon the background of the authors and their interpretations of the works of other geographers. Thus the student at this level has modest opportunity to become engaged in a reasoning process based upon the detailed backgrounds and methods that are the prerequisites of the generalizations presented here. For this reason, the student is encouraged to extend his background by reading selections in the periodical literature of geography.

GRAPHIC TOOLS

Probably no tool is so vital to the geographer as is the map. The map, a reduced representation of the earth or portions of the earth, is significant for field recording, laboratory analysis, and report presentation. By means of the map, small or large areas, even the entire world, can be brought down to laboratory scale for study and analysis. When the features of interest are placed upon maps in proper form and position, many relevant facts and relationships become apparent. When maps of several elements related to a particular phenomenon are compared, relationships that may be causal factors underlying the phenomenon often stand out. For example, a comparison of maps showing distribution of characteristics of surface, climate, and soils may lead to an understanding of the patterns of land usage. By use of maps, it is also possible to establish and delimit areas of similar geographic elements or relationships.

The globe. The globe is the only true representation of the earth. It shows correctly, with a minimum of distortion, the mathematical characteristics of the earth, the spheroidal shape, the relative sizes of areas, distances, and directions, and it exhibits a correct scale relationship with the real earth. It also indicates such concepts as meridians and parallels, great circles, and loxodromes in proper perspective. The globe is used primarily to illustrate the three-dimensional aspect of the earth. Aside from this use, the globe loses practicability because it is unwieldy to transport and store, expensive to construct, not serviceable for data plotting, and viewable in only one-half of its entirety at any one time. It should be repeated, however, that only the globe correctly represents all the earth's mathematical relationships; consequently, it should be a basic reference in a student's study of maps and projections.

The map projection. The disadvantages of the globe can be eliminated to a certain degree by projecting the spherical surface to the flat surface. By "projecting" the parallels and meridians mathematically to flat surfaces (see Figure 1.1), a variety of projections have been developed that serve as a framework upon which the earth's phenomena are mapped. The map projection is characterized by a number of advantages lacking in the globe: It can be handled, stored, and reproduced easily; it can be designed for a specific purpose; it can be readily used for plotting; and it can be "enlarged" to almost any size so that one is able to observe a city block or the whole world on one sheet of paper. Some projections retain the areas of the earth in true size relationships (equal-area), others preserve the true shapes of things (conformal), and others show the true distances or directions. There is, however, one limitation imposed upon all projections: the inability to display at once all the attributes of the earth from which it is projected—the one surface is round and the other flat; therefore some distortion is necessary upon projecting. (Imagine attempting to flatten a basketball without splitting or causing some distortion in it.)

Theoretically, there can be an infinite number of projections, and each may have its own properties, characteristic grid, use, value, distortion, and limitations. In the table of projections (pages 16–20) only frequently used types are described. Through examination and study of these, it will be seen that each one has been designed for a specific purpose and that for every need there is a proper projection. If areas of the earth are to be evaluated in size relationship with one another, a projection retaining an equal-area scale throughout should be used. Since equal-area projections cannot maintain true shapes, the choice of any one in particular will depend largely on how little it distorts the shape of the region being mapped. For example, if the United States is being mapped and equivalence is desired, an Albers equal-area projection would be a sensible choice; it shows areas equal throughout and also retains an almost true shape

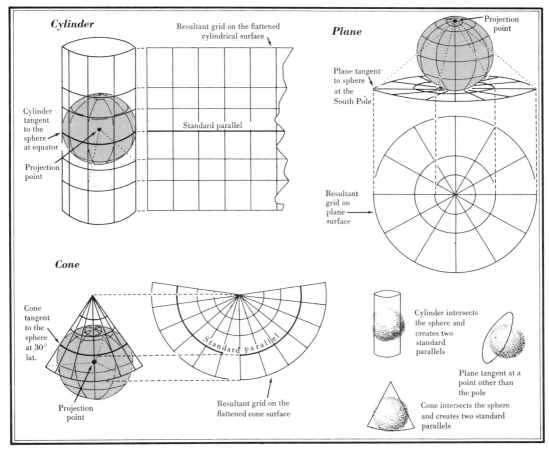

FIG. 1.1. Standard geometric forms which aid in visualizing the projection of the sphere to the flat surface. Notice that the point from which we project the earth's grid (the projection point) may change and that the geometric form may intersect the sphere in a number of different ways, producing an infinite variety of projections.

of an area with a great east-to-west extent (see projections). In mapping smaller areas of land on a large scale, it is possible to choose from a variety of projections without suffering too much shape or area distortion.

***Location on maps** (**Latitude and Longitude**)*. On the round earth, location is of prime importance; to establish a definite location, a grid has been superimposed on the earth. The equator, which bisects the earth, is 0° latitude. From the equator to either pole there are 90° of latitude—one-quarter of the earth's circumference. Circles, called parallels, become increasingly smaller as they approach the poles. Distance north or south of the equator is measured in degrees of latitude, each degree being equal to approximately 70 miles.

TABLE 1.1

Latitude	Statute Miles
0°	69.17
15°	66.83
30°	59.96
45°	49.00
60°	34.67
75°	17.96
90°	0.00

Measurement in an east-west direction is accomplished through the medium of the degrees of longitude. These are based upon great circles, or meridians, each bisecting the earth. The meridian that passes through Greenwich, England, was chosen as the starting line. This is called the Prime Meridian, or 0° longitude. In our numbering system, all meridians east of Greenwich are called east longitude and those to the west, west longitude. The meeting point of the east and west meridians is the line on the opposite side of the earth from the Prime Meridian. This line coincides approximately with the International Date Line (which is adjusted to land surfaces). The approximate distance for a degree of longitude at the equator is 70 miles; the degree distance becomes increasingly shorter toward the poles because of the convergence of the meridians. (Table 1.1).

Degrees of both latitude and longitude are subdivided: One degree is divided into 60 minutes ($'$); one minute is subdivided into 60 seconds ($''$).

Scale. Prior to a discussion of map and symbol types, it is necessary to have an appreciation of the scale of the map and to understand how this may limit or facilitate map-making. *Scale* is the size relationship between the map and the earth expressed as a ratio. For example, on a map with a scale of 1:1,000,000, one unit on the map represents one million of the same units on the earth's surface; expressed as a fraction $\frac{1}{1,000,000}$, presented graphically $\begin{array}{c} 0\ \ 10\ \ 20 \\ \vdash\!\!-\!\!+\!\!-\!\!\dashv \\ \text{miles} \end{array}$, or written, one inch on the map represents 15.78 miles on the earth. The topographic maps prepared by the United States Government indicate a great number of features such as cities, roads, railroads, and contours. They are detailed and are mapped on a *large scale* (1:24,000 to 1:250,000). The maps in this book, on the other hand, are *small-scale*, special-purpose maps (1:500,000 and smaller) depicting with relatively little detail one or two distributions on the earth's surface. (Small scale greatly limits the amount of detail as well as the degree of accuracy on a map.)

Map types and map symbols. Upon the globe and projection frameworks previously described, maps showing distributions and relationships can be developed. Some indicate different kinds of phenomena: land masses and oceans, vegetative types, or climatic types (Frontispiece and Figures 3.16 and 3.17 are examples). Other maps indicate kinds and *amounts:* population *numbers, inches* of precipitation, and *tons* of ocean-traffic flow (see Figure 2.2). Sometimes a number of these earth phenomena may be combined on one map. All map types show the material represented by means of greatly simplified symbols, which may be aptly termed the shorthand of maps. Examples of this "script" are colors, patterns, dots, lines, and shades of gray (Figures 3.3, 3.4, 3.7 to 3.9, and 3.16 are examples). The map symbols and the map types must show things greatly reduced from reality; therefore, all have been generalized and simplified in appearance in order to show more clearly the relevant material being mapped. The amount of generalization of maps and symbols varies inversely with the scale of each map.

Photographs. Photographs are highly useful in geographic study and instruction. The photograph gives a synoptic record as seen by the eye, thus implementing field notes and maps in a study. The photograph, moreover, presents detail as well as local relations that would require many words to describe. The photographs in the text possess such values and should receive careful study.

Aerial photographs. Photographs taken from the air at an angle of 90° to the earth's surface (vertical photos) or those taken at an angle less than 90° to the earth's surface (oblique photos) are now available for many portions of the earth (see Figure 1.2). Much is covered by overlapping vertical photographs, which, when viewed in pairs with stereoscopes, reveal features in three dimensions.

Because air photos record everything on the earth from a vantage point not normal to the eye, their interpretation requires a great deal of skill. With practice, however, skill can be acquired to enable the user of air photographs to observe landforms, drainage patterns, crops, soils, forests, roads, cities, and so on and then to use this information for soil, forest, crop, and settlement analysis as well as for city surveys, road building, military intelligence, and so on.

The term "remote sensing" is now used to refer to the various means of "observing from a distance." Improved hardware and spacecraft have extended the capacity of air photographs and added other kinds of sensors, such as radar, to the tools for analysis of the features on the earth's surface. For example, color infrared photography gives ability to record and visually use the near infrared portion of the spectrum and has considerable capacity to segregate green tones

FIG. 1.2. *Top:* a vertical aerial photograph. *Bottom:* an oblique aerial photograph.

of plants, improving the identification of species, disease, relative vigor, etc.

In addition to direct analysis, a most important use of air photos is as an aid in the construction of maps. Once, years of laborious and costly field surveying of an area were required: now airplanes equipped with special cameras take photographs which, when viewed through complex equipment, yield remarkably exact information about roads, cities, and other topographic data.

The map and the photograph are closely related tools in that both are used to reduce the earth to an observable and comprehensible size. Both can also be used to present in a clear and simple way material witnessed on the earth's surface, thus augmenting the otherwise difficult job of description through the written word.

Graphs. Graphs comprise the final category of research and presentation devices to be mentioned here. The value of graphs is based upon their provision of ready comprehension of the broad characteristics of the data they are designed to represent. Although only a few types are employed in this book, it should be understood that there are many others. Different purposes require different types. Essentially any geographic periodical will illustrate the variety of types and their uses.

Terms and Definitions Helpful in the Understanding of Maps and Projections

Area scale. Ratio of area on the map to the same area on the earth.

Azimuth. The angle, measured from north, that a great circle makes when intersecting a meridian.

Azimuthal. Projection upon which directions of lines from the center agree with directions, or azimuths, of the same lines on the earth.

Central meridian. The meridian which bisects a map projection.

Conformal (equi-angular). Term used to describe projections which retain true shapes of the earth's regions by having parallels and meridians intersect at right angles. The linear scale is the same along parallels and meridians from their point of intersection. (The scale, however, does not remain constant throughout the projection.)

Earth grid (graticule). An orderly system of imaginary (man-conceived) lines on the earth's surface intersecting at right angles. This grid enables accurate location of places and measurement of distances and directions.

Equal-area (equivalent). Term used to describe projections which retain a consistent area scale throughout. (Not conformal.)

Equatorial case (meridional). Projection centered on the equator.

Great circle. Formed by the surface of the earth intersecting a plane which bisects the earth. An arc of this circle is the shortest distance between two points on the spherical surface.

Latitude. The distance north or south of the equator measured in degrees.

Linear scale. Ratio between distance on the map to the same distance on the earth.

Longitude. The distance east or west of the Prime Meridian measured in degrees.

Loxodrome. A line of constant compass direction.

Meridian. The north-south component of the earth grid extending from pole to pole which enables the measuring of distances (180°) in an east or west direction from the Prime Meridian.

Oblique case. Projection centered on any latitude between the pole and the equator.

Parallel. The east-west component of the earth grid paralleling the equator which enables the measuring of distances 90° in a north or south direction from the equator.

Polar case (azimuthal). Projection centered on the North or the South Pole.

Prime meridian. Designated as 0° longitude and passing through the observatory at Greenwich, England. It is the reference meridian for all east and west distance measurement.

Projection. Result of transforming (projecting) the spherical earth's grid and surface onto a plane surface.

Standard line. A line on the map projection along which the scale is true.

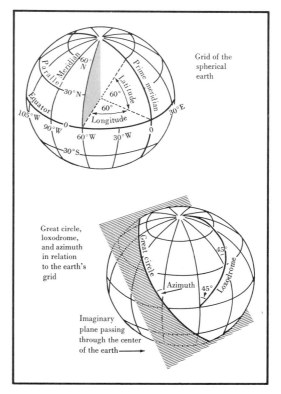

FIG. 1.3. Relationships which exist among the spherical earth, grid components, great circles, azimuths, and loxodromes.

Map Projections

The following map projections have been selected because of their frequent use and because many illustrate major differences between projection categories. They have been classified as to whether or not similar properties are possessed such as equivalence, conformality, etc., because a choice between projections is generally desired by the consideration of these special properties.

It should be reemphasized that only the globe represents all characteristics of the earth in true perspective. Therefore, when studying the following material, one should constantly refer to a globe in order to determine the ability of any projection to represent the earth realistically.

Equal-area projections*

Name and Appearance	Grid Characteristics	Attributes	Principal Uses
Molloweide	1. Meridians are elliptical and are spaced evenly on the parallels. 2. Parallels are straight lines and are spaced closer together as the poles are approached. 3. The equator is twice the length of the central meridian.	1. The deformation of shapes is extreme at the periphery. 2. Scale is true along the equator and along $42\frac{1}{2}°$ north and south of the equator. 3. Shows the earth in an uninterrupted manner. 4. Relatively easy to construct.	1. Showing world distributions emphasizing the mid-latitudes.
Goode's Interrupted Homolographic	1. Meridians are curved and converge on the poles. 2. Parallels are straight lines and are spaced more closely together as the poles are approached. 3. The poles are repeated.	1. The interruption improves the shapes of areas and helps maintain a relatively consistent linear scale. 2. Transoceanic relationships are lost because of the gaps.	1. Mapping of world statistical data depicted only on the land masses.

Copyright by The University of Chicago

* Consistent area scale retained throughout. (Areas directly comparable, but true shape is sacrificed.)

1. The meridians are equally spaced on the parallels.
2. The parallel spacing decreases as the poles are approached in order to retain the equal-area property.
3. Parallels and meridians are straight and meet at right angles.

1. The least average deformation of shape (angular deformation) when the standard parallels are 30°.
2. Extreme shape distortion near the polar regions.
3. Relatively easy to construct.

1. Good for mapping regions through which the standard parallels pass although seldom used.

1. On the polar case the meridians curve and radiate from the poles.
2. Spacing of the meridians and parallels is calculated to preserve equal-area property.

1. The shape and scale deformation is symmetrical around the center.
2. From the center azimuths are true.
3. From the center, equal great circle distances to points on the earth are represented by equal linear distances on the projection.
4. Limited to a hemisphere.
5. Relatively difficult to construct.

1. Mapping areas of equal north-south and east-west extent. (North America.)

1. Straight line meridians converge toward a point which is not the pole.
2. Two standard parallels.
3. The parallels are concentric circles and are spaced on the meridians to retain consistent area scale.

1. Little shape distortion around the standard parallels.
2. The scale is correct along the standard parallels.
3. The scale is greater within the standard parallels and smaller without.
4. Isoperimetric lines.

1. Mapping areas of greater east-west extent than north-south extent. (United States.)

Cylindrical

Lambert Equal-Area

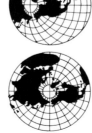

Albers

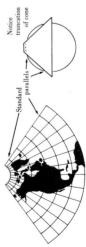

Notice truncation of cone

Standard parallels

17

Conformal projections*

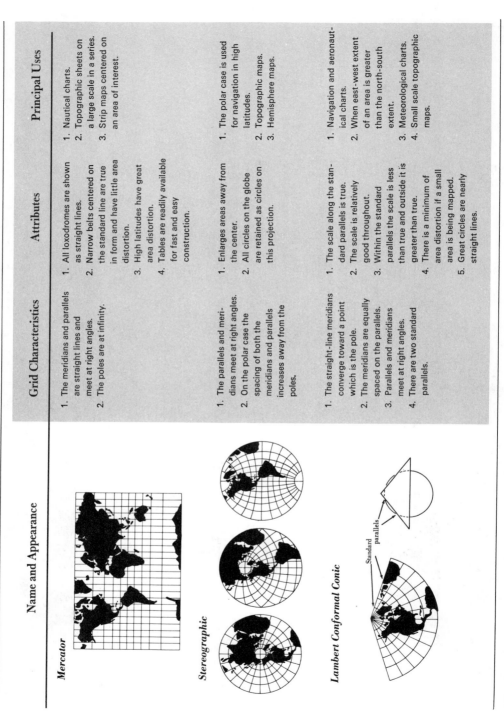

Name and Appearance	Grid Characteristics	Attributes	Principal Uses
Mercator	1. The meridians and parallels are straight lines and meet at right angles. 2. The poles are at infinity.	1. All loxodromes are shown as straight lines. 2. Narrow belts centered on the standard line are true in form and have little area distortion. 3. High latitudes have great area distortion. 4. Tables are readily available for fast and easy construction.	1. Nautical charts. 2. Topographic sheets on a large scale in a series. 3. Strip maps centered on an area of interest.
Stereographic	1. The parallels and meridians meet at right angles. 2. On the polar case the spacing of both the meridians and parallels increases away from the poles.	1. Enlarges areas away from the center. 2. All circles on the globe are retained as circles on this projection.	1. The polar case is used for navigation in high latitudes. 2. Topographic maps. 3. Hemisphere maps.
Lambert Conformal Conic	1. The straight-line meridians converge toward a point which is the pole. 2. The meridians are equally spaced on the parallels. 3. Parallels and meridians meet at right angles. 4. There are two standard parallels.	1. The scale along the standard parallels is true. 2. The scale is relatively good throughout. 3. Within the standard parallels the scale is less than true and outside it is greater than true. 4. There is a minimum of area distortion if a small area is being mapped. 5. Great circles are nearly straight lines.	1. Navigation and aeronautical charts. 2. When east-west extent of an area is greater than the north-south extent. 3. Meteorological charts. 4. Small scale topographic maps.

* Retain true shapes of regions (parallels and meridians intersect at right angles).
Scale is the same along parallels and meridians from their point of intersection, but scale changes from point to point.

Azimuthal*

Name and Appearance	Grid Characteristics	Attributes	Principal Uses
Azimuthal Equidistant	1. Meridians on the polar case are radiating straight lines and parallels are curved and spaced evenly on the meridians.	1. Distortion of angles and areas is great on the periphery. 2. Distance from the center is true in all directions. 3. Angular directions from the center are true. 4. Can show entire earth on the projection.	1. Isochronic (time factor) maps.
Gnomonic	1. The meridians are straight lines in the polar and equatorial cases with the spacing increasing away from the center. 2. The parallels are conic sections; the spacing and curvature increase away from the center on the equatorial and polar cases.	1. Great circles are shown as straight lines. 2. The scale is greatly distorted. 3. Shapes, distances, and areas are greatly distorted near the edges. 4. Less than a hemisphere can be shown practically. 5. Difficult to construct.	1. Used in conjunction with the Mercator in navigating. 2. Very, very small areas at very large scales such as harbor charts, etc.

* Azimuths from the center are correct.

Projections with no common unifying property*

Name and Appearance	Grid Characteristics	Attributes	Principal Uses
Van der Grinten	1. The complete sphere is contained within a circle. 2. The only straight lines are the central meridian and the equator.	1. Distorts both shape and area sizes. 2. Gives a fairly good representation of the world as it distorts polar areas less than the Mercator and the Mollweide.	1. In mapping world distributions where equal area is not primary.

* The following projections display neither equal area nor conformality and have no common unifying characteristic.

Name and Appearance	Grid Characteristics	Attributes	Principal Uses
Denoyer Semi-Elliptical 	1. The meridians are spaced less than the normal amount near the poles. 2. The parallels are straight lines and are spaced farther apart as they approach the poles. 3. The poles are one-third the length of the equator.	1. Not equal-area or conformal. 2. Has a pleasing appearance.	1. Used on many school wall maps.
Orthographic  	1. The meridians are arcs of ellipses on the oblique and equatorial cases. 2. Straight parallels only on the equatorial case.	1. It shows the earth as it might look when viewd from a distance. 2. Limited to a hemisphere. 3. Not equivalent, conformal, or equidistant.	1. When showing the earth as it might look from a distance. 2. Physiographic diagrams.
Polyconic 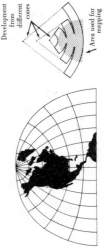	1. The meridians are spaced equally on the parallels. 2. The central meridian is a straight line. 3. The parallels are arcs of nonconcentric circles. 4. The parallels and meridians do not meet at right angles. 5. The equator, if shown, is a straight line.	1. The scale is relatively accurate throughout. 2. It is *close* to conformal and *close* to equal-area. 3. Equidistance is evidenced along the mid-meridian and all the parallels. 4. Tables for rapid construction are readily available.	1. The grid for the U.S.G.S. topographic sheets. 2. Mid-latitude areas elongated in a north-south direction. 3. Small areas on a large scale such as topographic maps in a series. 4. As a good compromise projection.

Selected Periodicals in English

UNITED STATES

Annals of the Association of American Geographers (quarterly).
Economic Geography, Clark University, Worcester, Mass. (quarterly).
Geographical Review, American Geographical Society, New York (quarterly).
Journal of Geography, National Council of Geography Teachers (monthly except June, July, and August).
The Professional Geographer, Association of American Geographers (bimonthly prior to 1971 and quarterly since 1971).

UNITED KINGDOM

Geographical Journal, Royal Geographical Society (quarterly).
Geography, Geographical Association (quarterly).
Scottish Geographical Magazine, Royal Scottish Geographical Society (three numbers a year).
Transactions of the Institute of British Geographers (two numbers a year).

OTHERS

Australian Geographer, Geographical Society of New South Wales (two numbers a year).
Canadian Geographer, Canadian Association of Geography (quarterly).
Geographical Review of India, Geographical Society of India (quarterly).
New Zealand Geographer, New Zealand Geographical Society (two numbers a year).
Geografiska Annaler, Swenska Sällskapet för Antropologi och Geografi.
For others, see Chauncy D. Harris, "Annotated World List of Selected Current Geographical Serials in English," *Research Paper No. 96,* 2d Ed. (Department of Geography, University of Chicago, 1964), see also Chauncy D. Harris and Jerome D. Fellman, "International List of Geographical Serials," *Research Paper No. 138,* 2d Ed. (Department of Geography, University of Chicago.)

two

MAN, CULTURE, AND THE EARTH

Modern geography views man and his environment as an interacting entity. In this view the earth, as the human habitat, is considered to be subject to the interpretations, decisions, and actions of man. Interests focus upon the character of this habitat, the processes—both physical and cultural—attending its modification, the relationships between the parts, and the impacts of the resultant changes. In the geographer's view, then, environment is not considered a static matrix that rigidly shapes the lives and activities of the human occupants. This is not to say that the physical environment has no influence upon man or that man can completely dominate or control all natural forces and conditions; it is rather to suggest that use of the environment is subject to human decisions and to the power of man to select, adapt, and alter, as well as to his increasing capability to minimize environmental restraints. Through the progression of time, in his relationship to the physical environment man has become an increasingly dynamic agent. To express these concepts some geographers have borrowed and modified a somewhat biological term "ecological dominant," as a shorthand expression of modern man's great capability to modify the environment.

The human occupants, however, have differed in these respects in the span of time, and they still differ from one part of the earth to another. Each society in each stage of growth interprets the values and utilities of the environment it inhabits, and it organizes its life and spatial unit accordingly. Interpretations or perceptions are cultural appraisals; they are made from the background of attitudes, experiences, and skills of the occupying society.[1] The society selects, organizes, and uses earth features to meet its needs; it may have had alternatives; a different society might have chosen different features; and the same society may change its selection with time, new needs, skills, and so on. Thus, as suggested above, the environment is not a rigid determinant; in a given situation it does, however, influence the scope of possibilities, and it can have influence upon the organization and uses of selections made. The roles performed can vary in time through changes in societal conditions or in the environment itself through human modifications both good and bad.

[1] Jan O. Broek has stated the concept well in suggesting "Each particular culture acts like a filter through which a people views its habitat." *Geography, Its Scope and Spirit* (Columbus, Ohio: Charles E. Merrill Books, Inc., 1965), p. 41.

Since the earth becomes meaningful and useful to man through perceptions conditioned by culture, it is essential that we develop at this point some basic notions about population and cultural differences. To accommodate our purposes, we will summarize these under three major groupings: population dynamics and distribution, societal organization, and technological development.

Population

Today (1972) approximately 3.75 billion people inhabit the earth, and the number is increasing. During most of the period of human existence on earth, the number of births did not significantly exceed the number of deaths; however, during the last few centuries, more particularly the last few decades, the balance has been drastically altered as the death rate has dropped.

Estimates have placed world population in the time of Christ at 250 million. During the next sixteen centuries the number expanded slowly, at a rate of increase of between 2.5 and 5 percent per century, and did not double until the end of the period.[2] By 1880 the world population had reached 906 million and by 1900 had expanded to about 1600 million, and the increase rate had reached nearly 1 percent per annum. Under the influences of advances in medical science and improved living conditions, the actual population and the increase rate have advanced astoundingly in the twentieth century. In 1969 the growth rate was 1.9 percent (see Figure 2-1).[3]

Although there are many and complex variables precluding accurate forecasts—such as continued lowering of infant mortality, further lengthening of the life span, and

[2] "The Future Growth of World Population," *United Nations Population Studies No. 28* (New York: United Nations, 1958).

[3] *Demographic Yearbooks*, United Nations, may be checked for recent figures. Also see the publications of the Population Reference Bureau (1755 Massachusetts Avenue, N.W., Washington, D.C., 20036).

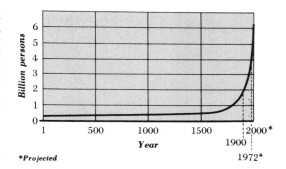

FIG. 2.1. Twenty centuries of world population growth transplotted for all years for which data were available. (U.S. Department of Agriculture.)

the nature of the birth rate—there is little doubt that an increasing number of people will live on the earth. The growth rate, whatever it comes to be, is applied to a constantly increasing population base.

SPATIAL PATTERN AND DIFFERENTIAL GROWTH RATES

There are several fundamental questions respecting populations that geography seeks to answer. Where are the people? Why are they where they are? What is the pattern of growth—is it even or uneven? What are the details of distribution or density? What are the movement—immigration and emigration—trends? We will call attention to only a few significant facts now; answers to most of these questions will be woven into the body of the analysis.

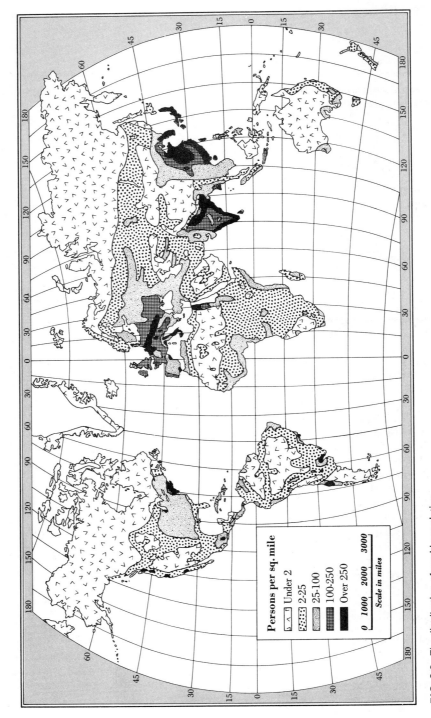

FIG. 2.2. The distribution of world population.

One of the most fundamental facts concerning the geographer is that population is very unevenly distributed over the earth's surface. This is true in the gross pattern, or macro view, as well as in the detailed, or micro view. The validity of the macro view is borne out by the generalized world-population map (Figures 2.2, 2.3) and by Tables 2.1 and 2.2, which show trends by broad regions. Asia is especially noticeable, accounting for over half of the total; yet on this landmass there are major variations in concentration. Outstanding blocks of concentrated populations occur in the east and south, whereas the interior and north are thinly inhabited; but even in these latter areas detailed maps would reveal differences —for example, small agricultural oases with relatively densely settled populations occur in a few places in the interior.

Other noteworthy population concentrations exist in Europe and the western part of the USSR and in the eastern part of North America. Several smaller concentrations occur in Latin America and Africa; and it can be reasoned from the map that the majority of the Australians live in the

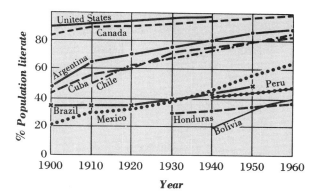

FIG. 2.3. Literacy trends in the Western Hemisphere. (U.S. Department of Agriculture.)

southeast corner of that continent. Also of major interest are the large areas, especially within the tropics and high latitudes, that presently support scant populations.

Population growth rates also vary markedly from one region to another. Rates have been higher than the world average in most of Latin America, Africa, and Asia, somewhat less than average in North

TABLE 2.1

World population by regions*

	World	Africa	Asia	North America	Latin America	Europe	USSR	Oceania
Mid 1969 Millions	3551	344	1990	225	276	456	241	19
Percent of Total	100	9.7	56.0	6.3	7.8	12.8	6.8	0.5
Growth rate	1.9	2.4	2.0	1.1	2.9	0.8	1.0	1.8
2000 Projection Millions	6130	768	3458	354	638	527	32	353

* "1969 World Population Data Sheet," Population Reference Bureau, April 1969.

TABLE 2.2
The Largest Populations, Mid 1969*

Country	Millions	Current Growth Rate
Mainland China	740.3	2.6
India	536.9	2.5
USSR	241.0	1.0
United States	203.1	1.0
Indonesia	115.1	2.4
Japan	102.1	1.1
Brazil	90.6	2.8
West Germany	58.1	0.4
United Kingdom	55.7	0.6
Nigeria	53.7	2.5
Italy	53.1	0.7
France	50.1	1.0
Mexico	49.0	3.4
Philippines	37.1	3.5

* "1969 World Population Data Sheet," Population Reference Bureau, April 1969.

America, and the USSR, and considerably less in Europe (see Tables 2.2 and 2.3). In other words, the growth rates are highest in the technically unadvanced countries of the world—in the very regions that can, at present, least effectively cope with rapid increases in population.

At present these countries contain about two-thirds of the world population; and current trends would indicate a sizable increase in their portion by the turn of the century. A significant feature of this growth pattern is its impact upon the prospects for economic advancement—especially in countries such as China and India which currently have huge populations. Under these circumstances, the expanding yearly requirements for basic needs limit capital accumulation for investment in more diversified and productive economies.

POPULATION DENSITY

The factor of population density is of special significance. In general usage, density

TABLE 2.3
Annual population growth rates in selected countries, Mid 1969*

Africa		Latin America		Asia		Denmark	
Libya	3.6	Costa Rica	3.8	Israel	2.9	Denmark	0.9
Sudan	3.0	Bolivia	2.4	Kuwait	7.6	Ireland	0.5
UAR	2.9	Ecuador	3.4	Saudi Arabia	1.8	Belgium	0.1
Ghana	2.5	Peru	3.1	Burma	2.2	Switzerland	0.9
Kenya	3.0	Argentina	1.5	Thailand	3.1	Bulgaria	0.6
Rwanda	2.7	Chile	2.3	South Korea	2.8	Hungary	0.3
Chad	1.7					Greece	1.2
South Africa	2.4					Yugoslavia	1.1

* "1969 World Population Data Sheet." See also Table 2.2

refers to the occurrence of people within a common unit, usually a square mile; it is expressed as a ratio between population number and this standard unit. Average density for the world is approximately 60 persons per square mile. It has already been suggested, however, that the situation is vastly different in reality. In eastern and southern Asia and in Europe, settlement is most dense, with many areas supporting over 1000 persons per square mile—and the broader regions exceed the world average several times over (see Table 2.4). Southwest Asia, Africa, the Americas, the Soviet Union, and Oceania are less densely settled; only in Middle America is the world average exceeded. Averages for broad areas can be somewhat misleading because they fail to recognize the actual distribution of people: for example, well over half the population of Indonesia is concentrated on the island of Java. In the United States and Western

TABLE 2.4

Population density by major world areas and selected countries, 1968*

Region	Persons per square mile	Region	Persons per square mile
Europe	161	South America	26
France	238	Argentina	22
East Germany	409	Bolivia	9.0
Greece	172	Brazil	26
Italy	452	Chile	32
Netherlands	907	Ecuador	54
Spain	166	Peru	26
United Kingdom	832		
Asia	117		
Burma	100	Africa	28
China	198	Chad	6.9
India	422	Congo (Kinshasa)	18
Iran	42	Kenya	45
Japan	704	Nigeria	175
South Korea	791	South Africa	40
		Tanzania	34
North America	33		
Canada	5.4		
United States	54	USSR	28
Central America	75		
Guatemala	115		
Mexico	61	Oceania	5.6

* Compiled from United Nations data.

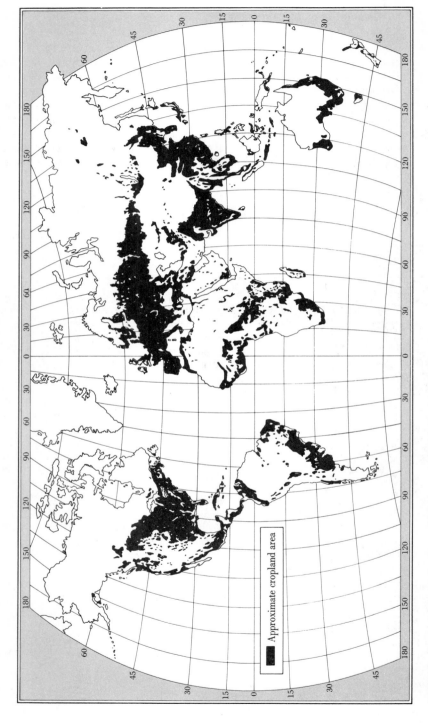

FIG. 2.4. Approximate cropland area. Partly because sufficiently detailed data on land use are not available for some countries and partly because the map is small, the shaded portions include scattered areas of land not used for crops and the unshaded portions scattered cropland areas. (*U. S. Department of Agriculture.*)

Europe more than two-thirds of the population lives in the urban centers, that is, upon a small portion of the total area. In these instances a significant share of the populations lives in environments that have been drastically altered and, not uncommonly, seriously degraded.

TABLE 2.5

Cultivated land and per capita quantities*

	Billions of Acres	Acres per Person
Africa	0.39	1.3
Asia	1.28	0.7
Australia and New Zealand	0.04	2.9
Europe	0.38	0.9
North America	0.59	2.3
South America	0.19	1.0
USSR	0.56	2.4
World	3.43	1.0

* The World Food Problem, A Report of the President's Science Advisory Committee, II, Report of the Panel on the World Food Supply, May 1967, 434.

It is generally more meaningful to consider density in terms of various productive measures, such as cropland per capita for predominantly agricultural populations; or in more diversified economies the population/resource or income ratios may be more meaningful (see Table 2.5 and Figure 2.4). The usefulness of the density concept lies in its possible value as an indication of population pressure upon space, resources, and productive output.

MIGRATION

Spatial movements of population can be a factor influencing density trends. The reader will be aware of the historical role of migration in the colonization of new lands. Although freedom of movement has been somewhat restricted by the political compartmentalization of the earth, migration remains a significant feature in density and growth relationships in many situations. On the international level most countries have regulations, especially with respect to admitting immigrants; such regulations commonly concern numbers, ages, skills, health, and races. At present some countries are more open than others: for example, Australia, with a small population, encourages white migrants, especially from British Commonwealth nations or the United States; Japan, on the other hand, possessing a dense population, does not actively encourage new people from foreign areas; and the communist countries, in general, have such restrictive regulations as to discourage migration.

Except in the communist or totalitarian countries, where internal movements may be restricted, peoples can generally move freely within countries. The impact of this mobility is illustrated by the United States, where people may and do move from one locale to another—especially for improvement of economic opportunity. Much of the relatively rapid population growth in the West Coast and Southwest in recent years has come as the result of migration (Figure 2.5).

Figure 2.6, showing percentage change in the population of the United States between 1960 and 1970, indicates an uneven growth rate for the nation. Heavy growth occurred where urbanizing influences were especially strong, whereas the predominately farming states were below the national average. The Dakotas actually showed losses for the period.

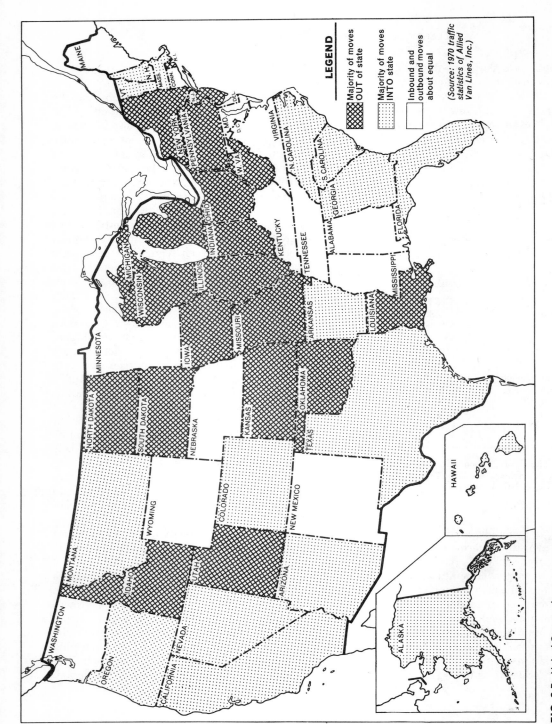

FIG. 2.5. United States migratory movement. (1970 traffic statistics of Allied Van Lines, Inc.)

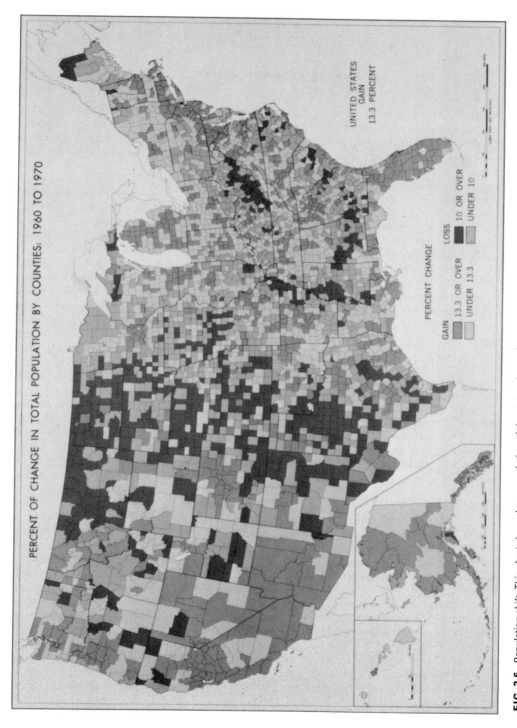

FIG. 2.6. Population shift. This chart shows how population of the states changed from 1960 to 1970, as revealed by the official census as of April, 1970. Seven states were heavy gainers and only the Dakotas showed losses. (United States Department of Commerce Bureau of Census.)

Social Organization

Several qualities of societal organization are of prime concern to geography, since they have spatial dimensions and bear upon man-earth relations within the spatial units. Although these tend to be complexly interrelated, there is value in categorizing them here and noting a few of the more prominent interests and possible relations. For this purpose, they may be recognized in institutional groupings, namely: political, economic, educational, and ideological. Although this coverage can only hint at the idea, it should be recognized that a hierarchy of organizations and influences exists in each of these institutional categories.

POLITICAL

Since the earth is divided into territorial units, each of which is under some form of political system, all people live under or in contact with the influences of some form of government. The significance to geography of governmental systems stems from their differing philosophies, organization, policies and legal and honored traditional arrangements regarding the control of land and other resources (tenure or conditions of possession and property rights or conditions governing discretionary use and disposition of gains). Also, it will be remembered, there is significance in the fact that people cannot in most cases move freely from one national unit to another. In the most extreme situations, the political processes of a government may largely dictate the life of people living under it, directing their economic activities, controlling national productive wealth and individual expenditures, and determining the patterns of settlement and development. In such cases, the ownership of land and other resources and essentially all the decision-making authority concerning man-earth relationships rest with the central government. Delegated down through the hierarchy of control, this central authority affects each individual.

Under democratic forms (as understood in the free world), the ideal role of government is to do for the people collectively what they cannot do for themselves individually; thus populations living under democracies have considerable individual freedom. They can, as individuals, own and buy or sell property and make the decisions attending its use—or, in a sense, "govern" their property. Even under democracies, however, the roles of government—through several categories of regulations, taxation, eminent domain, policing power, education, public and special group services, and development promotion—bear actively to some degree upon every citizen. In fact, the relationship is highly complex. In the United States, for example, the governmental hierarchy includes influences stemming from the federal, state, county, and city levels. And a given citizen may be under the further influence of a zoning district, a fire district, an irrigation district, a sanitary district, and so on.

In some of the less advanced countries, the national governments have tended, in the past, to exert rather passive influences. At the local level in the least advanced tribal societies, traditions, varying somewhat from one situation to another, have been the main kind of government influence. The investiture of decision making and property ownership differs somewhat from group to group. In certain groups authority rests with a council, or the like. In some instances property is controlled by the governing authority to parcel out as it desires; in some, productive activities and land use are communally oriented; and in others there is essentially private ownership. In most of the "primitive" agricultural societies there is a unique (and, perhaps, valuable) creed respecting the ownership of land: It is regarded as belonging to all the people—a heritage from those past, a trust for those present, and a legacy for those unborn.

On the extranational level, governmental policies and activities are strongly related to the propensity to exchange either ideas or goods as well as to the actual establishment of linkages between nations. Most students are aware of the various political alignments existing at the international level and recognize the influence of these on international relations, including economic exchange systems.

ECONOMIC

Economic organization is a second major category of cultural variables with spatial dimensions and relationships. Economic organization especially has bearing upon productive activities and the linkages between activities. There are many facets.

All people must have some means of earning a livelihood. Some may be considered primary producers—raising crops, rearing animals, harvesting natural biota, or extracting useful earth materials; others may be considered secondary producers—processing or manufacturing useful items from the raw materials of the primary producers; still others can be classed in an even more dependent category—moving and trading valuable goods and providing the several forms of services. Together these comprise the economic activities of man. The third forms, which obviously can exist only when supported by the first two, are most numerous in the highly advanced economies in which "commercialization begets specialization" and efficiency in profit making is the prime motive.

In the broadest order we can recognize, at the extremes, that some livelihood forms are characterized by little or no dependence upon exchange, whereas others have a high degree of dependence upon exchange. Nonexchange economies are, then, essentially self-sufficient, with the individual producers being also the prime consumers; therefore, they have little or no dependence upon the outside and, conversely, are relatively little influenced by the outside. They are, thus, essentially closed systems. In the commercial, or exchange, economies, on the other hand, there is a strong relationship between producers and consumers and formal linkages between the two groups are essential; the "rules of the game" are institutionalized with cost, price, profit, and capital availability—highly important considerations among management concerns. Japan, for example, supports a significant share of her large population by obtaining profits from processing, handling, and selling products. To keep her factories running she must import a major share of the necessary raw materials; and to maintain her economy she must export the finished products. In commercially oriented economies, profits are essential and are the motivators of change, technological innovation, and efficiency in the management of production; producers must be aware of the demands of the market and be able to adjust to changes. In such economies the existence of capital credit systems is essential.

Economists and economic geographers, viewing economic development through the perspective of time, suggest that there are evolutionary stages in the advancement from traditional, self-sufficient economies to highly developed, sophisticated ones. In reality, most countries are at a transitional point somewhere on a scale that would include some of the traditional, self-sufficient societies of parts of the tropics on one end and the United States and several of the highly commercial countries of Western Europe on the other.[4]

Obviously, different forces operate during the different stages. In the traditional, self-sufficient societies, basic family needs and available, time-honored techniques are the major conditioners of production and the use of the natural environment; whereas in commercial economies markets and many production and distribution features bear upon the organization and use of the environment. In the most sophisticated forms, where very high mass consumption exists, overt aims and wants of the people become significant.

Although the concept of stages of growth serves as an aid in understanding of the development process, in reality few large countries are entirely in one stage, and, moreover, some have never gone through all stages. In the United States, for example, the traditional, self-sufficient stage, if it ever truly existed, was fairly short-lived.

EDUCATION

Knowledge is the foundation of all progress. The nature of educational systems has direct influence upon the knowledge level of a given people and has much to do with their attitudes toward the purposes and organization of life as well as their attitudes toward the environment and their abilities to use it. Among the societies on the earth, there are vast differences in the types and levels of educational systems. Many of the rudimentary agricultural societies of the tropics are without formal education systems, and the level or degree of literacy is low. Their knowledge is accumulated and passed on through the living experience.

[4] See, for example, W. W. Rostow, *The Stages of Economic Growth* (Cambridge: Cambridge University Press, 1962).

They have changed relatively little throughout history.

In contrast, the countries in the advanced stages of economic development have well-organized education and research systems and a high level of literacy (see Figure 2-3). These systems include not only formal education and research institutions but also many means of informing the public such as newspapers, magazines, radio, and television. In such societies new ideas are usually more readily understood, and, because the means for rapid diffusion are present, changes can be rapidly induced. Literacy, in general, broadens the range of contacts and expands the variety of stimuli to which individuals are exposed. Ideas move very slowly when dependent only upon oral media; moreover, it is difficult under such circumstances to transmit complex ideas accurately.

IDEOLOGY

Variations in the scheme of concepts and value systems about human life or culture have spatial dimensions and relations that are of marked significance in the man-earth system. Some aspects operate through the institutions previously discussed; however, there are many other cultural traits in this category that affect human attitudes and, thus, human behavior and social acceptability. Examples are religious beliefs, food preferences, pride in racial, regional, or national background, or even voting behavior. One example will suffice to illustrate the point: The discouraging by the Hindu religion of the killing or eating of cattle has resulted in a large cattle population in India but not in a significant cattle industry; in fact, the existence of large numbers of animals results in many seemingly adverse relationships, especially in matters of land competition and management.

Development of Technology

The body of technology possessed by a people is strongly related to its ability to alter, organize, and use the earth's physical-biotic system. Two aspects are especially significant, "resource-converting techniques" and "space-adjusting techniques."[5]

RESOURCE-CONVERTING TECHNIQUES

This category includes those means by which features of the natural environment are brought into the service of man. The term "techniques" or "technologies" in this sense is used broadly. For example, the techniques of land use include in this total view all the developed skills, tools, methods, and materials (including selected and improved plants and animals, chemical and other soil treatments, and disease and pest controls)—in other words, all of the arts, sciences, and technologies—basic to bringing land into functional service. Other facets of resource-converting techniques include a similar range for each of the other elements of environment or environmental complexes that are used by man. Thus they include the arts, skills, sciences, and methods underlying the use of minerals, water, and so on. Through these techniques man brings about the greatest changes in the physical-biotic system. Since there are spatial distributional differences in the techniques possessed by the peoples of the world and since techniques have changed in the time span, geographers are deeply concerned with these features in a space-time continuum.

[5] Edward A. Ackerman, "Geography as a Fundamental Research Discipline," *Research Paper No. 53* (Department of Geography, The University of Chicago, 1958), p. 26.

SPACE-ADJUSTING TECHNIQUES

"The space-adjusting techniques either shorten the effective distance of travel and transportation or permit intensification of space employment beyond that possible on the land surface provided by nature."[6] The arts, skills, sciences, and technologies of construction are especially noteworthy. The building of a canal to bring water to dry land, the construction of a highway bridge or the development of an airline to provide connectivity between potential producers and potential markets, the terracing of slopes for residential or agricultural use, the draining of swamp or water-covered land such as the polders of the Netherlands, and the filling of water fronts to create new land are obvious examples.

The growth of great cities may be suggested as a further example; certainly they represent an extreme intensification of space use—and they cast their influence far beyond the space they occupy. They have been dependent upon the advance of a variety of engineering and architectural techniques to provide for the necessary circulation, sanitation, and building design. Recently the new field of development planning has been added to the space-adjusting techniques. This brings the fruits of the social, earth, engineering, and architectural arts and sciences to bear upon the rational ordering of the occupance and growth patterns of cities and regions.

Throughout history, with every major change in space-adjusting techniques have come alterations in space relations. The development of railroads, the engine-powered

[6] Ibid.

ships, the automobile and superhighway, the airplane, and other forms of communication have been basic to the ever broadening spheres of contact between people and places and, thus, also to increased or intensified spatial interaction. They have been, moreover, the underlying factor in the modern mass concentration of people in large urbanized areas (for example, the almost contiguous population concentration is the megalopolis on the eastern seaboard of the United States).

Summary

Although we have briefly examined some of the partitive aspects of culture, we do not mean to imply that there are not other features of culture of interest to geography. Our intention, here, has been to crystallize the concept that man is the active agent of environmental organization and use, that he is conditioned by his culture, and that he differs in numbers and culture in both space and time. Moreover, we stress again that the several categories discussed are, in reality, interrelated. It is, thus, common in general geography to consider culture as a system consisting of the traditional ideas, values, and patterns of behavior of human groups. It may be suggested that cultural systems are products of human action as well as conditioning elements of further human actions. In their relations to the man-earth system, cultural systems and subsystems are of concern to the discipline of geography.

THE PHYSICAL EARTH AND MAN

Man lives and works against the background of the physical earth, the features of which differ widely from one setting to another. (Variations characterize both the individual features and the combinations that exist in complex interrelationships. Upon and amidst the various conditions, man has established and is in the process of establishing his habitats.) The earth's surface provides the stage for man's occupance, and the components of the natural environment provide the bases and substances of his support. Some aspects of the natural environment, however, may be in the form of obstacles or resistances calling for modification or circumvention; others may be more or less neutral.

The American viewing the earth would reason that some natural environments offer comparatively wider opportunities for economic activity and greater possibilities for organization and use than others. Each society, however, appraises the potentials of its environmental setting through the "filter" of its culture. Thus the tribesman of tropical Africa, under his culture, is quite "at home" in an environment that presently is unattractive to most persons who are products of American culture. Moreover, in two similar natural environments, two societies may find quite different assets and limitations and thus will be likely to develop quite different occupance patterns and activities. Also, any group in a given area will be likely to evolve different patterns and uses with passage of time. Man thus is recognized as a dynamic agent in the context of both space and time. For this reason some of the more fundamental and variable aspects of man and culture have been examined in chapter 2. There follows a brief consideration of the earth's natural system and subsystems.

The Frame of Reference

In the broader scope geographers have several interests in the physical earth. These can be summarized into several related clusters: (1) the development of understanding of the spatial distributions, relations, processes, and variations of and in the physical-biological system and its subsystems; (2) an interpretation of the physical-biological world, in spatial units, as the foundation for human occupance and activities; and (3) the development of an understanding of the changes and processes of changes in the physical-biological world, resulting from man's modification of his habitats. Although individual geographers may have a central interest in one aspect of the physical earth, they work with an awareness of all. In this brief introductory text in which our principal mission is to develop an understanding of the broad outline of and variation in the vast system on the earth's surface comprised of man and the natural environment—or, stated in another way, understanding of the earth as the home of man—we must generalize our consideration somewhat. Thus, in this chapter, we will concern ourselves primarily with an identification of: (1) the elements of the physical-biological system, (2) some of the major relationships among the elements, and (3) the principal characteristics of particular concern to man. Our objective at this point is primarily to establish a modest background for the regional chapters that follow.

Although the elements of the natural environment exist together in complex relationships, in fact, in interrelated systems, our analysis will be aided by first examining each independently.

THE EARTH IN SPACE

Earth, the home of man, is a whirling mass of matter in the boundless expanses of space. In the scheme of our solar system, earth obeys the orderly laws of the universe rotating on its axis and following a set orbital path around its incandescent star, the sun. Its movements and distance from the central star make it unique among the planets in its physical and biological complexities. The sun radiates an almost infinite amount of energy into space. The minute portion intercepted by the earth is the source of life.

The earth, an oblate sphere, has a very slight bulge around the equator and a slight flatness at the poles. Its equatorial diameter is 7,926.68 miles and its polar diameter is 7,899.99 miles. The circumference at the equator is 24,902.45 miles and around the poles is 24,818.60 miles. The area of the earth is 196,940,400 square miles.

The earth has two primary motions: rotation and revolution. The earth rotates eastward about an axis through the poles. The time required for one rotation in respect to the sun is 24 hours and has been established as clock time. The second movement is revolution. The rotating earth moves around the sun in an elliptical orbit. The time required for a complete orbit is $365\frac{1}{4}$ days which determines the number of days in the year or calendar time. Leap year occurring every four years is the adjustment to the quarter day.

The combination of the rythmic movements of the earth, the fixed position of the inclined axis, and the different angles at which the sun's rays strike the earth at different latitudes produce the seasons and the changes in the length of day and night. In addition these factors influence air pressures, winds, temperatures, precipitation, storms, and ocean currents.

SOLAR ENERGY

Except for the energy received from the sun and some lost back to space, the earth is essentially a closed system. The solar energy, however, is central to the functioning of the atmosphere and powers all life on the earth. From the sun comes an endless stream of shortwave radiation, which is the basis of earth's light and heat. Only one two-billionths of the sun's energy is intercepted by the earth. An average of 34 percent of the intercepted energy is lost directly to space, through

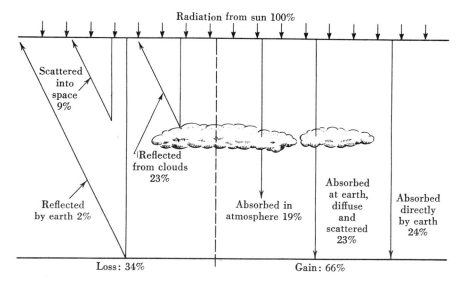

FIG. 3.1. Heating of the earth and its atmosphere by insolation. (Critchfield, *General Climatology* [Prentice-Hall, 1960], p. 16.)

reflection from clouds, earth, and so on; 19 percent is absorbed by the atmosphere; and the remaining 47 percent is received by the earth's surface. Upon absorption of the incoming solar radiation, the earth in turn becomes a radiating body, "broadcasting" energy of longer wavelength. Thus the earth's blanket of atmosphere is largely heated from the earth's surface.

The amount of energy received by the earth is equal to the amount lost to space; consequently, the mean temperature of the earth remains essentially constant. The even exchange of energy lost and received is called the *heat balance*. All parts of the earth's surface, however, do not receive equal amounts of incoming radiation. The low latitudes receive the greatest amount of radiation per unit surface area, and a greater accumulation of heat occurs in these areas than in higher latitude regions. The balance is achieved, however, by winds and ocean currents, which transfer heated equatorial air to deficient higher-latitude areas.

THE ELEMENTS OF THE PHYSICAL-BIOLOGICAL SYSTEM

Weather and climate. Weather and climate are basic components of the physical environment. No other natural elements play so decisive and important roles in man-earth relationships. Landforms, soils, water, and vegetation feel the impact of the authority of climate. Although weather and climate are inseparable, there does exist a difference in meaning and function. Weather is the day-to-day condition of the atmosphere—the observable, sensed phenomena at the moment. It is the warmth or chill, rain or snow, sunshine, clouds, fog, wind, or storm. Climate, not merely an average of weather components, is the composite of weather, including extremes, frequencies, and the annual march of events in the atmosphere. Weather is like the pages of a book, each page a new, varied experience; climate is the sum total, the volume.

Man on every part of the globe is subject to the action of weather and the seasonal regimen of climate, and man in every walk of life is cognizant and concerned with atmospheric conditions. The rice farmer of India prays for the life-giving rains that accompany the summer monsoon; the wheat farmer fears the hailstorm that may destroy his crop; the orchardist is alerted to frost warning; the yachtsman rejoices in clear skies and fair wind; the logger, merchant, football coach, astronaut, and vacationist keep posted on the predictions. Plans for events and possible outcomes are subject to weather whims. Man constructs shelters and designs clothing to combat the heat, cold, moisture, and wind. Man feels, observes, predicts, and measures the ingredients of weather. It is a daily item of television programs, the radio, and newspapers. In summary, weather has an integral and intimate role in man's environmental relationships.

An understanding of weather and climate of the world is achieved through the knowledge of the ingredients or basic elements: (1) air temperature; (2) atmospheric moisture including clouds, fog, humidity, and precipitation; and (3) pressure and winds. The varying occurrences and proportion of these elements provide a variety of climates ranging from the monotony of the tropics to the extreme diversity of the middle latitudes.

Air temperature. The concepts of heat and cold originated as a part of man's sense organs. Sensory measurement, however, has a wide range of unreliability and quantita-

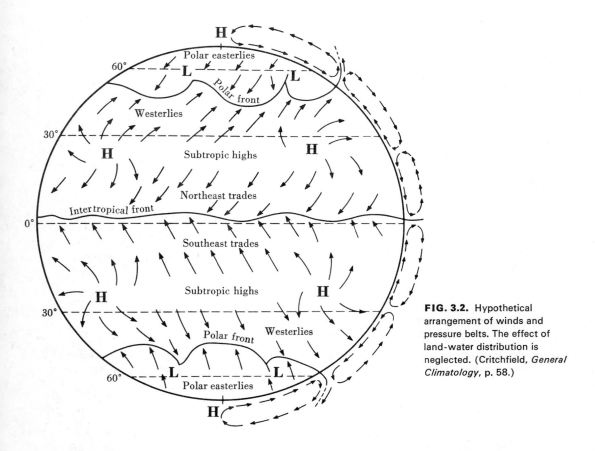

FIG. 3.2. Hypothetical arrangement of winds and pressure belts. The effect of land-water distribution is neglected. (Critchfield, *General Climatology*, p. 58.)

FIG. 3.3. The world's wind and pressure system, generalized for January.

41

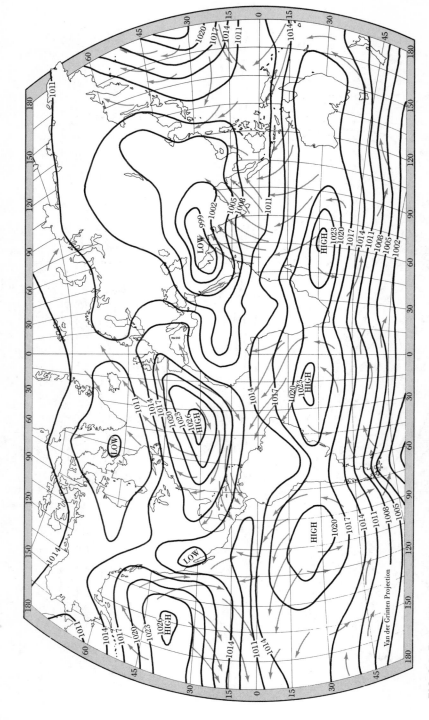

FIG. 3.4. The world's wind and pressure system, generalized for July.

tive methods are necessary for temperature standardization for apprasial and comparison. For this purpose, temperature information is obtained from a properly placed thermometer, and the readings provide the basis for a number of significant figures. The most common is the daily mean temperature, which is obtained by averaging the diurnal maximum and minimum readings. The mean monthly temperature is obtained by the addition of the daily means of the month divided by the number of days of the month. These data are used in the construction of climatic graphs found in the text. The term "annual range" refers to the difference between the mean temperatures of the warmest and coldest months. Climatic variability, however, cannot be detected in this general data.

Temperatures, as a result of reports from recording stations all over the world, may be shown on the map. The device employed is the isotherm, or line of equal temperature. On the world map the isotherms used are for the months of January and July. If distance from the equator were the only consideration for isotherm plotting, the lines would closely parallel the equator. But many factors—such as, differential heating of land and water, mountain barriers, and elevation —or controls disturb an orderly arrangement of isotherms. These climatic controls will be discussed later in the chapter.

Pressure and winds. Wind is air in horizontal motion—movement paralleling the earth's surface. Pressure difference is the immediate cause of wind; unequal heating of the earth's surface is the basic cause of pressure differentials. When air becomes heated (as over a landmass in summer) it expands, forming a low-pressure area; conversely, air pressure increases with decreased temperature (as over a landmass in winter). The most accurate instrument used in measuring pressure is the mercurial barometer. Pressure zones may be mapped, as is the case with precipitation and temperature. The plotting of pressure is accomplished by the use of isobars or lines of equal pressure.

The air, being free to move, tends to equalize pressure differentials. Wind is thus created as air flows from high-pressure areas to areas of lower pressure. Winds blow from "highs" to "lows." Winds do not flow directly north or south toward the low-pressure zones but follow a curved path, due to the earth's rotation. This deflection is caused by the *Coriolis force.* This is not actually a force but an effect resulting from the rotational movement of the earth and the movement of the air in relation to the earth. As a result of the Coriolis force, winds in the northern hemisphere are deflected to the right, and those of the southern hemisphere are deflected to the left, looking downwind.

Winds receive their names from their source-areas; therefore, a north wind comes from that direction and blows toward the south, an east wind is from the east and blows westward, and so on. Wind direction may be ascertained by observing cloud movement, smoke, waves, vegetation, and so forth. A weather vane is used for more accurate information, and wind speed is usually determined by an anemometer (consisting of several hollow hemispheres mounted on a freely moving vertical axis). Gears similar to those on an automobile speedometer move a device which indicates speed.

Distributed over the earth is a pattern of pressure cells and resulting winds, consisting fundamentally of a series of alternate high and low pressures and alternate east and west wind systems. Near the equator, where the earth receives a maximum of incoming solar radiation, there is a broad, discontinuous zone of low pressure variously known by several names: the *equatorial low, doldrums,* the *belt of equatorial convergence,* and the *intertropical front.* This belt is generally characterized by rising air, calms, variable winds (except along coasts, where sea breezes are important), and local thunderstorms, which may be extremely violent.

Two belts of high-pressure cells, known as the subtropical highs, circle the earth at approximately 20° to 35° north and south

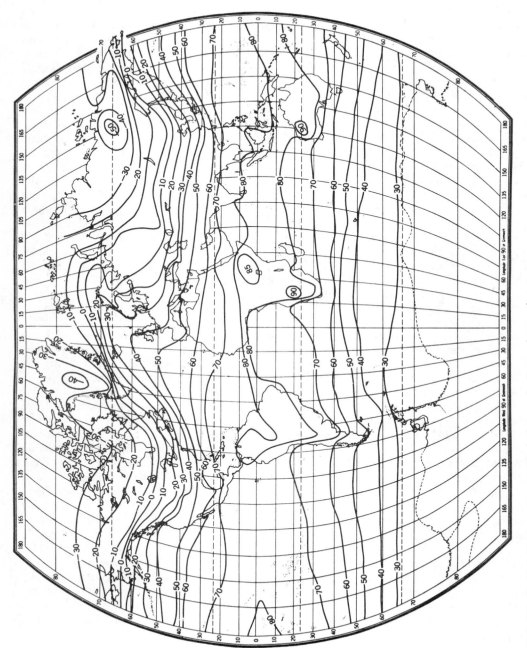

FIG. 3.5. World distribution of temperature in January. Isotherm values are in Fahrenheit degrees. (Courtesy of A. J. Nystrom & Co., Chicago.)

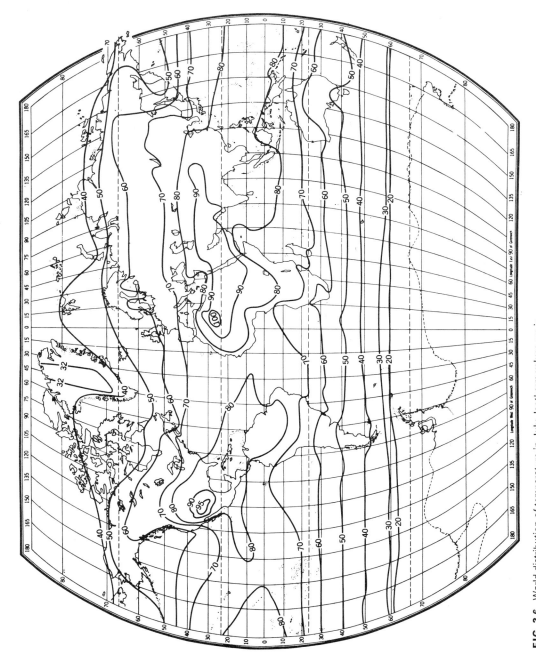

FIG. 3.6. World distribution of temperature in July. Isotherm values are in Fahrenheit degrees. (Courtesy of A. J. Nystrom & Co., Chicago.)

latitude. These belts, sometimes called "the horse latitudes," are characterized by descending and warming air. Winds in these belts are variable, and clear, dry weather prevails. The trades, blowing equatorward from the subtropical highs, are known as the steadiest winds on earth, especially when they blow over water surfaces. Low-pressure cells in the vicinity of 55° to 65° north and south latitude are known as the subpolar lows. These are areas of storminess, particularly in the winter season.

The cyclonic westerly winds, originating on the poleward margins of the subtropical highs, flow toward these low-pressure zones. The cyclonic westerly winds in the southern hemisphere are known by seamen as the "roaring forties," since no great landmasses disrupt their movement and they are strong and persistent. Landmasses in the northern hemisphere cause the cyclonic westerlies to be more variable, both in strength and direction, than their southern-hemisphere counterparts. The north and south polar ice caps are source-areas for easterly winds that blow equatorward toward the subpolar lows.

A convergence of winds occurs locally in the low-pressure belts—the trades in the equatorial low and the cyclonic westerlies and polar easterlies in the subpolar lows. These meeting places of the winds are characterized by storms and capricious weather. Where the surface winds are flowing equatorward out of high-pressure belts, particularly in the subtropical high, air descends from aloft, giving a warming and drying effect and usually producing fair weather.

The entire pressure and wind system is extremely complex. The belted pressure pattern of the northern hemisphere is broken into a series of separate high- and low-pressure centers, due to the powerful influences of the great landmasses of North America and Asia and their associated ocean bodies. A second highly significant factor is that all the winds and pressure belts shift seasonally with the apparent seasonal motion of the vertical rays of the sun.

Atmospheric moisture. Heat transforms vast quantities of the earth's water into vapor, which is present in varying amounts in all parts of the lower atmosphere. Living things, especially plants, also contribute to atmospheric moisture by transpiration. This water vapor later condenses and returns to the earth through precipitation as rain, snow, hail, or sleet. There is a constant exchange between the atmosphere and the earth's water supply. The land receives water by condensation and precipitation from the atmosphere, the sea by precipitation and runoff, and the atmosphere receives water vapor by evaporation from the sea and the land and through transpiration by plants. Heat is required to evaporate water. The heat is not lost but locked within the water vapor as latent heat of condensation.

The air's capacity for water vapor depends largely upon its temperature; as temperature decreases, capacity also decreases—and vice versa. When air can hold no more water vapor at existing temperature and pressure, it is said to be saturated. The amount of water vapor actually present in the air compared to the amount it could hold if saturated at that same temperature and pressure is referred to as *relative humidity*. The ratio is expressed in percentages. As air is cooled it eventually reaches the saturation temperature, or "dew point." When air is cooled below the dew point, it releases excess water vapor by condensation. On the earth's surface, condensation takes place on vegetation, sidewalks, and other such objects in the form of dew. If the temperature is below 32° Fahrenheit, sublimation produces frost. In the atmosphere, condensation takes place around hygroscopic nuclei—minute particles in the air that have an affinity for water. Hygroscopic particles are salts from ocean spray and chemical compounds from industrial and domestic smoke. This condensation is visible as clouds and fog. During the process of condensation, the latent heat stored in the water vapor is released.

Clouds may form pleasing or ominous, as

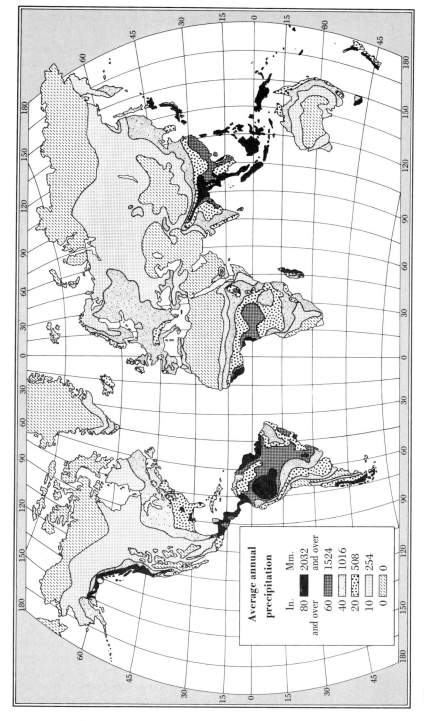

FIG. 3.7. Average annual precipitation. (U.S. Department of Agriculture.)

well as telltale, patterns in the sky. To the weather observer, clouds are clues to the action in the atmosphere and bases for weather forecasting. Certain types are hazardous to aircraft and navigation, and others may have a direct effect on such industry as electric-power producers. The presence of a cloud cover does not necessarily mean precipitation; clouds indicate potential, more than actual, precipitation. The actual precipitation from clouds takes place when many minute particles of water or ice crystals coalesce to form a mass too heavy to remain suspended. Rain, snow, sleet, or hail then falls to replenish the earth's supply of moisture lost through evaporation, and the cycle is completed (Figure 3.7).

Precipitation mechanics may be divided into three categories: *convectional, orographic,* and *cyclonic.* Convectional precipitation (Figure 3.8) is the result of the interaction of a set of physical laws—the earth absorbs and is warmed by incoming solar radiation. The humid surface air becomes heated, causing a decrease in air pressure. The warm air, expanding and rising, is further lifted by an influx of cooler air into the low-pressure area. The warm air, filled with a high water vapor content, ascends and cools, and the vapor condenses in the upper altitudes, forming numerous billowy cumulus and cumulonimbus clouds with anvil-shaped crowns. The maximum of heating is usually in the early afternoon; at this time the cycle of convection is often completed by the release from the cumulus clouds of torrential rainstorms, often accompanied by thunder and lightning. Hail is also a form associated with convective precipitation.

Orographic precipitation (Figure 3.9) is due to the rising and cooling of air because of topographic barriers such as mountain ranges and plateau escarpments. The slope in the path of the winds is known as the windward side, and warm, moist air forced up the slope cools and produces some of the greatest annual precipitation totals in the world. The leeward, or protected, side of the barrier where the air is descending and warming is much drier and is called the dry (or rain) shadow. The amount of precipitation received on the dry-shadow side is dependent upon the altitude and the continuity of the mountain barrier. High, continuous mountains reduce the precipitation to arid proportions in the lee areas.

Cyclonic or *frontal* precipitation (Figure 3.10) has its greatest development in the middle latitudes, which are the scenes of conflict between polar and tropical air masses. The boundary zone between these contrasting air masses is known as the polar front and represents the meeting place of cold air from the poles and the warmer air masses from the tropics. Along this zone of contact, waves are formed that develop into great eastward-moving eddies of air. Air

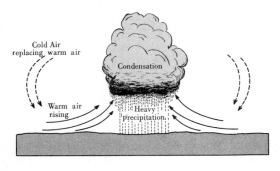

FIG. 3.8. The mechanics of convectional precipitation.

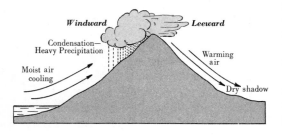

FIG. 3.9. The mechanics of orographic precipitation.

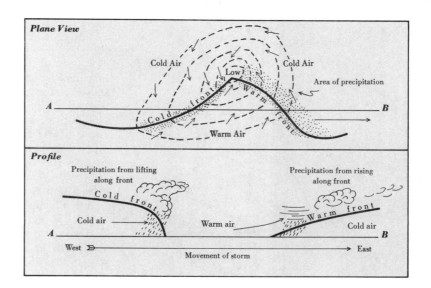

FIG. 3.10. The cyclonic storm.

whirls converging into a low-pressure center are known as cyclones or "lows." Anticyclones or "highs" develop when relatively dry stable air dominates and winds blow out from the high-pressure center. Highs bring cold, clear weather in winter or hot, dry conditions in summer, whereas lows are responsible for cloudy, wet weather. Cyclones and anticyclones vary in size, but they usually cover thousands of square miles. The storms move in a general eastward direction across North America and Eurasia at speeds between 20 and 30 miles per hour. A winter journey of a low across the United States takes from three to five days. Cyclonic activity is most prevalent in winter when the sun is more effective in the opposite hemisphere and the polar front has moved south into the middle latitudes. High velocity winds in the upper troposphere (the first layer of the atmosphere), occurring in the zones of the westerlies and known as jet streams, are believed to act as steering devices for cyclones and anticyclones. Jet streams are well developed and have their greatest velocities during the winter season.

When air masses of different temperatures, pressures, and moisture contents meet they do not usually mix but develop boundaries, or zones of discontinuity called fronts. Usually the boundary moves along the surface of the earth as one air mass moves and another replaces it. If colder air is replacing warmer air, the advancing edge is termed a cold front; if warmer air replaces colder air, it is termed a warm front. No two fronts are alike in all details, but major characteristics of cold fronts or warm fronts are sufficiently similar for helpful generalizations. When cold air advances, friction retards the movement of the surface air but does not interfere with the cold air aloft. It advances with a steep or blunt front (the cold front), displacing the warmer air ahead and forcing it to rise quickly, often causing heavy precipitation and sudden change in wind direction. The activity is concentrated in a rather narrow belt, 50 to 100 miles in width but extending hundreds of miles in length, usually in a northeast-southwest direction. The cold front moves at the average rate of 20 to 35 miles an hour. After the passing of a cold front, the weather is clear and the air cooler and drier.

When a warm front advances, the warm, humid air rises gently up over the wedge of colder air lying ahead of it. In the lifting process its temperature is lowered, condensation occurs, low clouds form, and drizzle begins. As the warm air continues up the slope, widespread rain develops, and the

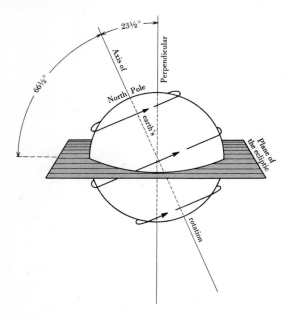

FIG. 3.11. The inclination of the earth's axis to the plane of the ecliptic.

help answer such questions as: Why is there snow on the equator? Why does the northwest coast of North America receive such heavy precipitation? Why is the St. Lawrence River closed to navigation each winter when the ports of northern Norway, many degrees farther north in latitude, remain ice-free? Why are there great deserts in Africa and Asia?

Of all climatic controls, latitude, or sun, control is the most significant. Sun control is related to seasonality, changing lengths of day, and the intensity and duration of solar radiation. All these in turn have far-reaching effects on the physical and cultural phenomena of the earth. Basic to an understanding of sun control is a study of earth motions and positions in relation to the sun.

The earth's eastward rotation on an axis accounts for day and night in each 24-hour period. A second movement is the revolution of the earth around the sun in a slightly elliptical orbit each $365\frac{1}{4}$ days. The earth's axis has an inclination of $23\frac{1}{2}°$ from a vertical to the plane of the ecliptic, an imaginary plane passed through the sun and extended through all points of the earth's orbit (see Figure 3.11). The angle of the axis remains constant (this is called parallelism of the earth's axis). The combination of rotation, revolution, inclination, and parallelism accounts for differences in the distribution of solar energy over the earth, lengths of day and night, and the seasons.

The sequence of the seasons follows the revolution of the earth around the sun. In the following discussion, the earth's orbital journey and its position in respect to the sun is described with reference to the northern hemisphere. A study of Figure 3.12 will be helpful in interpreting the description. March 20 or 21, the vernal or spring equinox, is used as the starting point.[1] On this date the

cloud cover becomes more complete, with high clouds extending as much as 500 miles in advance of the actual front. Warm fronts generally move more slowly than cold fronts; consequently, warm fronts are frequently overtaken by cold fronts.

Whereas isotherms are used to show the distribution of temperature, a similar device, the *isohyet,* or line of equal precipitation, is used to show areal distribution of precipitation on the world map. The isohyets may be drawn, for example, through places having the same annual average precipitation. Notice that in general the precipitation is heaviest in the moist, tropical, equatorial areas and least in the colder, higher latitudes where the air has a small capacity for moisture.

Climate controls. The climate of a particular place is the result of complex factors which influence the atmosphere. These factors, known as climate controls,

[1] The difference in dates occurs because a calendar year has 365 days, but $365\frac{1}{4}$ days are required to complete a revolution. (The quarter-days are accumulated and added as an extra day every fourth year.)

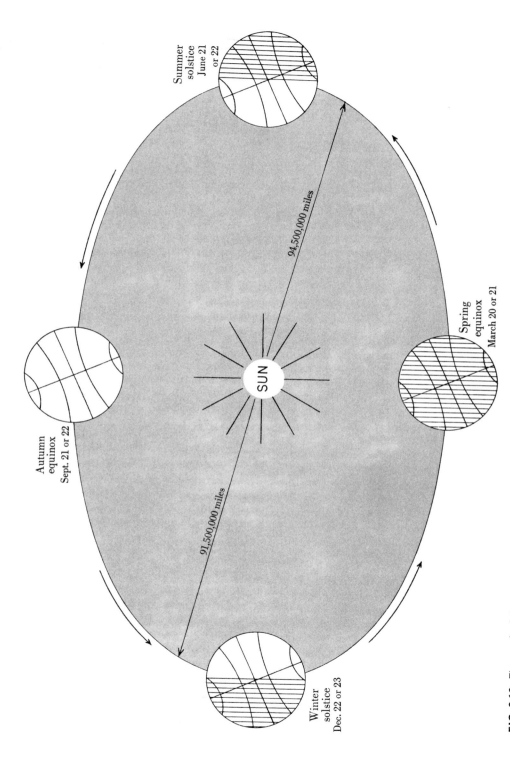

FIG. 3.12. The march of the seasons.

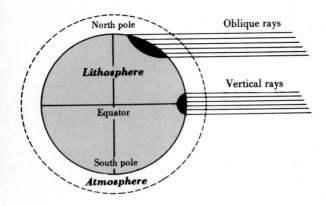

FIG. 3.13. Vertical and oblique rays of the sun.

noonday sun is directly overhead at the equator and the length of day and night is equal for all parts of the earth. This gives rise to the term equinox—equal day and night. The earth has completed one-fourth of its revolution by June 21 or 22, the summer solstice and longest day of the year. The sun is now directly overhead at noon on the Tropic of Cancer ($23\frac{1}{2}°$N.). The entire area within the Arctic Circle experiences 24 hours of daylight. The next quarter-turn finds the earth in a position (on September 22 or 23) in which the sun is again directly over the equator—the autumnal equinox. Day and night are again equal all over the earth. During the six-month period (March to September) the North Pole receives continuous sunlight. When three-fourths of the journey has been completed (on December 22 or 23, the winter solstice and the shortest day of the year), the sun is directly overhead at the Tropic of Capricorn ($23\frac{1}{2}°$S.). At this time the entire area south of the Antarctic Circle ($66\frac{1}{2}°$S.) is bathed in light and experiences 24 hours of continuous daylight, and the sun is no longer visible north of the Arctic Circle ($66\frac{1}{2}°$N.). During this six-month period (September to March) the South Pole has continuous sunlight, while at the North Pole the sun is completely out of sight. The annual revolution is completed on March 21 or 22. The vertical rays of the sun, during the circuit, appear to be migrating north and south between the Tropic of Cancer and the Tropic of Capricorn, the two extreme positions of the sun's vertical rays. Days begin to lengthen from the time of the winter solstice to the summer solstice date; conversely, days shorten from June 21 or 22 to December 21 or 22. Only in areas between the two tropic lines are days and nights of almost equal duration throughout the year. The northern hemisphere's spring and summer occur in the middle and higher latitudes when the sun's vertical rays are north of the equator. Areas between the tropic lines, constantly under the influence of the overhead sun, do not develop these temperature fluctuations (winter and summer). The seasons are reversed in the southern hemisphere.

The amount of solar energy, or *insolation*, received by any portion of the earth's surface is chiefly dependent upon the hours of sunlight and the angle of the sun's rays to the surface. Between the Tropics of Cancer and Capricorn the sun's near-vertical rays provide direct heating. Poleward of the tropic lines the angle of the rays with the earth's surface becomes increasingly oblique. Oblique rays are less effective than direct rays, since heat energy is spread over a larger surface and since the rays must also pass through a thicker layer of absorbing atmosphere (see Figure 3.13). Thus the earth-heating effect of a middle-latitude winter sun is less than that of a summer sun.

Light and heat are two determining factors in plant growth. The number of frost-free days and the amount of heat accumulated during the frostless period are important considerations in agricultural production. Although many plants are able to withstand short cold snaps and some are able to winter through prolonged subfreezing weather, damage to plants from cold is still the chief hazard throughout many agricultural areas. The tropical regions, except for high elevations, enjoy a continuous frost-free season with no interruptions in the periods of growth. Frost generally becomes increasingly more prevalent toward the poles, culminating in the perpetual frost of the polar regions. Areas in high latitudes having short frost-free periods or growing seasons are somewhat compensated in summer by 17 to 18 or more hours of sunlight each day.

The differential heating of land and water is another important climatic control. Land and water surfaces may intercept an equal amount of insolation yet have unequal rates of heating and cooling. Since water is translucent, solar heat penetrates to considerable depths; the circulation of water, vertical and horizontal, is extremely important in governing the distribution of energy through a large mass. Water requires about five times more heat to raise its temperature one degree Fahrenheit than does relatively dry land. Water not only requires more time to heat than land but, due to the greater volume of stored heat, takes longer to cool. Land is opaque and is not subject to tides, currents, and similar movements; consequently, its heating is concentrated in a small surface volume. The consequence of this differential heating and cooling is exemplified by the fact that when it is continental or land-controlled, climate is characterized by large seasonal, as well as daily, extremes of temperature, whereas marine or ocean-controlled climate is more moderate.

Winds of local significance are a result of the differential heating of land and water and the consequent pressure differences. A common phenomenon associated with coastal areas is the land and sea breeze. On clear summer days both land and water receive equal amounts of incoming solar radiation. The land, however, heats faster and develops a lower pressure. As a result, there is a circulation of cooler air from the water to the land called the sea breeze, with the

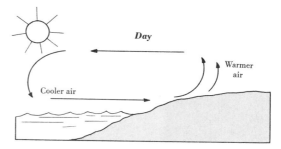

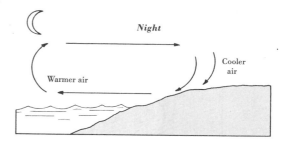

FIG. 3.14. The sea and land breezes.

effectiveness greatest in the afternoon. Sea breezes seldom penetrate the interior for more than thirty miles. A reverse situation develops at night when the more rapidly cooling land develops high-pressure characteristics in contrast to the slower cooling water with its consequent lower pressure. The flow of air is now toward the sea and is known as the land breeze (see Figure 3.14). The sea breeze is of special importance to tropical areas and brings welcome relief from the hot, humid weather.

In addition to the land sea breezes other localized winds include the diurnal movement of air up and down mountain and valley slopes, the mountain and valley breezes. Another *katabatic wind* is caused by cool air draining off high-plateau or mountain areas.

Continental winds developed through the principle of differential heating of land and

FIG. 3.15. The pattern of world ocean currents.
——— warm currents
- - - - - cool currents

water and subsequent pressure difference are the monsoons.[2] These winds are associated with large land and water masses. As the land becomes heated by the summer sun low pressure develops, and moisture-laden air moves in from the sea. Associated with the summer monsoon are cloudy weather and heavy rainfall. When the landmass cools in winter and high pressures prevail, the air movement is reversed and flow is toward the low pressures of the sea. Associated with the winter monsoon is dry, clear weather. The best example of monsoon winds is found in south and east Asia.

Winds are a third major climatic control because of their direct and indirect affect on temperature and precipitation. Their two most important functions are (1) the movement of heated air from low latitudes to higher latitudes—without this circulation there would be a constant increase of temperature in the low latitudes and a decrease in the middle and high latitudes, and (2) the transport of water vapor over land, where it may condense and replenish the earth's water needs. Both local and regional changes in temperature are associated with winds. In the northern hemisphere a south wind usually means a rise in temperatures. Winds sweep into the air salt particles that serve as nuclei for the condensation of water vapor. The direction and movement of ocean currents and drifts are also affected by winds.

Ocean currents are a significant climate control. The world system of ocean currents transfers warm water from low to higher latitudes; conversely they import cold water from the polar areas (see Figure 3.15). The warm waters of the North Atlantic Drift make winter temperatures in countries of northwest Europe remarkable mild for the latitude. Cold water paralleling tropical coasts not only cools these areas but causes aridity. Thick fogs are associated with converging cold and warm currents, such as those found along the northeast coasts of North America and Asia.

Other controls are the cyclonic storms. (discussed in the preceding section on precipitation), which are of particular significance in the middle latitudes. Mountains have a decided effect on the amount of precipitation received on lee- and windward sides and act as barriers to the movement of cold air. The Rocky Mountains of the United States are effective in restricting the westward invasion of polar air from Canada. Locally, temperatures are modified by the orientation of mountain slopes. South-facing slopes (in the northern hemisphere) receive considerably more insolation than north-facing slopes. Another important control is the decrease in temperature with elevation. The average rate of decrease is about 3.3° per thousand feet of altitude. A visible evidence of this control is the permanent snow on high mountain peaks located on the equator.

Lithosphere. Geographic interest in the subsurface centers around several possible kinds of relations. The nature of the physical composition of the bedrock and its resistance to weathering and erosion have direct bearing upon the surface forms of the land and the nature of soils. These characteristics, as well as the depth of the solid rock from the surface, may have influence upon the kinds of buildings that can be constructed on the surface and upon the cost of providing adequate foundations; in a similar manner, bedrock influences the construction of transportation lines, since it affects the ease and cost of establishing satisfactory grades, weight-supporting standards, and the accessibility of ballast materials. The arrangement, slope, composition, porosity, permeability, and depth of bedrock strata may have bearing upon the quantity and quality of underground water and the cost of bringing it to

---

[2] The mechanics of the Monsoon, particularly the Indian, are much more complicated. Current research indicates the influence of the jet stream located immediately south of the Himalayan wall.

the surface of the earth. The concentration of minerals useful to man is a prime concern of economic geography with the subsurface of the earth. The utility of a given mineral deposit depends upon such physical characteristics as type of mineral, quality, form, depth, and other aspects of accessibility, and upon such human characteristics as need or market, capital, technical ability, and ease of assembly of other minerals.

The earth's crust is composed of chemical elements. Out of more than ninety possible elements, only eight make up about 98 percent of the weight—oxygen, 45.60; silicon, 27.72; aluminum, 8.13; iron, 5.00; calcium, 3.63; sodium, 2.83; potassium, 2.59; and magnesium, 2.09. About a dozen more comprise the bulk of the remaining 2 percent. The elements form minerals; geologists define a mineral as "a naturally occurring inorganic substance with a characteristic internal structure and with chemical composition and physical properties that are either uniform or variable within definite limits." Most are compounds of two or more elements, but a few are single elements —such as native gold, graphite, and sulfur. Coal and petroleum are of organic origin. About two thousand minerals have been identified, but most are rare. The earth's crust is composed chiefly of nine, and these (common rock-forming minerals) exist in physical mixtures known as rocks. There are many possible combinations, thus many kinds of rocks.

For our purpose, three general groups of rocks, based upon mode of origin, may be recognized: igneous, sedimentary, and metamorphic. Igneous rock, the chief component of the earth's crust, has been formed from the solidification of molten matter that originated within the earth and pushed upward toward or onto the surface. Two subdivisions are recognized: (1) intrusive igneous rocks, which solidified beneath the surface and are generally completely crystallized and coarse-grained (for example, granite); and (2) extrusive igneous rocks, which were formed by flows and/or ejections from vents. The quiet flows produce lavas, of which the most common is dark-colored, fine-grained basalt; and explosive processes frequently produce fragmental rocks such as tuff and breccia.

Most sedimentary rocks are formed from the weathered waste of older rocks. Down through the earth's history, weathering processes have tended to break up the surface rocks. Action of gravity, water, ice, and wind may move fragmented materials from places of origin. These materials may be laid down on the floors or shores of water bodies and later be compacted by pressures of increasing weight and/or the deposition of some cementing agent. Organic materials are sometimes compacted with the sediments or sometimes separately. Some rocks, such as rock salt and gypsum, are evaporites from saltwater bodies. Sedimentary rocks range from fine-grained limestones and shale, through medium-textured sandstone, to coarse, irregular conglomerate that resembles concrete. Sedimentary rocks commonly exist in layered arrangements; this is known as *stratification*.

Metamorphic rocks are made by the transformation of previous igneous or sedimentary rocks through changes in either mineral composition or structure, or both. The common causes of transformation are high pressure, high temperature, and hot fluids within the earth. Every common igneous and sedimentary rock has a metamorphic equivalent: for example, metamorphized sandstone becomes quartzite, shale becomes slate, and limestone becomes marble. The metamorphic rocks have the least surface distribution of the three groups.

Igneous rocks originally formed the bulk of the earth's surface (upon solidification). Today they are the most common class a few hundred feet below the surface and constitute virtually all the rock below about 1500 feet. In addition, surface outcrops are present in many areas of the world. Sedimentary rocks and loose sediments, however, immediately underlie approximately three-fourths of the area of the continents and much of the ocean floors.

This section should not omit a discussion of a few common relationships between rock types and other elements of the physical system. Since rocks vary in their resistance to weathering and erosion, in humid lands soft rocks such as shale (the most common of the sedimentary class) commonly give rise to gentle slopes, plains, and subdued relief features. Those such as granite, being more resistant to deformation, weathering, and erosion, tend to produce more enduring and bold landforms. Underground water is most easily accumulated and carried in sedimentary rocks, a layer often being an aquifer (water bearer). Lignite, bituminous coal, petroleum, natural gas, lime, and salts are associated with sedimentary rocks; and anthracite coal, slate, and marble are associated with metamorphic rocks. Metallic minerals, because they may be concentrated by means of a variety of processes, may be associated with any of the three. The relationships between rocks and soils will be brought out later.

Landforms. The irregularities that characterize the surface forms of the land are significantly related to other elements of the physical system as well as to the distribution of people and the character and pattern of human activities. Two opposing groups of forces—tectonic forces, which originate within the earth, and gradational forces, which originate outside the earth—are constantly at work changing the surface, producing landforms, and altering their shapes. The tectonic forces, deriving energy from within the earth's interior, tend to build up the relief of the land. Variations in elevation are continuously being produced by forces that bend, warp, and break the earth's surface, depressing some portions and elevating others. These same results have been accomplished in some areas by the extrusion of molten material or lava onto the surface. The gradational forces, in direct opposition to the tectonic ones, are working to bring the earth down to a uniform level by the processes of degradation, which tear down the elevations, and aggradation, which fills the depressions; this change is accomplished through the forces of gravity, water, wind, and ice. The surface irregularities resulting from these processes appear in great variety and in countless combinations—ridges, peaks, valleys, plateaus, canyons, hills, knolls, flat plains, rolling plains, floodplains, and so on. Yet it is possible, on the basis of broad similarities, to generalize and classify them to bring order and understanding to the world patterns and distributions.

The earth has a surface area of nearly 200 million square miles, of which approximately 71 percent is covered with oceans. More than three-fourths of the land is contained in the large masses of Eurasia, Africa, and the Americas. The northern hemisphere contains the dominant share and for this reason is sometimes called the land hemisphere. Geographers classify the major forms, which give broad order variety to the surface of the land masses, into four groups: plains, plateaus, hills, and mountains. The bases for this classification are: (1) elevation, or height above the sea level; (2) local relief, or the difference between the highest and lowest elevations within limited areas; (3) degree of slope; and (4) configuration, or the proportions and relationships of sloping land to level land.

Plains are generally, but not always, areas of low elevation. The principal criteria for setting plains apart from other landforms are that they be distinguished by local relief of a few hundred feet or less and that they have a predominance of relatively smooth surfaces. Thus, the Great Plains of North America, with elevations of more than 5000 feet, and the Atlantic Coastal Plain, for the most part under 500 feet, both fit the classification. Actually there are many kinds

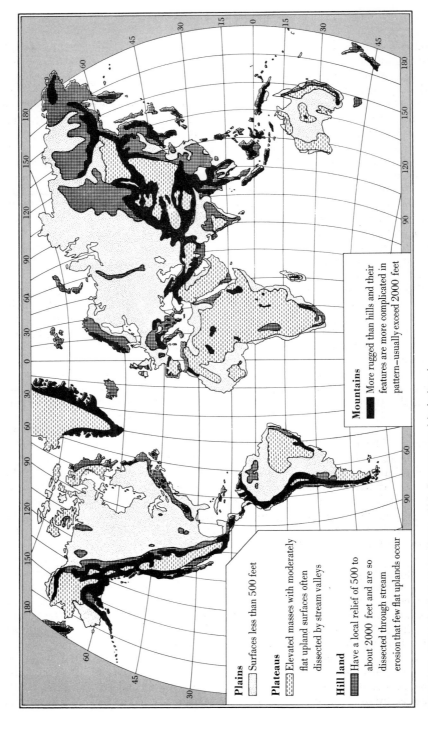

FIG. 3.16. Principal classes of landforms. (U.S. Department of Agriculture.)

TABLE 3.1
Landforms of the continents

	Plains	Plateaus	Hills	Mountains
Asia	32	23	23	20
Europe	67	8	21	4
Africa	25	71	1	3
North America	52	24	11	13
South America	56	25	9	11
Oceania	48	24	19	9
Total*	41	33	14	12

* Excluding Antarctica

and sizes, ranging from small floodplains that have been formed by the flood action of streams to large expanses (such as the Atlantic Coastal Plain) which, within the limitations of a few hundred feet of local relief, display considerable variety in the shapes or forms of the details of the surface.

Plateaus are similar to plains in that their surfaces are predominantly smooth to slightly irregular. They differ from plains, however, in that they lie at higher elevations and their local relief is considerably greater. Often streams have cut deep canyons, and small features rise above the general surface. Plateaus commonly are bordered on at least one side by steep slopes. The terms "hills" and "mountains" are used to identify landforms that are characterized by a high proportion of their areas being in appreciable slopes. They are divided on the basis of elevation and local relief. The "hill" classification generally is restricted to areas of moderate elevation with local relief ranging from about 500 to 2000 feet; whereas "mountain" is applied to massive areas of higher elevation with local relief exceeding 2000 feet and perceptible altitudinal zonation in temperature and vegetation. In local parlance, however, identification is often influenced by relativity: for example, the plains people may refer to a slight rise as a hill and name it.

The distribution of the four main groups is shown in Figure 3.16. The percent of occurrence by continents has been approximated as shown in Table 3.1.[3]

A comparison of Figures 2.2 and 3.16 will reveal a close correlation between the distribution of plains and the distribution of world population. Although there are many exceptions because other elements of the physical system may be adverse, man tends to prefer plains as dwelling places. Where people are found in the highlands they usually choose the valleys, basins, and other areas of relatively smooth surface. Most of the major population supporting economic activities—and/or resource-converting and space-adjusting techniques—are more readily adapted to the more gentle terrain. Transportation facilities and other linkages, settlements, farming, and many other facets of land organization are favored where extremes of relief are absent.

[3] Frank A. Pearson and Floyd A. Harper, *The World's Hunger* (Ithaca, N.Y.: Cornell University Press, 1945), Table 16.

There are relationships between landforms and other elements of environment (as well as cultural relationships) that may draw man to the highlands. Landforms function as a noteworthy climatic control. Exposed highlands frequently receive higher precipitation totals than neighboring lowlands. Within the highlands the *aspect* of slopes has marked bearing upon temperature: For example, in the northern hemisphere southern exposures receive more sunshine than do northern slopes. Locally, lower valley slopes may be seasonally more frost-proof than neighboring valley floors, because cold air, like water, tends to flow downslope and settle in low places. Slope influences soil accumulation: The forces of gravity, running water, and moving ice cause down-slope movement of particles. Thus in many highland localities soils are shallow and in some places completely absent. There is also a relationship between highlands and minerals. It is in these areas that weakness in the earth's crust has allowed magmas, formed in the depths, to intrude near the surface, later to be exposed by weathering and erosion. These magmas in many cases have brought minerals in sufficient concentration to be classed as ores that can be profitably mined.

These few illustrations of environmental interrelationships are sufficient to lend understanding to the following generalizations. Highlands in various instances can be more attractive than lowlands for settlement because of more moderate temperatures (in the tropics) or more favorable precipitation (in the drylands). They are important to lowlands for water supply, nurturing many streams and commonly sending their snowmelt to the lowlands during the warm seasons. Highlands favor water control with deep valleys for dam and reservoir construction and gradients for hydroelectricity development. They are the major forested areas of the middle latitudes, because of limited competition from agriculture. Some provide significant grazing grounds. Temperature, scenery, and snow have allowed some highland areas to become significant recreation sites. Also, mountain areas of the world are significant sources of metallic minerals.

Vegetation. Natural vegetation is intimately related to other elements of environment. All life on the earth depends upon energy from the sun. Chlorophyll-bearing organisms—green and purple bacteria, blue-green algae, phytoplankton, and the vast array of higher plants—utilize solar energy for photosynthetic reduction of carbon dioxide from the atmosphere and soil water to form organic compounds and molecular oxygen. Of the requirements for life only solar energy is supplied by a continuous extraterrestrial source. Other needed substances such as carbon, nitrogen, potassium, phosphorus, sulfur, water, etc., must continue to cycle in the natural system on the earth.

Vegetation responds especially to the detailed characteristics of climate, soils, and drainage and vegetation in turn influences the nature of soils, soil moisture, and erosion. Vegetation also is the primary level of food supply for the complex web of life on the earth which is bound together in intricate food chains.

In response to differences in the related elements of the physical system, the character of vegetation differs significantly from one part of the earth to another. Under some conditions it is comprised of stands of merchantable timber, under others of vast expanses of grass, and under still others of desert shrub or tundra that have little utility. In general, trees require more moisture than grasses. Hence the more humid parts of the earth tend to be forested; and subhumind, semiarid, and long-season drought parts tend to be grass-covered. There are, of course, many exceptions—some grasses can tolerate excessive water and persistent low temperatures better than trees, and poorly drained or relatively cold humid areas may be grass-covered.

In general, the vegetation of drylands either tolerates the vagaries of low and erratic precipitation or has facility for evading the dry periods. Seed-producing annuals may hasten through their entire life cycle in a few weeks after a rainstorm and the moistening of the soils, following which the seeds may lie in a dormant state for many months or even several years until the next rain. The most apparent vegetation types of the drylands are shrubs that are physiologically equipped to withstand drought (xerophytes), with special features for accumulating and storing moisture and/or resisting transpiration such as widespread root systems, minimum leaf surfaces, spongy interiors of stems, and so on. Desert shrubs are able to survive by the ability to become dormant before wilting occurs; many microflora, such as lichens and mosses, have similar properties. The succulents, such as cacti, are able to survive on the basis of water storage in their tissues, widespread root systems, minimum leaf surfaces, spongy interiors of stems, and so on. Some perennial grasses are also found in the drylands, especialy in those portions with a regular season of precipitation but low totals, surviving through their ability to become dormant.

Although variations occur in species associations, density, and size, it is possible to group and map at world scale the vegetation in three general classes: forest and woodlands, grasslands, and shrubs (see Figure 3.17). Each of these is a response, in the main, to interrelated conditions of temperature and moisture; and, in turn, each of the classes shows both temperature and moisture-gradient responses. For example, the northernmost forests, just south of the tundra, are characterized by evergreen conifers of low species diversity; often one or two species of spruce and fir form pure stands. In the warmer, moist middle latitudes deciduous forests with greater stratification and diversity are characteristic. Pines may be found in both forests, commonly in a seral or developmental stage. A third forest occurs in the tropics, ranging from broad-leaved evergreens in the portions that have well-distributed rainfall to deciduous broad-leafs where there is seasonal drought. The tropical rain forest has the greatest species diversity of all the forests. The temperate rain forest of the Pacific Northwest coastal area of the United States and the chaparral woodlands of California and maquis woodlands of the Mediterranean Basin may be cited as examples of moisture-gradient forests, representing the opposite ends of the scale. In all the forests, variations in micro environment, including soils, drainage, and climate, may produce vegetation differing from the general pattern. Altitudinal zonation is discussed in the Tropical Highlands chapter.

Similar temperature and moisture responses occur in the other two vegetation classes. Grasslands vary from the tall, dense savannas of the tropics and prairies of the middle latitudes to steppes of shorter and more sparse grasses in both zones. The complexes formed in the shrublands vary with the different temperature conditions of the cold deserts (tundra), the seasonally cold deserts, and the continuously warm and hot deserts.

Native animals. The nature of the animal life of the earth varies from area to area in accordance with the tolerance of species and variances in food supply, cover, water, and human interference. In some ways the distributional patterns are more complex than other facets of the physical-biotic system. Animals are mobile and thus not so likely to have fixed ties; migrations are a natural feature of the life patterns of many. Insects are the most abundant category of the earth's animal life. Although some are useful, man is constantly at work to control those that spread disease, attack crops, and cause human physical discomfort. Mammals, fish, birds, reptiles, and amphibians are higher orders of animal life. Of these, the various species of fish are the most valuable. Some primitive peoples still depend on native animals for subsistence. In man's commercial economy, fish and fur-bearing animals are the most important; but in a number of countries game animals for meat supply are significant.

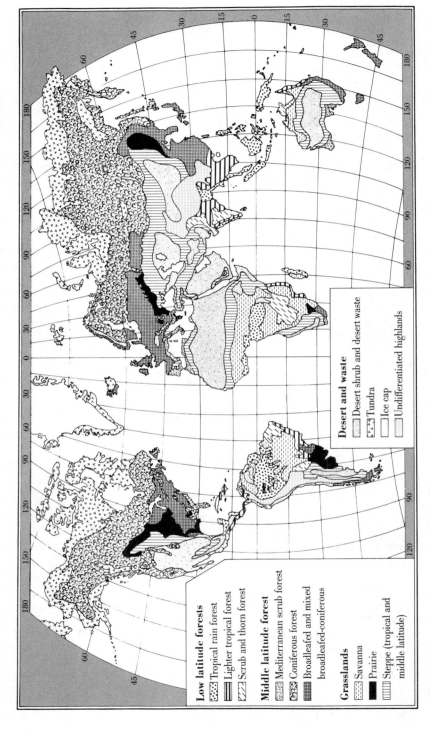

FIG. 3.17. Natural vegetation. (U.S. Department of Agriculture.)

The recreation, or sport, facet of the man-animal relation is likewise of importance in many areas. In regions of relatively dense population, man is hardly cognizant of the fauna in his physical environment; but in world areas less exploited by man, native animal life still forms a vital, colorful part of the natural landscape.

In general terms it may be stated that the greatest varieties and densities of native animals occur where the natural conditions are most favorable and the imprint of man is least in evidence. For example, the savannas or tropical grasslands of Africa are famous for their varied and colorful big-game animals as well as for their myriad rodent and insect populations. In these areas the pyramid of animal life is well illustrated in many complex food chains: the grazing food chain which includes in stepwise progression grass, herbivora, and carnivora; and the decay food chain whereby organic debris may be totally consumed by bacteria, fungi, and small animals of decay that release carbon dioxide, water, and heat which are in turn recycled in the ecological system, or organic debris or partially decayed debris may enter complex food webs involving large animals.

A fundamental ecological fact is that only a fraction of the energy entering a population is available for transfer to the population that feeds upon it without major disruption of either. The amount of energy transferred undoubtedly varies widely. It has been suggested that in the grazing chain probably 10 to 20 percent of the energy fixed by the plant community can be transferred to herbivores, and 10 to 20 percent entering the herbivore can be transferred to the first level of the carnivores, etc.[4]

[4] Woodwell, George M., "The Energy Cycle of the Biosphere," *The Biosphere* (San Francisco: W. H. Freeman and Co., Publishers, 1970) p. 29. This publication is a reprint of the September 1970 issue of *Scientific American*. The entire publication is recommended as a readily understandable source on the several subsystems and cycles of the natural environment.

In many of the lands that have been densely settled, such as parts of China, the alteration of the environment through human use has radically changed the food supply, limited cover, and interfered with the natural balance. In these areas native animals, unless protected, have been reduced to successional types, mainly rodents and other small, wily species. In the cold and drylands species diversity is naturally low; however, especially in the cold regions, the existing animal life has been an important base for human support. Concern for dwindling native animal populations has stimulated active protection and redevelopment programs in many regions of the world. These have taken the form of refuges, hunting and fishing regulations, artificial propagation (especially of fish), and integrated planning for land-use–wildlife development. In many countries active informational programs have been developed to foster favorable public awareness.

Soils. As an environmental element, soil is of significance primarily because it has an influence upon the actual and potential plant growth of a locale. Since soil is the upper, biologically molded portion of the earth's surface, it is subject over a period of time to the action of varying combinations and sequences of physical and chemical forces. The principal soil-forming factors are: the nature of the bedrock or parent material; the characteristics of the climate, which influence the kind and rapidity of weathering and the kinds of biota; the biota, which contribute organic material, influence nutrient cycling, and aid in mixing; the slope of the site, which influences erosion, drainage, and accumulation; and age, which relates to the length of time a particular soil has been under the influence of other forces. Man, also, may be added to this list, since cultivation, crop production, fertilization, and other management techniques have a marked bearing upon the development of agricultural soils. Over a long period of time cultivated soils become significantly different; culturally

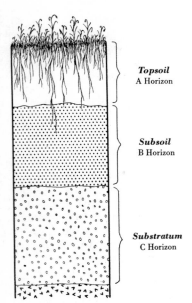

FIG. 3.18. Soil profile. A mature soil with a smooth topography is a natural active body, having characteristics different from the parent material. A vertical cut through a mature soil displays a profile with distinct layers called horizons. The A Horizon contains weathered solid, mineral material, and a certain amount of organic matter in various stages of decomposition. Some materials have been carried by leaching into the B horizon, which is also weathered but contains little or no organic matter and is therefore lighter in color. There are accumulations of iron, aluminum, and fine clays leached from the surface layers. In arid regions there are also accumulations of salts in this layer. In the C horizon the modified soil material of the B horizon merges with the unconsolidated parent soil material, which in turn covers the solid rock.

induced characteristics replace natural characteristics.

Variation in the soil-forming factors in turn result in variations in soils—in their depth, texture, structure, fertility (or plant foods), pH (acidity-alkalinity) rating, water permeability and holding capacity, and drainage. These variations in turn influence soil productivity, or capacity to yield crops under management. On the farm level the details of these characteristics on a field-to-field basis are important considerations and influences, and classification and evaluation systems have been devised for most of the advanced countries. Very few, however, have been completely mapped in detail.

For our general view of the world, it is possible to utilize a broader-scale classification —the great soil groups.[5] The great soil groups, based upon the general characteristics of the soil profiles (a vertical slice from the surface to the unmodified parent rock; see Figure 3.18), comprise three broad categories: azonal, zonal, and intrazonal. Azonal soils are youthful, not having developed the layered arrangement typical of mature soils, and they reflect their geologic backgrounds. Azonal soils include the lithosols of highland slopes, the fresh (alluvial) sediments deposited on land by streams, and beach and other deposits of sand.

The zonal groups are the most significant. They are mature soils that have lost their youthful geologic characteristics and have developed characteristics induced by climate and vegetation. Over broad areas of the earth, with varying bedrock but with good drainage and similarities in climate and associated vegetation, soils occur with similarities in the general characteristics of their profiles. The distribution of the major groups is shown on Figure 3.19. The relevant details of these soils will be discussed in the context

---

[5] Although soil scientists are in the process of modifying classification, new world maps are not yet available. The new classification, among other improvements, considers the impact of cultivation.

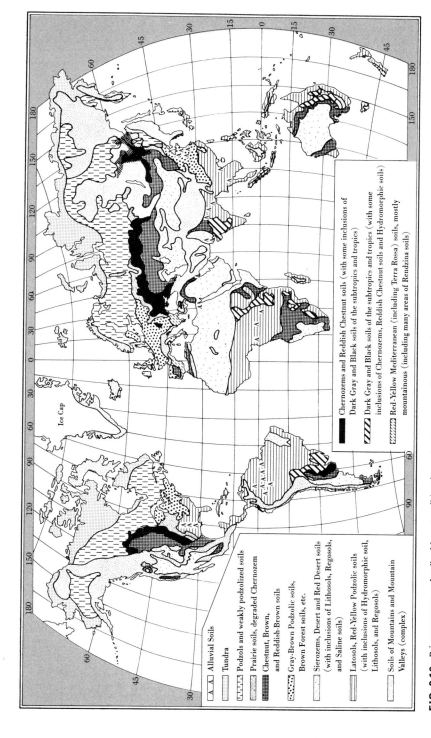

FIG. 3.19. Primary groups of soils. Many small but important areas not shown on the map occur in all parts of the world. (U.S. Department of Agriculture. Adapted from map by Dr. Kellogg.)

of our regionalization of the man-earth system.

Intrazonal soils have more or less well-developed soil characteristics that result from predomination of such influences as drainage and parent material over the normal effects of climate and associated vegetation. In humid areas, for example, bog soils form in low places of poor drainage where thick deposits of plant remains accumulate; and in arid areas saline and alkaline soils form in similar low positions where salts accumulate. In other cases dark, nonacid soils develop in humid areas where calcareous parent material remains dominant. Any one of the intrazonal soils may be associated with several zonal groups. Azonal and intrazonal soils together comprise a relatively small portion of the world's soils and are scattered through the zonal groups: Thus the generalized world map indicates only major occurrences.

It is difficult to produce a definitive rating of soils in terms of their value to man. This is true because both natural plants and cultivated crops vary in their soil requirements and also because soil management is constantly improving. Therefore, any evaluation system would demand consideration of man's needs, desires, and techniques for both groups, as well as for individual plants and crops. Although tropical red and yellow soils produce luxuriant natural vegetation and the podzols produce valuable coniferous trees, they are both relatively poor agricultural soils under present levels of resource-converting techniques. From the standpoint of general crop production, the chestnut, chernozem, prairie, gray-brown, subtropical red and yellow soils of the zonal groups, and the alluvial and rendzina soils of the intrazonal groups have the most favorable agricultural characteristics. It will be brought out later, however, that many other conditions, both natural and cultural, influence the utility of soils.

Water. Water rates a high position on the list of environmental elements important to man—it is absolutely essential to life of all kinds. And its importance and utility go far beyond physiological requirements. Technically advanced societies have developed many industrial processes and products that utilize water in large quantities. In addition, man has extended croplands and the range of crops in many areas through irrigation. He also uses water for the generation of power, for transportation, and for waste disposal; he gains food and raw materials from water bodies; and he has acquired a liking for many water-based recreation activities.

In many ways it can be said that water is the most complex of all the physical elements because its supply, characteristics, and availability are intricately interwoven with other natural elements—indeed, with the total physical-biological system. The water of the earth can be divided into two general classes: saltwater occurring predominantly in the oceans and freshwater on the continents. Man, a land-dweller, makes far greater direct use of the latter. Yet the two classes are related—freshwater has been and will be again saltwater. Through the action and energy of the sun and gravity, water moves endlessly from the sea to the land and back again. Water is evaporated wherever it occurs at the surface (primarily from the sea), is transported by the atmosphere, and is released by condensation and precipitation, some on the sea and some on the land. With its salt left behind at the point of evaporation and with its state having changed from gas back to liquid during condensation, water is in its most useful form when it reaches the land. Although man may temporarily trap some for his use, water inexorably makes its way back to the sea, partially by surface streams, partially by seepage from the groundwater reservoir where exposures occur on the continental shelf, but mostly by the same medium that brought it—the atmosphere. This restless movement, with no beginning or end, is known as the hydrologic cycle (Figure 3.20).

Three temporary natural reservoirs on the land are of significance to man: surface streams and lakes, the soil, and groundwater. The first and third may be tapped for bulk uses; the second is the basis for plant growth, both natural and cultivated. The areal differences in supply of all three categories are related to other environmental conditions. Precipitation is the keystone in the arch of water supply. In general, humid lands are more fortunate than those receiving low annual and seasonal amounts of precipitation. Yet, many other features can affect the local supplies in any of the three reservoirs. Temperatures influence both evaporation and transportation rates, as does the density of vegetation and its foliage; slope and vegetation influence runoff and soil moisture absorption; soil texture, structure, organic content, and depth influence storage capacity; and all of these factors, plus the conditions of the pore spaces and cracks in the rocks below the soilwater zone, influence the quantity in the groundwater zone. Moreover, groundwater, like streams on the surface, moves under the force of gravity. If slope exists, a permeable stratum, especially if there is an impervious stratum below it, may carry a groundwater supply a considerable distance beyond the intake source area.

Because of the various and complex influences on water in the three reservoirs, it is as yet impossible to produce a world map that would show the areal characteristics and patterns of all. Surface sources alone are readily mapped, and there are in existence relatively accurate maps for essentially all the earth. A correlation of precipitation, soil, and

FIG. 3.20. The hydrolic cycle—the restless, endless movement of water.

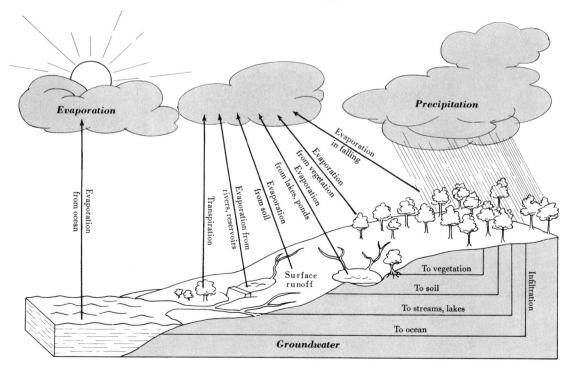

vegetation maps gives a fair evaluation of the soil-moisture reservoir. But very little work has been done in preparing groundwater maps of the world; for example, only a highly generalized groundwater map has been prepared for the United States.

PHYSICAL–BIOLOGICAL SYSTEMS— NATURAL ENVIRONMENTS

The point was made in the introduction to this chapter that the biological and physical components of the natural environment exist together in varied combinations over the earth's surface; however, these combinations are not without order. They exist in functional entities or systems in which each element is an interdependent, interacting part of the whole. These functional entities are now popularly recognized as ecosystems. From the point of view of human utility or habitat, each element can have an influence or measure of veto power over other elements within a complex. For example, some combinations produce well-watered plains with fertile soils, fruitful climate, and forest vegetation, that are underlain with useful minerals. Some produce plains with fertile soils and subhumid, short frost-free season climates, and grass vegetation; some produce well-watered mountains with poor soils and forest vegetation. Others produce dry, essentially barren mountains; some produce dry, rolling surfaces variegated by patches of sand, thin soils, and rock exposures; and others produce dry, fertile floodplains with streams close at hand. Indeed, the combinations at the local level are almost infinite.

From the point of view of viability of human occupance systems it is essential that modifications of the earth recognize and be in harmony with the realities of natural systems. Changes in one component usually affects others. Failure to monitor, moderate, adjust, and compensate these changes can impair the base of human existence.

Natural Resources, a Cultural Evaluation of Environments

The major significance of environmental characteristics and variations from the geographic point of view lies in the realities of their interrelationships with man. As noted previously, the physical environment provides not only the stage for human occupance and economic activities but also the natural resources which are the basic supports of these economic activities. Environments and societies vary throughout the earth; therefore, an understanding of both groups of variables and of relationships between the groups must precede any geographic explanations.

Let us now emphasize a theme introduced in chapter 2 and reemphasized at the beginning of this chapter, that physical environments are evaluated within a cultural frame of reference. Man is the dynamic agent of environmental interpretation and its organization and use—he differs in numbers, wants, ingenuity, techniques, and social organization both in time and space. The possibilities and the resistances offered by a particular environment, therefore, differ with social groups and with changing technology. Within this cultural concept of environmental utility, we may define natural resources as those elements and combinations of elements of nature that function in the service of man—that have human utility or human value.[6] Thus, although land may be defined in physical terms—as the surface expression of the sum of the characteristics presented by the earth's physical-biological system at a given site—the addition of the term "resource" (that is, land resources), identifies the functional utility of land to man. Thus we speak of cropland, forest land, grazing land, urban land, and so on. In the varied combinations of man and the earth any natural feature—sunshine, rock, mineral, soil, surface, water, grass, and so on—may have a value or usefulness. As implied earlier, however, in a general sense the occupants of a given environment may variously assess these groups of natural qualities as natural resources, natural resistances, and neutral features. The first are useful, the second must be circumvented or overcome, and the third have little or no influence. Ingenuity has been depicted as man's greatest resource; it is the mother of all others. In large measure man applies his resourcefulness to the natural environment, thus creating resources to overcome resistances in the processes of improving his habitats and conditions for living.

Any or all of the societal features discussed in chapter 2 may influence resource evaluation in a given circumstance. In the technically advanced societies, the cultural conditions influencing the value of the elements of nature are quite different from those in the less advanced or the primitive societies. On the one hand, evaluations are made under the influence of rapid growth in demand, advancing scientific and technical implements, rising standards of living, and—in commercial economies—profit motives. On the other hand, in contrast, traditional uses and methods are common, and existence,

[6] For amplification of these ideas, see Richard M. Highsmith, Jr., J. Granville Jensen, and Robert D. Rudd, *Conservation in the United States*, 2nd ed. (Skokie, Ill.: Rand McNally & Co., 1969), chapter 1.

rather than monetary profit, is the chief concern of the less advanced and primitive societies. In the former there are growing demands upon nature for more quantities and varieties of resources; in the latter the demands tend to be strongly upon the arable land and, perhaps, biotic resources.

Because of the dynamics of the agency of man, the resources of the earth cannot be considered either finite or stable. Environments can be altered by human action; neutral features and even former natural resistances may become resources; some resources can be exhausted through use, and others can be diminished in utility through imprudent management and/or failure to recognize environment relationships and connections. Forests can be cleared, land can be drained, crop plants can be substituted for natural plants, slopes can be terraced, water can be brought to some dry land, soil can be loosened or compacted (with erosion as a result), mineral deposits can be mined out, and surface waters can be restricted in utility by pollution. The utility of a particular site may be altered with improved transportation or changes in need or demand. A mineral deposit heretofore too lean for mining may become useful and made valuable with improved technology or changes in economic circumstances; and previously unused elements of environment may become useful as the result of scientific discoveries. Thus it may again be concluded that the resource values of an environment are a cultural evaluation of a particular group at a given time—and rates and patterns of changes are cardinal problems in geographic consideration.

RESOURCES AND ECONOMIES

Man supports himself by using the natural resources available to him. Both his evaluation and use of these basic components of environment are conditioned by the sum of the characteristics that comprise his culture. All areas of the earth are by no means equally attractive. In some environments the limitations are far greater than the attractions. The economic well-being and opportunities of peoples bear a strong relationship to the resources available to them and their ability and desire to put them to use. Areas endowed with abundant and varied resources and inhabited by technically advanced people are likely to support varied activities and complex, advanced economies; whereas areas with few and restricted resource potentials, populated by less advanced people, are more likely to support simple activities and subsistence economies. There are, of course, many exceptions: Mainland China, by present evaluation standards, ranks high in world inventory of land suited to cropping, forests, and grazing and is well endowed with water, coal, and several metallic minerals; but a long history of preoccupation with tilling the soil resulted in a subsistence agricultural economy which had little regard for other resources. Today that nation, following a political revolution, is in the midst of a revolution in the organization of her economy with major attention being given to industrial resources. Although she may rank high in amounts in the world inventory of a number of these resources, the development of her huge population during the long history of subsistence agriculture has resulted in a relatively low per capita resource base. This relationship is likely to place permanent limitations on major improvements in the standard of living of the Chinese people; the possibility of a rising standard of living is made less likely by the current population increase rate. In contrast, Japan, with an equally low or, perhaps, lower per capita resource base but with an advanced technology and a century of concerted development effort, is economically more advanced. Restricted in area and in agricultural and industrial resources, she supports her large population in large measure by selling her "resourcefulness" to the world in the form of manufactured products. External linkages are essential to her well-being and advancement; she must import a large share of her raw materials and 15 to 20 percent of her food requirements, and she must export a large share of her manufactured products to sustain her economy.[7]

In reality, trade is essential for all modern economies. No area and no nation contains within its borders all the resources required by modern man, or at least, technically advanced man. Differences in resource endowments and in per capita production are the fundamental bases for trade and they are a major reason for the need for world peace and friendly relations among nations. The resource demands of the world's nations today cause many to reach far beyond domestic borders to procure supplies. Industrial nations seek raw materials and perhaps additional foods; agrarian and developing nations make foreign purchases of manufactured goods. The constantly increasing demand for resources, especially for minerals, has continued to alter former limitations of isolated location and of resource combinations unfavorable to local development. The countries of the Middle East, for example, lacking other resources to support industrialization, now supply an important share of the world's petroleum—finding both markets and development capital readily available. Technically advanced nations, for both political and economic reasons, are engaging in programs of technical aid to many underdeveloped countries, with noteworthy changes in the basis for resource evaluations as a result.

The nations in the best position to make selective use of their natural endowments are those with varied and abundant resources, advanced technology, and high living stan-

---

[7] A brief summary of the modern economy is provided by Robert B. Hall, Jr., *Japan: Industrial Power of Asia* (New York: Van Nostrand Reinhold Company, 1963).

dards. They may be able to concentrate on the highest-quality land for biotic production and hold in reserve low-quality resources that would be utilized by nations less well endowed. They may be in better positions to reap commercial benefits from distinctive resource qualities. One could note, for example, the intensive commercial agriculture of California that is intimately geared to the available production and marketing technology and to the high standard of living of the United States and, in contrast, the extensive use of similar land in Chile where the resource values are restricted by limited domestic markets, historic organizational forms, and distance from foreign markets.

Technically advanced nations may make resource evaluations that usually cannot be made in less advanced nations: For example, outdoor-recreation resources, at least on a large scale, are an evaluation category of advanced nations with relatively high standards of living. Recreation may give special utility to rugged highlands, snow, water, or sunshine. Similar features would not be so likely to receive recreation evaluation in either the less-developed or lower-standard-of-living areas. On the other hand, reverse considerations may prevail: Coconuts and seashells are used as utensils by natives of the Central Pacific Islands, but they have no such utility in advanced economies.

four

INTRODUCTION TO WORLD REGIONS

The concept of *region* has long served as a device for generalizing areal likeness and differences and as a tool for area analysis.[1] Courses in world regional geography date from the beginning of college and university instruction in the United States. In fact, a major share of scholarship had a regional orientation almost until the time of World War II. This is understandable, of course, in view of the relatively small number of practitioners and the sizable task of appraising the basic characteristics of the earth-space order. During this period topical instruction and research was commonly restricted to elements of the physical system, especially in the sub-disciplines of geomorphology or landforms and climatology. Commercial geography gained some significance late in the last century, and the broader interests of economic geography gained increasing importance following World War I.

Early regionalizations, for the most part, were based upon similarities in physical or natural features—landforms, climate, vegetation, or soil types. This orientation was compatible with the leading writers' dominating concern for "environmental influences," at least until the 1920s. Some efforts were also devoted to natural composite regions. By the 1930s other criteria were also being employed, and geographers were beginning to profit from and contribute to work along these lines in other disciplines. Moreover, during the 1930s, the rising federal interest in regional planning offered opportunities for some geographers to turn their attention to applied problems. In the academic field, geographers began to employ the regional concept to the human uses of the earth and to economic activities.

In the post World War II period, the increasing numbers of practitioners and the increasing specialization in the topical dimensions of the field have led to the use of the regional concept in many kinds of studies about various aspects of the earth and the man-earth system. In recent years a number of geographers have used cultural regions as a device for presenting introductory-level world courses for the college liberal arts

---

[1] The thinking of American geographers on this topic is well summarized by Derwent Whittlesey, "The Regional Concept and the Regional Method," in *American Geography, Inventory and Prospect*, ed. Preston E. James and Clarence F. Jones. Published for the Association of American Geographers (Syracuse: Syracuse University Press, 1954), pp. 21–68.

student. That format gained a considerable following with the appearance of a textbook, *Culture Worlds,* in 1951.[2] Since then, several other textbooks based upon forms of culture regions have appeared.

Most geographers today would agree that there is no single best form of regionalization but that there are many criteria upon which regions can be based and that selections can and should vary with purpose. Most, furthermore, would agree with Derwent Whittlesey's summation of the region

...as a device for selecting and studying areal groupings of the complex phenomena found on the earth. Any segment or portion of the earth's surface is a region if it is homogeneous in terms of such an areal grouping. Homogeneity is determined by criteria formulated for the purpose of sorting from the whole range of earth phenomena the items required to express or illuminate a particular grouping areally cohesive. So defined, a region is not an object, either self-determined or nature-given. It is an intellectual concept, an entity for the purpose of thought, created by the selection of certain features that are relevant to an areal interest or problem and by the disregard of all features that are considered to be irrelevant.[3]

Thus, the regional concept has been both refined and broadened in recent years. Despite the increasing attention given to the topical dimensions of the field, the region has not lost significance in geographic study. The organizing principles of spatial distributions and spatial relations also pervade topical fields of study. In other words, here, too, attention centers upon the subject's areas of distribution and its associations and interconnections with other subjects within these areas. As Whittlesey suggests, "The study of a topical field in geography involves the identification of areas of homogeneity which is the regional approach; the study of regions that are homogeneous in terms of specific criteria make use of the topical approach, because the defining criteria are topical."[4]

World Regional Approach

Subject matter may be organized in several different ways for introductory courses. The most common formats are either "regional" or "systematic," in which the material is grouped into convenient categories (segments of the earth's surface, or topics such as climate, vegetation, population, settlements).

[2] Richard Joel Russell and Fred Bowerman Kniffen, *Culture Worlds* (New York: The Macmillan Company, 1951), 620 pp. "Through the device of recognizing seven culture worlds, the geography of the inhabited parts of the earth is presented in an orderly manner. Each culture world is a reasonably unified subdivision of the earth's surface occupied by people who are strikingly alien to inhabitants of other culture worlds. Each has its individual culture traits, and its inhabitants have similar ways of changing landscapes." (From the Preface.)

Some instructors are also experimenting with more thematic approaches, seeking an emphasis that will develop understanding of processes shaping the spatial distribution and association of features of the man-earth system; others are emphasizing the methods, concepts, and theories of the discipline, sometimes focusing upon theoretical problems. Each approach has its place; no one approach is equally adaptable to all situations.[5] Students

[3] James and Jones, *American Geography,* p. 30.
[4] Ibid., p. 31.
[5] A report of The Geography in Liberal Education Project (Association of American Geographers, 1965), *Geography in Undergraduate Liberal Education* discusses a number of these courses along with their desired aims.

74 THE STUDY OF THE MAN-EARTH SYSTEM

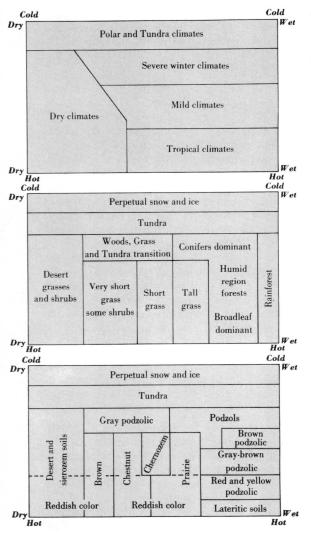

FIG. 4.1. These schematic diagrams show the generalized world distributions of climates, vegetation, and soils. Study will reveal significant relationships.

courses varies from one college to another as well as from one curriculum to another.

This book is designed for students with limited backgrounds in geography and with curricular patterns that permit little additional course work in this discipline. It thus attempts, first, to introduce the nature, content, major concepts, and major themes of geography and, second, to provide a survey appraisal of the earth as the home of man. The authors believe that the first aspects can be consolidated and the second accomplished through a world regional approach.

It is also believed that for such courses balance can best be achieved by employing, with modifications, a traditional world regionalization based upon similarities of climatic characteristics. One of the more significant advantages is the provision of a definite and unifying theme, which permits the instructor, if he desires, to concentrate upon human aspects of the man-earth system since the text provides the student with a logical index or reference of geographic knowledge. A second advantage results from the influential role of climate in physical-biotic processes. As pointed out in chapter 3, vegetation, soil, and water, especially, respond to the characteristics of climate, and there are close correlations in the broad regional patterns of each (see Figure 4.1). Thus climatic regions tend to be, at least broadly speaking, "ecosystems," or areal associations of interrelated natural processes; they therefore provide one of the easier avenues toward comprehension of harmonies and disharmonies, similarities and differences, and spatial distributions in the natural environment.

It is, furthermore, the belief of the authors that for introductory survey courses regionalization of the earth on the basis of climate provides a highly satisfactory framework for demonstrating the role of culture in man's selections and modifications in habitat development. The format, as presented in this

vary in their high school backgrounds, and the level of introductory college geography book, allows some degree of secondary classification and/or regionalization of the earth on the basis of culture considerations, thus showing that men of different backgrounds have organized and changed similar environments in strikingly different ways and at different rates. Thus the paramount concept of man as the agent of environmental organization and use, or man as the "ecological dominant," can be emphasized within this format. Moreover, at the same time, the development of a sound understanding of the earth's physical-biotic system and its subsystems is permitted.

The Presentation Format

The presentation in each of the regional chapters follows a somewhat standard organization. Each proceeds from a general introduction to location, to a consideration of climate, and other elements of natural environment. The aim is to provide a background for understanding the raw stage, on the one hand, and the nature and relations of the physical and biotic elements and processes, the generalized ecosystem, on the other. The consideration, then, turns to occupance. For this analysis, the climatic regions are subdivided on the basis of broad occupance types. For each, the pertinent characteristics and processes are considered.

PERTINENT TERMS AND CONSIDERATIONS

At this point it will be of value to identify several terms that will be used frequently: occupance, cultural landscape, settlement, and economy type. As commonly used in geographic parlance, *occupance* refers to the total of the components of the man system on the earth. It thus includes the characteristics of the organization of earth territory by human communities, of their ways of life, and of their creations of man-made or modified features. To identify the spatial associations of man-made features, the concrete evidences of human occupance, the term *cultural landscape* is generally employed. The occupance of the earth displays varying scales of intensity—ranging from situations in which there is simply political (and/or ownership) control of areas of the earth and few evidences of the presence of man to those in which dense populations and human modifications have obscured or drastically altered the physical environment.

Settlement is used in two different ways by geographers. It may refer to the colonization of an uninhabited land, thus having a connotation of process; or it may refer to appurtenances by which man attaches himself to the land and to the grouping of people and their dwellings and buildings into communities, hamlets, villages, towns, and cities. This second use, in contrast to the first, is a classification system for the actual patterns. We will concern ourselves with the second meaning in the present discussion.

Economy type, as we have previously suggested, refers to a system of making a living and/or producing wealth. In singular form, it is specific and in common usage is normally employed with an indentifying prefix, such as: agricultural economy, mining economy, fishing economy, manufacturing economy, and so on. The term may also be used in a somewhat broader sense to refer to characteristics of exchange, that is, subsistence economy or nonexchange economy and commercial economy or exchange economy; or it may be used to refer to an areal-, regional-, or national-economy complex. As suggested in chapter 2, these designations convey or imply some specific meanings and characteristics.

From the foregoing definitions it will be recognized that occupance is the more inclusive term, including features of territorial control, economy, and settlement.

Several aspects of occupance are of interest; in the main these relate to degree or intensity, distribution, and associated forms. Implied also are interests in the underlying processes by which the components came into existence, the relations among the components and between the components and the physical environment, and the interactions between the occupance type and other types or areas.

Since a pervading consideration in man's selection of living space is related to a means of deriving a livelihood, it follows that many of the major characteristics of occupance show close correlations with economic activities. In the broadest order of relationships it may be noted that both the distribution and the density of mankind over the earth are associated, to a significant degree, with the opportunities for winning a living. On the individual basis the latter truth holds for the teacher, machine operator, or merchant as well as for the farmer, miner, or logger.

The long background of study of the man-earth system and its subsystems has revealed that no two places on the earth are exactly alike. Within the range of the many variables possible in both nature and culture there is room for an almost infinite variety, even in discrete characteristics. And this variety exists in reality! Nevertheless, basic similarities (within certain levels of variance) do exist in nature, as well as in the several facets of culture and cultural response.

Here we should emphasize that geographers use many kinds of classifications that allow the grouping of like phenomena at various levels and degrees of similarity. These permit generalization and are an aid to comparisons and to understanding. Although we are employing climate as our major basis for regionalizing the earth, we will also be using other classifications to aid understanding of differences and similarities within the general framework.

CLIMATE CLASSIFICATION

Climate has been the object of classification since the days of the ancient Greeks (Figure 4.2). In more modern times a number of systems have been devised, each of which has merit for particular purposes. Most of those employed in introductory college geography textbooks in the United States are modifications of a system developed by Dr. Wladimir Köppen, working from his post at the University of Graz, Austria, during the latter part of the second decade of this century. Köppen employed an empirical quantitative approach; he defined each climate according to fixed values of temperature and precipitation, computed as averages of the year or of individual months. He then based the boundaries upon known vegetation patterns.

The Köppen system employs a letter code, designating major climate groups, subgroups, and seasonal characteristics of temperature and precipitation. The main elements of the system are as follows:[6]

Five major groups are designated with capital letters A, B, C, D, E. Four of these,

---

[6] See W. Köppen, *Grundriss der Klimakunde* (Berlin: Walter De Gruyter and Company, 1931).

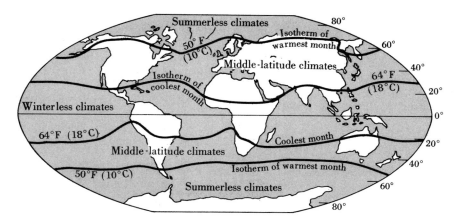

FIG. 4.2. Three major groups of climate on the basis of broad similarities of temperatures. (After Strahler.)

A, C, D, and E are identified by temperature values, and B, the dry climates, is identified by low precipitation. Three are mainly tree climates—A, C, and D.

- A Tropical Forest Climates. These climates have no winter. Average temperatures of every month exceed 64.4°F (18°C). Annual rainfall is high, exceeding annual evaporation.
- B Dry Climates. In these climates evaporation exceeds precipitation.
- C Warm Temperate Forest (mesothermal)[7] Climates. The coldest month in these climates is between 64.4°F (18°C) and 26.6°F (−3°C); at least one month, however, exceeds 50°F (10°C), thus C climates have both summer and winter seasons.

[7] "Temperate" is not considered an appropriate designation of the extremes characteristic of either the seasonal or daily weather pattern in these regions. "Mesothermal," thus, is widely substituted to indicate the intermediate position of temperatures, between the extreme heat of the tropics and the extreme cold of the polar lands. "Microthermal," meaning "small heat," likewise is preferred.

- D Snow Forest (microthermal)[7] Climates. In these climates the temperatures of the coldest month are under 26.6°F (−3°C) and those of the warmest month are over 50°F (10°C). The poleward boundary, 50°F average temperature for the warmest month, is approximately the poleward limit of tree growth.
- E Ice Climates. The average temperature of the warmest month in these climates is below 50°F (10°C), thus they have no true summers.

Within the five major groups, subgroups are identified with a second letter as follows:

- S Steppe Climate. Exact precipitation boundaries in these semiarid climates are determined by a formula: for example, in the low latitudes the amounts would be between about 15 and 30 inches (applies only to the B group).
- W Desert Climate. Precipitation in these arid climates, usually less than 10 inches annually, is also determined by a formula (applies only to the B group).

T Tundra Climate. Used as a subgroup of E climates. Warmest month between 50°F (10°C) and 32°F (0°C).
F Perpetual Frost Climates. Used as a subgroup of E climates. No month averages above 32°F.
f Moist at all seasons (applies to A, D, and C groups).
w Dry season in winter or low-sun months.
s Dry season in summer or high-sun months.
m Rain-forest climate despite short dry season (applies to monsoon type of the A climates).

A third letter added to the code gives further refinement of weather elements.

a Hot summer. Warmest month over 71.6°F (22°C) (applies to C and D groups).
b Warm summer. Warmest month below 71.6°F (22°C) (applies to C and D groups).
c Cool, short summer. One to three months

TABLE 4.1

The main climatic types of the Köppen system*

Af	Always hot, always moist climate
Am	Always hot, seasonally excessively moist climate
Aw	Always hot, seasonally droughty climate
BSh	Semiarid, hot climate
BSk	Semiarid, seasonally cool or cold climate
BWh	Arid, hot climate
BWk	Arid, seasonally cool or cold climate
Cfa	Mild winter, always moist climate with long, hot summers
Cfb	Mild winter, always moist climate with short, warm summers
Cfc	Mild winter, always moist climate with very short, cool summers
Cwa	Mild winter, moist summer climate with long, hot summers
Cwb	Mild winter, moist summer climate with short, warm summers
Csa	Mild winter, moist winter climate with long, hot, droughty summers
Csb	Mild winter, moist winter climate with short, warm, droughty summers
Dfa	Severe winter, always moist climate with long, hot summers
Dwa	Severe winter, moist summer climate with long, hot summers
Dfb	Severe winter, always moist climate with short, warm summers
Dwb	Severe winter, moist summer climate with short, warm summers
Dfc	Severe winter, always moist climate with very short summers and excessively cold winters
Dwc	Severe winter, moist summer climate with very short, cool summers
Dfd	Severe winter, always moist climate with short summers and excessively cold winters
Dwd	Severe winter, moist summer climate with short summers and excessively cold winters
ET	Polar climate with very short period of plant growth
EF	Polar climate in which plant growth is impossible

* Based upon Henry Madison Kendall, R. M. Glendinning, and Clifford H. MacFadden, *Introduction to Geography,* 3d ed. (New York: Harcourt Brace Jovanovich, Inc., 1962).

over 50°F (22°C) (applies to C and D groups).

d Very cold winter. Coldest month below −36.4°F (−38°C) (applies only to the D group).

h Dry-hot. Mean annual temperature exceeds 64.4°F (18°C) (applies only to the B group).

k Dry cold. Mean annual temperature under 64.4°F (applies only to the B group).

The Köppen system permits recognition of twenty-four climate types, as shown in the summary on Table 4.1.

For the purposes of regionalization in this book, this classification is considered too detailed. Thus, groupings have been made to yield fewer types. Moreover, boundaries have been modified somewhat to accommodate other considerations, and descriptive names have been employed. Climate criteria used for regionalization are identified within the chapters. The rough equivalents to Köppen classification are as in Table 4.2.

TABLE 4.2

Text Type Name	Köppen Type
Rainy Tropics	Af
Wet-Dry Tropics	Aw and BSh
Monsoon Tropics	Am and Cwa
Tropical Deserts	BWh
Dry Summer Subtropics	Csa and Csb
Humid Subtropics	Cfa and Cw
Humid Continentals	Dfa, Dfb, Dwa, and Dwb
Dry Continentals	Bsk and Bwk
Marine West Coasts	Cfb and Cfc
Subarctics	Dfc, Dwc, Dfd, and Dwd
Polar Lands	ET and EF

PART TWO

THE TROPICAL REALM

Our examination of the earth as the human habitat begins with the tropical realm, which centers on the equator and extends poleward to 25° or 30° north and south latitude (see Figure 4.2). Climatically it is identified as the winterless portion of the earth. In terms of vegetation, it is the habitat of hot-weather plants; where soil moisture is sufficient, plants grow the year around. The realm, however, is by no means uniform in physical-biotic conditions. Parts are characterized by rainfall throughout the year and by dense, luxuriant forests; parts have seasons of drought of varying length and a responding vegetation that ranges from woodlands to vast stretches of grass; and parts are deserts that receive exceedingly low amounts of rain in a highly erratic pattern and thus support little vegetation. The nature of the surface is also varied, ranging from low, marshy plains along some streams and littoral positions to plateaus and mountains—the highest peaks of which contain snowfields.

The nature and intensity of human occupance is also varied. Large portions of the realm are entirely without permanent occupants, and other large portions are sparsely inhabited. There are, however, great contrasts in density. India, Pakistan, Java, the Nile Valley of Egypt, and several islands of the West Indies rank with the most heavily populated areas of the earth. In general, the realm today is considered to be the last major stronghold of primitive peoples. Small groups (in isolated situations) support themselves by hunting, gathering, and fishing; somewhat larger numbers are dependent upon a migratory form of grazing on the desert margins; but most have advanced to a dependence upon farming. The majority of the farmers are very low on the world scale of techniques; many practice a rudimentary shifting form. A larger number, however, depend upon a fixed land base; some are essentially gardeners.

There are great differences in the cultural backgrounds of the peoples of the tropics. Dr. George H. T. Kimble in his study on Tropical Africa reports that "Tropical Africa is an area composed of 600-odd groups, with differing origins, customs, and aspirations...." Similar differences are to be found elsewhere.

The occupance continuity of some areas apparently extends back to the earliest period of man on the earth. The early flourish of civilization and the impressive accomplishments in the Nile Valley, the Near and Middle East, and other parts of tropical Asia are well known. During the last several hundred years these accomplishments, however, have been eclipsed by the accomplishments of Western man and Western civilization, which have been aided by a continual advance in resource-converting techniques. Although contacts, in places, have been relatively continuous for at least five centuries, the impress of Western culture is modest. This is true in spite of the fact that much of the realm was held in colonial status by European nations for one hundred to three hundred years. Western man, however, has been in the realm primarily as a political administrator, a trade firm representative, an exploiter of tropical resources, or a missionary, and, outside of Latin America, not mainly for the purpose of establishing a permanent home.

The Western contact, nevertheless, did result in a certain veneer of Western culture, most notable in commercial centers, education systems (where they exist), and political systems. Political stability, the work in health and sanitation, and such developments as irrigation projects established during the colonial period had much to do with population growth in some areas. The provision of market linkages and money systems also encouraged some commercial orientation of native economics. By and large it may be said, however, that the Western gains have been greater than those accruing to the indigenous peoples, most of whom continue to live by their traditional livelihood forms.

In many places within the tropical realm far-reaching changes have been under way for several decades. Especially noteworthy has been the rise of nationalism and the establishment of independent states. The leaderships of most of these new nations recognize their underdeveloped status, and development planning is under way in many. In this, the aid of advanced technology is being sought. The achievement of equality in economic and social advancement is, however, some distance in the future, and the course is beset with many problems. These are, of course, greater in some areas than others.

In order to recognize both physical and cultural differences, we will divide the realm into four climatic regional types—the Rainy Tropics, the Wet-Dry Tropics, the Monsoon Tropics, and the Tropical Deserts—and then sort out the Tropical Highlands for independent appraisal.

five

RAINY TROPICS

The Rainy Tropics, viewed from space, appear as an irregular although broken ring of green encircling mid-earth. The world's densest vegetative cover masks the details of the landforms in this green girdle and man-made clearings commonly are lost in the sea of forest. Warm, wet, and humid are the key words to describe the weather and climate. Temperatures are constantly high, and showers drench the land almost daily. The Rainy Tropics cover about 10 percent of the earth's surface but contain only 5 percent of the earth's population. Surprising population paradoxes exist, however. The small island of Java teems with people, whereas man is negligible in the vast Amazon basin.

The Rainy Tropics are the principal strongholds of primitive cultures, although varying degrees of Western influence, present in some areas for many years, have resulted in islands of Western culture. In Southeast Asia there exists an overlay of Chinese culture, along with Moslem and Hindu cultures in parts. Much of the total area of the Rainy Tropics has at one time or another been under colonial rule. In the past, outsiders have usually directed their energies toward the exploitation of resources, rather than the social and economic development of the natives.

During this century there has been a rising wave of nationalism, resulting in a decline in colonialism and a growth of independent nations, especially since the end of World War II. Many of these newly independent states are seeking to improve their people's living standards; many are laying plans for social and economic advancement and are seeking assistance from the technically advanced nations. For the most part, aid focuses upon improving education, health and sanitation, resource inventory, and resource-converting techniques.

With such new interests, growth in world population, and the increasing food and industrial requirements, there appears to be little doubt of increasing development in the Rainy Tropics. The optimum climate for many plants and the enormous spaces devoid of people provide the regions with great agricultural potentialities, the development of which is dependent in part upon increased research in tropical soil management and a scientific approach to production. Acreage in crops currently grown on a commercial scale could be expanded many times, and the commercial production of other native plants usable for food, sap, or fiber awaits needs and markets. The rain forests offer one of the major hopes for filling the increasing wood demands of the world; utilization depends primarily upon the advancing field of wood technology to make practical the logging of the highly mixed stands. Mineral shortages in other regions have caused intensification of critical surveys; the list of strategic minerals supplied by tropical areas is growing. The hydroelectric potentials have been scarcely touched; Africa alone has almost four times the potential of North America. Although the industrial wave is just beginning to lap upon the Rainy Tropics, trends point to a growing world significance of these regions.

Undoubtedly, much capital and direction will continue to come from northern-hemisphere industrial nations, but there is increasing interest by the local people in the shaping of their own destinies. Most of the countries with areas in the Rainy Tropics are awaking to their possibilities and beginning to move ahead with scientific programs for development.

To many, the Rainy Tropics conjure up visions of mysterious jungles peopled by savages—lands of pleasant climate where nature is so lavish in her gifts that every need is easily supplied. In reality, the majority of natives are farmers, forests dominate the vegetation, climate is monotonous, and nature in general presents many problems. Frontiers are gradually being invaded, but great areas are still question marks on the map. The Rainy Tropics remain lands of mystery, their veil of secrecy being only partially lifted.

Location

The Rainy Tropics straddle the equator in an irregular latitudinal belt of from 5° to 10° on either side with an extension of from 15° to 25° on windward coasts. This places the major portion in the doldrums, or belt of equatorial calms, a belt of low pressure encircling the earth near the equator where the noonday sun's rays are never far from

FIG. 5.1. The Rainy Tropics.

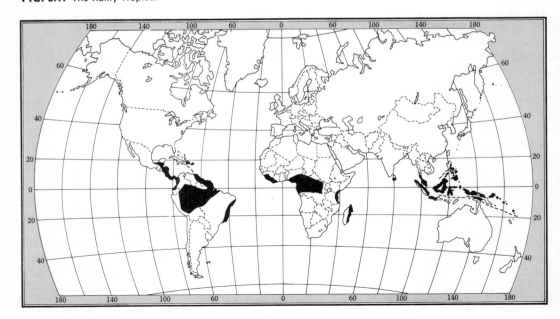

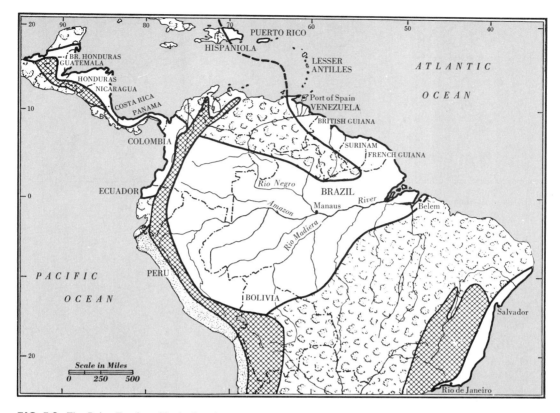

FIG. 5.2. The Rainy Tropics of Latin America.

vertical. The Rainy Tropics attain their greatest areal extent in the northern part of South America, the heart of Africa, and the islands of Southeast Asia. The regions are not continuous, as boundaries in all areas are often disrupted by highlands (Figure 5.1).

The largest Rainy Tropics region is in the Amazon Basin of South America. On the northeast, the region extends along the coast of the Guianas. The Andes Mountains isolate a small part from the main region. A narrow strip also extends along the trade wind coast of Brazil.

The onshore trade winds cause the Rainy Tropics to extend into Caribbean America, where the region is restricted to the eastern periphery of the mainland by a mountainous backbone. Also included are the windward sides of many islands of the West Indies.

Like the South American region, the bulk of the Rainy Tropics in Africa lies chiefly in the basin of a large river, the Congo. A smaller section is located along the coast of the Gulf of Guinea, and a narrow strip is found along the east coast of the continent, including the offshore islands of Pemba and Zanzibar. The last area is the windward coast of eastern Malagasy.

The Malay Peninsula is the only section of the Asiatic mainland in the Rainy Tropics. Most of the region consists of equatorial islands that include the eastern one-third of the Philippines, most of Indonesia, southwest Ceylon, and many islands of Oceania.

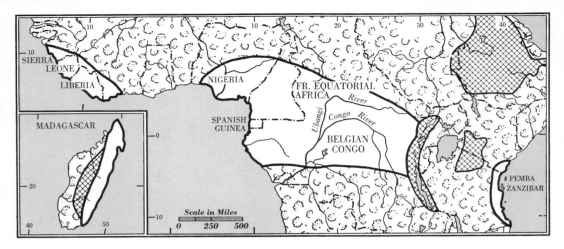

FIG. 5.3. The Rainy Tropics of Africa.

FIG. 5.4. The Rainy Tropics of Southeast Asia.

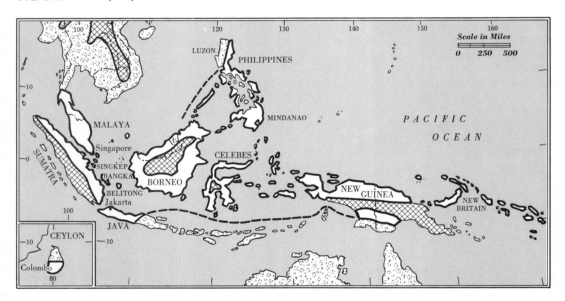

Physical Environment

CLIMATE

The climate of the Rainy Tropics results chiefly from the combination of conditions produced by location in the belt of equatorial low pressure. In these low latitudes the sun's noonday rays are always nearly vertical. Consequently, insolation is uniformly high, and days and nights are of approximately equal length all year. The sun rises and sets at about the same time throughout each month, and the periods of dusk and dawn are never prolonged. The alternation of a season of long summer days with a season of short winter days is lacking. January is quite like July—it is inappropriate to divide the year into seasons of summer, winter, spring, and fall. There may be a wet season and a less wet season, but throughout the year a uniformly warm temperature is prevalent.

The northeast and southeast trade wind air masses converge to form the intertropical front in the equatorial low, causing air to move upward from the earth. But surface wind conditions are complex; squalls alternate with light variable winds and periods of dead calm. On the whole, these regions are poorly ventilated. Clear days are rare, and billowy cumulus clouds have usually formed in the sky by noon. Thunderstorms are frequent and heavy showers a common occurrence. Relative humidity is constantly high. Each day is like the next; through the years, weather and climate are synonymous. The unvaried length of day and night, absence of seasons, tedious heat, and almost daily rain produce a monotonous day-to-day, month-to-month climatic environment.

Temperature. The vertical rays of the overhead sun that dominate the equatorial belt throughout the year produce a constant supply of heat—a year of continuous summer. Temperatures have three dominant characteristics: (1) they are uniformly high throughout the year; (2) there is a small range between the coolest and warmest month; and (3) there are no great extremes. There is an unending procession of months with thermometer readings hovering between the high 70s and 90s; typical yearly averages are close to 80°. The difference between the average monthly temperatures of the warmest and coolest months seldom exceeds five degrees. Manaus in the interior of the Amazon Basin has a yearly average of 81.3°, ranging from a March average of 80.4° to an October average of 82.9°. Average monthly temperatures at Nouvelle Anvers in the Congo Basin range from 76.3° to 80° and at Singapore on the tip of Malaya from 79° to 81°. Daily extremes as high as 100° are exceptional. In contrast, summer recordings in New York have reached 102°, in Chicago, 105°, and in Omaha, 114°. Although several other climatic regions have seasonally higher temperatures, no region can match the combination of consistently high temperatures and small yearly range of the Rainy Tropics. Temperature is not excessive, but the air is constantly charged with water vapor. The heat and the high humidity produce a high *sensible* temperature that creates considerable body discomfort.[1]

The diurnal range, or the difference between the warmest and coolest daily thermometer readings, averages about 15°. Night-time temperatures usually fall to around 70°. This has led some writers to speak of night as the "winter of the tropics." A high percentage of cloudiness prevents greater diurnal ranges; the cloud layer, plus humidity, retards heat radiation. During the less wet season, however, some areas experience clearer nights and temperatures as low as 65°. Day temperatures tend to be higher during this period. Island or coastal locations have lower average temperatures than inland stations. Islands and coasts are favored by a cooling sea breeze (in places referred to as the "doctor") that tempers the heat of the day, making these areas much more attractive for living than the interiors.

Frost is unknown, except in areas of higher altitude, and there is no cessation of growth. Harvest can occur at any time of the year, being determined by the time of sowing and the length of time required for a crop to mature.

Precipitation. The average yearly amount of precipitation is from 60 to 120 inches; there is no definite dry season, but there is often a less wet season, especially in those portions of the regions more distant from the

---

[1] Sensible temperature does not refer to thermometer readings but to the heat or cold one actually feels.

equator. During the season of heaviest precipitation, rain falls almost every day; whereas in the less wet period showers are lighter, and there are a few days with no rain.

The equatorial low, frequently known as the *convectional* belt, is an area of rising air currents where the resulting rain is seldom of long duration. This precipitation is chiefly the convectional type.

The convectional belt has an average of 75 to 150 thunderstorms a year. More thunder is heard in the islands of Southeast Asia than in any other part of the world. Bogor, Java, records 322 days a year with thunder, usually in the afternoon. In some localities several thunderstorms occur during one day. Showers are usually of short duration, and the sun often appears after the storm has passed. Sometimes rain continues to fall into the evening, but as the heat of the day diminishes, there is a tendency for skies to clear. There have been instances, however, when rain has continued for 24 hours or more. Besides the usual daily occurrence of the showers, there is also a somewhat clock-like regularity to their arrival. These precipitation conditions are usually associated with large landmasses.

The daily rainfall regime over island areas or coastal fringes tends to reach its peak toward evening. Bodies of water heat more slowly than land and reach their maximum temperature later in the day, producing the convectional condition that causes showers to occur early in the evening or in the night.

The rainfall of the windward coasts is less influenced by the convectional belt. Heavy precipitation in these areas, the *orographic* type, is due to the continuous trade winds rising along the highlands paralleling the coast. The steady trades, blowing over wide expanses of warm water, take up great quantities of water vapor. These winds, charged with vapor, rise along the windward slopes of the highland barriers. The forced ascent causes cooling in the higher altitudes, and heavy precipitation falls on the windward sides (Figure 5.5).

SURFACE FEATURES

Climate is one of several factors that influence the nature of landforms. Rocks are decomposed rapidly by the chemical action of the atmospheric elements. In the Rainy

FIG. 5.5. Typical climatic graphs of Rainy Tropic stations.

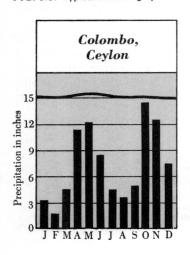
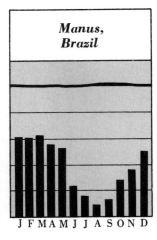
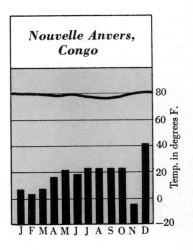

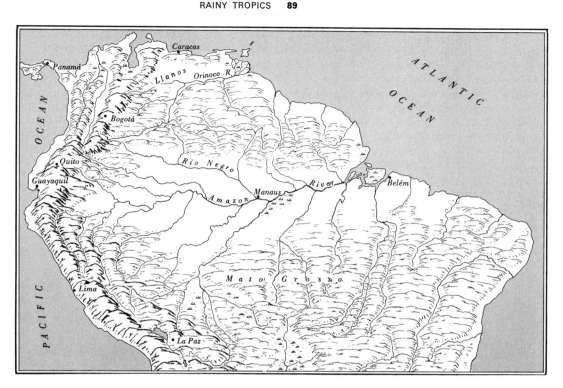

FIG. 5.6. Physiographic diagram of northern South America.

Tropics, prevailing high temperature, heavy rainfall, and organic acids generated by both living and decaying plants and animals combine to produce rapid chemical weathering. Thus, one of the characteristic features of these regions is a deep layer of *regolith,* or weathered material overlaying the solid bedrock and completely obscuring it. The heavy mantle of vegetation covering this unconsolidated mass gives a rounded form to the land surface.

Numerous perennial streams result from the heavy and persistent rainfall. Two of the longest rivers in the world, the Amazon and the Congo, are found here. The concept of extensive swamps covering much of the area is erroneous. Rapid runoff and evaporation do not permit this condition, except along low-lying positions on flood plains, deltas, and coastal fringes.

Amazon basin. The basin drained by the Amazon system is one of the major lowlands of the world. In an area so large, it is apparent that there would be a variety of surface features, but away from the streams extend vast stretches which are little known; detailed maps and information are lacking.

The Amazon lowlands, which extend westward in a *dendritic,* or tree-like, pattern, have been formed by stream deposits and comprise one of the world's major alluvial plains. In the lower course of the main stem this floodplain is about 50 miles wide, but it decreases in width toward the headwaters. Narrower floodplains extend finger-like out from the Amazon and up the tributaries. The rise in elevation progressing up the river plain is slight, and the river has a remarkably low gradient all the way from the mountain front to the sea; this is a major factor producing the seasonal widespread inundation. The floodplain is usually constricted by bluffs, and the *interfluves,* or areas between the streams, are generally considered to be low, undulating uplands.

Since the earliest penetrations by man, the Amazon system has been the transport artery of this great basin. Rivers are still the most important means of access. The largest of oceangoing vessels can sail regularly to Manaus, about 1,000 miles from the open sea, and smaller craft can go virtually to the foot of the Andes and hundreds of miles up many of the tributaries of the Amazon. The magnitude of this giant river is difficult to comprehend (Figure 5.6). Earl Parker Hanson in his *New Worlds Emerging* states:

In length, the Amazon River is exceeded only by the Nile, and very little by it; in volume of water, by no three of the world's other rivers combined. Three thousand nine hundred miles long, it has its western-most sources high in the Andes, within a hundred miles of the Pacific Ocean, and its mouth in the Atlantic. It is estimated that one-fifth of all the world's running fresh water is carried by the Amazon; at low water its mouth, a hundred and fifty miles wide, pours some 60 billion gallons per hour into the sea to turn the ocean from salt to brackish for over a hundred miles from shore. It would take a score of Mississippi Rivers to equal that low-water flow, which is vastly exceeded in the rainy season.

Nobody has more than a vague idea of the Amazon's annual high-water flow. Near Manaus,

FIG. 5.7. The Panama Canal. This man-made artery is one of the most significant transportation routes in the world, reducing ship transit time by many days. The canal, from Colon on the Caribbean to Balboa on the Bay of Panama (Pacific), is 40.3 miles from shore to shore and 50.7 miles from deep water to deep water. It has a minimum depth of 41 feet and its width varies from 100 to 300 feet. Six pairs of locks lift and lower ships over the elevation of 85 feet. (Dorothy A. Heintzelman.)

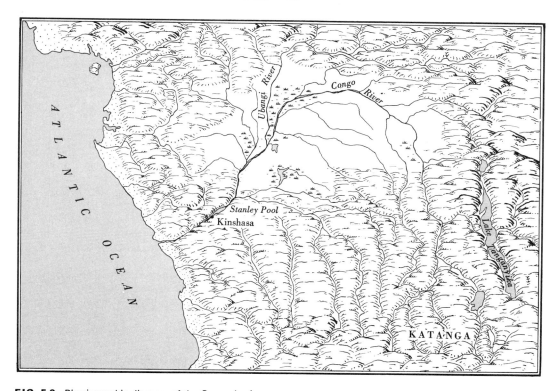

FIG. 5.8. Physiographic diagram of the Congo basin.

a thousand miles upstream, it has a yearly rise, between the dry season and the rainy, of up to 60 feet. At its peak, the river increases hundreds of miles in width at several points, flooding thousands of square miles of forest-covered land.

There are some eleven hundred known tributaries of the Amazon, not counting thousands of brooks. Nine or ten of these tributaries exceed the Rhine in length and carry more than double its volume of water. Seven of them are over a thousand miles long, and one, the Madeira, is nearly three thousand miles from source to mouth. Yet the main stream often receives the waters of these subject rivers without showing any appreciable increase in either width or current.[2]

[2] From Earl Parker Hanson, *New Worlds Emerging* (Des Moines, Iowa: Meredith Press, 1949), pp. 109–10. By permission of the publishers and author.

Middle America. In Middle America, the Rainy Tropics occupy a relatively narrow coastal belt, backed by the slopes of the rugged cordillera, which is part of a continuous zone of highlands stretching from Alaska to Tierra del Fuego. This coastal zone, largely an undulating plain and in proximity to United States markets, has been favored for the establishment of numerous banana plantations. Many of the West Indies included in this region have interior highlands with narrow coastal plains (Figure 5.7).

Congo basin. In contrast to the lowland drained by the Amazon River, the basin drained by the second great river of the Rainy Tropics, the Congo, is 1,000 feet and more above sea level. In fact, the entire continent of Africa may be thought of as a

large plateau with steep edges bordered by a narrow coastal fringe of lowland of varying width. The Congo basin is an immense shallow depression within the surface of this plateau (Figure 5.8). The rim of the plateau forms a highland border, which becomes more pronounced where the East African Highlands form a distinct barrier, displacing the Rainy Tropics climate with Tropical Highland climate. The lowest portion of the basin is in the north where the drainage divide between the Ubangi tributary of the Congo and the Shari River draining northward into Lake Chad is almost imperceptible. Drainage of the basin is provided by the Congo River system; the main stem enters in the southeast, swings in a broad curve through the basin, taking up tributaries from the north and south, and leaves via the southwest, where the Congo cuts through the rimming plateau and descends nearly 1,000 feet to the coast in a series of wild rapids. Inability to navigate the lower Congo was a major factor retarding exploration of the heart of Africa. Above the falls the main stem is navigable for river steamers from Stanley Pool to Stanley Falls; many miles of the major tributaries are also navigable. The western slopes of the plateau are drained by many short, swift-flowing streams.

Southeast Asia and adjoining islands. The Southeast Asia realm comprises the southern two-thirds of the Malay Peninsula and several thousand islands of Indonesia and the eastern Philippines. The surface features are characterized by mountainous interiors and intricate patterns of lowlands and valleys. The major plains are coastal; many are deltaic, formed by the accumulation of sediments at the

FIG. 5.9. Jungle scene near Belem in the Amazon Basin. Note the variety, luxuriance, and impenetrable character of the vegetation. This type is common along streams where light is available near the surface. (Brazilian Government Trade Bureau.)

mouths of the short, swift streams draining the mountain slopes. Many of the Pacific islands display similar characteristics, although low, flat islands are also found; the nature of these will be discussed in the following chapter.

NATURAL VEGETATION

The high temperatures, heavy precipitation, and high humidities of the Rainy Tropics produce a climatic optimum for luxuriant vegetative growth. Few areas compare with its abundant plant life; the many varieties defy easy classification. Studies made in Malaya and the island of Trinidad indicated that several thousand different species may be found in a relatively small area. In all regions many botanical specimens still await scientific study and nomenclature.

Natural vegetation of the Rainy Tropics is grouped into (1) the selva or rain forest, (2) jungle, and (3) coastal types. The *selva* is a broadleaf evergreen forest. The many varieties are mixed and do not grow in pure stands like the forests of Douglas fir in the North American Pacific Northwest. The trunks, with little branch structure, stretch 100 to 150 feet into the air, bursting suddenly into branches and foliage. Each tree carries on a struggle for light. The separate crowns coalesce and form a thick canopy above the forest floor. Above the canopy tower the forest giants, the *emergents*, forming a jagged skyline. Weaving around the trunks and intermingled with the crowns is an intricate network of vines (lianas), colorful epiphytes (airplants), and parasitic plants. The canopy prevents sunlight from reaching the forest floor; this limits the growth of grass and other low vegetative cover. The interior of the forest is gloomy and the light is subdued even at midday; the lianas hanging in garlands from the canopy give an eerie effect. Travel is not restricted, but it is easy to become lost. There is neither guiding sun nor landmarks, only the tree columns and the green roof overhead.

On steep slopes, along drainage ways, and in other places where light can penetrate to the forest floor, the jungle, a dense thicket of low trees, shrubs, and vines, takes command (Figure 5.9). If a clearing has been made by man, storm, or fire, it becomes a battleground with every green thing competing for the life-giving light. The jungle is an impenetrable barrier that hinders development and exploration and has relatively little economic value. Because waterways were the chief means of penetration, the heavy jungle growth overhanging these streams erroneously gave the early impression that vast areas consisted of nothing but jungle.

Heavy stands of saline tolerant mangroves fringe mud flats, lagoons, and marshes and cluster around river mouths entering salt water. During high water, the mangroves appear as a thick wall of greenery; as the water subsides, their exposed stilt-like roots present a weird, impenetrable maze. Coastlines with sandy shores are frequently rimmed with stately coconut palms. Coconut seeds are carried by ocean currents; the palm is found on nearly all islands of the humid tropics.

There is no holiday for vegetation in the Rainy Tropics—growth is rapid, uninterrupted, and continuous.

NATIVE ANIMAL LIFE

The native animal life matches the natural vegetation in abundance and variety. Over large areas man has little altered the virgin conditions. All the regions are characterized by: (1) an immense number and variety of insects, (2) a majority of aboreal fauna, (3) few grazing and carnivorous types, and (4) abundant aquatic life. Similarities, however, stop with these generalizations; each region has its own special species.

The count of insect life is endless. There are clouds of brilliant butterflies, endless termites, mosquitoes, spiders, ferocious driver ants, ticks, and gnats. Many are stinging, blood-sucking, and disease-carrying. The warm, humid climate affords ideal breeding conditions for insects, and the absence of frost prevents any interruption in their life cycle. Nature partly compensates for the gloom of the selva by the brilliant colors of its bird life, which varies from bee-like hummingbirds to gaudy parrots. Reptiles, camouflaged by color and resembling lianas, vary from deadly vipers to thirty-foot anacondas. Ranging through the canopy of foliage, which provides a variety of food such as fruit, nuts, sap, bark, and leaves, are monkeys, sloths, bats, and other arboreal creatures. There are few herbivorous animals in the selva, due to the lack of forage. The carnivores are also uncommon. Characterizing the ground types are the anteaters with tongues nearly two feet long, ground cats, wild pigs, and (in Africa) lowland gorillas and the strange okapis. Occasionally animals from the bordering savannas penetrate the fringes of the selvas. Streams abound in fish and water fowl, and crocodiles are often seen dozing on the mud banks. Herds of hippopotamuses, found only in Africa, feed on aquatic plants in the quiet stretches of tropical rivers. In the warm ocean bodies of the tropics, animal life is as abundant and varied as that found on the land.

SOILS

Soils are generally low in fertility, and removal of the natural vegetation soon exhausts the meager plant foods available for crops. This situation appears anomalous in view of the vigorous vegetative cover, but the climatic conditions that produce the heavy vegetation also account for infertile soils. Soil fertility refers to the cultivated crop-producing capacity of soils rather than to natural vegetative-producing capacity. The fertility elements most often needed for plant growth are nitrogen, phosphorus, and potash. Several additional elements are recognized as essential plant nutrients: Calcium, magnesium, sulfur, and iron are supplied by the soil, and from the air and water come carbon, oxygen, and hydrogen. Various types of malnutrition in plants are corrected, and better yields sometimes are obtained, when small traces of minerals such as boron, copper, and magnesium are available.

In the Rainy Tropics four features tend to reduce the quantity of fertility elements in the soil. (1) Prevailingly high temperatures and heavy rainfall promote exceedingly rapid chemical decomposition. (2) Leaching is extreme and continuous, as there is no frost season. Because of abundant rainfall, much water soaks into the ground. Water percolating through the soil tends to dissolve the soluble minerals as well as silica, removing them in the drainage and leaving mainly the insoluble residue that is largely iron and aluminum oxides. (3) The water soaking down through the soils tends to carry fine particles, concentrating them at lower levels out of the reach of plants. This process, *eluviation,* tends to form a soil with a coarse surface layer and a layer of fine accumulation at some depth. (4) The climate conditions favor the existence of an extremely active bacteria life on the forest floor; this rapidly destroys the limited organic matter by converting it to a leachable form, and, as a result, soils are very low in organic content. The soil-forming process common to the Rainy Tropics is known as *laterization;* the soils, often reddish in color, are termed *lateritic.*

There are significant exceptions to the general infertility of soils in the Rainy Tropics. Two are especially noteworthy. The first, and most widespread, is the young alluvial soils found on the floodplains and deltas of streams, where they have been carried from all parts of the individual drainage basins and deposited during periods of flood. The alluviums are usually relatively productive, and, moreover, their fertility is replenished by periodic flooding and new silts. The second exception is found where the soils have been derived from young materials, high in mineral-content; parent materials in these instances have not been exposed to the degrading processes long enough for normal deterioration of fertility. The most notable example is found on the island of Java, where rich, young soils developed from volcanic materials are a large factor in the island's ability to support a dense population.

In general, with increased work in scientific survey, it is being recognized that there are major differences evident in the quality of the lateritic soils.

Occupance in Rainy Tropics

On a population density map, the Rainy Tropics stand out as thinly settled regions of the world. Over extensive areas of the Amazon Basin, the population density is under two persons per square mile. Similar conditions prevail over much of the Congo Basin as well as on many of the large islands of Indonesia and the Philippines. Throughout the major portion of the Rainy Tropics, man has yet failed to become the conqueror, and the native vegetation cover, although in places altered, still dominates.

Somewhat denser populations tend to follow the pattern of the river valleys, where alluvial soils and transport possibilities are attractive. Coastal areas, too, sometimes provide these conditions as well as offering a more favorable climate. The trade wind coast of Brazil and the Guinea Coast, especially Nigeria, illustrate these conditions.

A great contrast to these sparsely-populated regions of Africa and South America are the oriental lands of Southeast Asia, where, in some areas, man and his works dominate. The island of Java, 50,745 square miles in area, has a population of about 75 million, making it one of the most densely populated lands in the world.

Occupance in the Rainy Tropics is strongly related to rural activities: farming, gathering useful forest products, and mining. Commercial, industrial, and service activities have a minor role in population support. For the purposes of this chapter we can recognize two broad categories of occupance—indigenous and nonindigenous.

INDIGENOUS OCCUPANCE

Among the native peoples that constitute the greater part of the population of the Rainy Tropics, there exists great diversity in race, customs, and stages of development. In many cases though, man of the Rainy Tropics would be classed near the bottom of modern commercial-industrial society's scale of culture. In contrast to man in the United States, Canada, or Europe, his living is primitive, and his needs and wants are few. Work is performed by hand, dwellings are simple structures, clothing often satisfies only minimum requirements, and transportation is on foot or by crude crafts. Modern trappings of industrial societies would not, for the most part, fit into his traditional scheme of living nor be adaptable to or compatible with his present mode of life. Many factors, encompassing his entire scale of values, resource-converting techniques, restricted spatial interactions, etc., account for his simple scale of living and slight resource development. The year-round harvest season removes the necessity for storing food and, at the same time, the uniformly warm, humid conditions promote food spoilage in simple storage; together these have tended to curb the practice of producing surplus. Owing to isolation and limited contact with the industrial world, the indigenous cultures were late in being affected, and material needs remain small. Where he has come in contact with European and American tropical exploiters, however, his mode of life has been altered and a veneer of their culture superimposed, not always to his advantage.

An understanding of the conditions, characteristics, processes, and variations in indigenous occupance will be improved by examining major types.

GATHERING

The oldest means of gaining a living was dependent upon the raw products of nature. Early man's needs were simple, and his resource-converting techniques were comprised mainly of crude implements fashioned from sticks and stone that were dependent upon his own hands and muscle. He met his modest needs by gathering useful products from his immediate locale, such as berries, seeds, nuts, ainmals, fish, vines, leaves, wood, and water. His only mastery over nature was his ability to move when local supplies of these items were diminished. This form of occupance obviously could support only very modest populations—the man-land ratio necessarily had to be low. Man's relationship to nature was essentially parasitic.

Although the number is very small and dwindling—to perhaps a few hundred thousand people—there are yet today some societies that are little removed from their Stone Age ancestors. For the most part they are localized in isolated situations where they have had, and continue to have, little or no regular contact with significantly more advanced groups. The major occurrence of these residuals of the past is within the tropics, with small numbers in parts of both the rainy tropics and the dry margins. Up to this time their environmental loci have restricted other forms of occupance.

The pygmies of the Congo Basin, certain primitive Indians of the upper Amazon, and a few of the tribes in the selvas of Asia live by hunting and gathering. By and large the primitive people have made little imprint on the natural environment other than long-term vegetation modifications; they continue to live off the land, employing simple traditional tools and weapons, on a day to day basis geared to the whims of nature. Most have no permanent abodes but fashion crude temporary shelters, because their direct dependence upon nature continues to dictate a nomadic existence as well as functional groups small in number. They do not wander aimlessly, however; they have remarkable knowledge of the potentials of their environment and follow systematic patterns in their movement. They do not own land in the usual sense; but, under the practice of *usufruct,* their right to use the land of their locale is recognized by others.

Some carry on a small amount of trade with other groups of nomads or rudimentary farmers; and some carry on a modest exchange with trading posts representing the outside world—bartering products of the forest for such items as steel knives, cloth, tobacco, and trinket jewelry. In this manner their traditional culture has been altered.

The Semang of the Malay peninsula, now amounting to but a few thousand, present an example of this way of life. The Semang wander in bands of 20 to 30 people, depending upon edible roots, stalks, leaves, fruits, and larvae for nourishment. A little fishing is practiced and small game is hunted with blowguns and poisoned darts. The Semang are true "jungle nomads"; their wanderings, however, are not without pattern. Most groups remain within a fairly fixed area with their right to its use under the practice of usufruct respected by other groups. The Semang have developed a small amount of dependence upon trade, bartering products of their gathering at trading posts for such items as machetes, tobacco, and cloth. Altogether the gathering people represent but a small segment of the total population of the Rainy Tropics. Most groups have advanced beyond this stage.

Actually, gathering, hunting, and/or fishing are common adjuncts to many other economy forms and thus are elements in the occupance pattern of many people. For example, many of the rudimentary farmers of the tropics engage in these activities as a sideline, and many people the world over supplement their income or food supply by fishing.

There are also purely commercial facets of each of these activities. A great variety of nuts, vines, leaves, roots, saps, gums, and so on are gathered from the tropical forests for middle-latitude industrial markets. As an occupance form, however, the imprint is minor, commonly represented by an occasional assembly station strategically located to facilitate shipment out and to serve a large area as well as the scattered modest homes of gatherers sometimes grouped into small communities. It is of particular importance in these instances to note that the perception of utility, the market, and thus the stimuli for these activities come from beyond the region. The development of economic and transportation linkages is basic to their existence.

SHIFTING AGRICULTURE

Shifting systems of agriculture, variously known as *swidden, milpa* in Latin America, *fang* in parts of Africa, and *ladang* in Indonesia, as well as by many other local names, are widely distributed through the humid tropics. Estimates by the Food and Agriculture Organization of the United Nations suggest that as many as 200 million tropical people are dependent upon them. The common denominators among the several forms are: (1) fields are used for crops for only one to three years and then are allowed to return to natural vegetation; (2) fire is used in the clearing process; (3) crops are produced with the primary intent of meeting the cultivators' subsistence needs; and (4) tools are elemental or very simple, usually consisting of the digging stick, machete, and perhaps a form of hoe—neither draft nor mechanical power is employed. Generally, cultivated land per family unit ranges between one to three acres. In locale of the activities, the fields appear as patches within a scene dominated by natural vegetation.

With the aid of a machete, a long, thick-bladed knife, these small islands are partially cleared in the selva by cutting down the smaller trees and girdling and trimming the limbs of the larger ones. The slash is burned, and crops are planted among the stumps and charred logs. Normally this process involves the simple dropping of seeds, tubers, or root cuttings into holes gouged with a pointed (dibble) stick; following the planting, the crops usually are given little care until the harvest. Fortunately, under the reliable climatic conditions, crop failures are rare. Crop emphasis varies from place to place, especially in staples—Cassava (manioc) in Latin America, yams in Africa, upland rice in Asia, and taro in the Pacific islands. Plantain, a member of the banana family, is grown widely. Plantain can be boiled, fried like potatoes, or ground and used as flour. Cloth may also be woven from the fibrous parts of its leaves. Other crops may include a variety of vegetables, usually grown as interplants. As a rule, few domestic animals are kept, but some groups do have dogs, small numbers of chickens and ducks, and perhaps a few swine or goats that are reserved for payment of debts, purchase of wives, or sacrifices. Much of the farm work is done by the women while the men are hunting, fishing, and gathering.

In addition to staples, there are many

FIG. 5.10. A small native village in central Sierra Leone. It is three miles from the nearest road and is reached by a narrow footpath through the forest. The people practice slash and burn agriculture and grow mainly cassava, rice, millet, and some bananas. (Anne Hollingshead.)

other variations in the details of the societies and their resource-converting techniques, such as societal and land organization, population density, size of functional dependent groups, crop associations, the care given to crops, and the settlement pattern. Some groups live communally and are little removed from the gathering stage. Others have evolved essentially to a permanent occupance. The space available to them is fixed; their villages are more or less permanent; and they employ a regular field-forest (or brush) rotation which provides for a given piece of land to be in cyclic use for crops every five to twenty-five years.

Villages are frequently located on a stream or within a short distance of a water supply. A stream not only furnishes domestic water but also fishing opportunities and a transport artery. The village consists of a small cluster of houses, often without plan; occasionally there is a linear pattern with houses lining a central street. Construction is simple; available natural materials are used. A framework of poles supports a thatch roof; walls are of woven grass, palm fronds, mud, or bark. In some settlements the houses are elevated on poles—this provides better air circulation and some protection from insects, marauding animals, and the dangers of high water (Figure 5.10). Furnishings are simple. A household will usually have an iron kettle, and other kitchen utensils are made of wood, coconut shells, or sea shells. Bamboo is used extensively in Southeast Asia for construction, and also for making containers for cooking food and carrying water. There is little to indicate permanence in the village or the clearing. It is a relatively simple matter to abandon one site and move to another.

The several forms of shifting agriculture differ in external contacts and dependence. The most rudimentary, generally in the most isolated situations, may have little or no outside contact or exchange. But the more advanced have been influenced by peoples of other cultures, usually through government administrators and trading companies, and have developed a small amount of exchange dependence: For example, a significant share of the palm oil of world commerce comes from the groves, largely natural, of the cultivators of the Guinea Coast of Africa.

In some areas, especially where population is increasing as in Nigeria, the system has been modified. Here migration is limited due to lack of space, the villages are permanent, and a fairly definite rotation system of field use-brush fallow has developed. Each family has an area of 7 to 10 acres out of which about 2 acres are cleared each year and 5 to 8 acres are in fallow. Fields are cropped rather intensively for a period of from 1 to 2 years and are fallowed for 3 to 6 years. During the fallow period brush grows to provide the ash to improve fertility. Yams, a high-yielding root crop, provide the staple, although many other crops are grown between the yam hills.

Not uncommonly each family has a garden

plot adjacent to the dwelling. These are used as a permanent base for vegetables and fruits, fertility being maintained with household refuse.

In these shifting forms the relations between man and nature are relatively direct and intimate. Man of simple wants and needs with minimal resource-converting technique has responded to a physical-biotic world featuring a climate which provides, on the one hand, favorable growing conditions throughout most of the year but, on the other hand, soils inherently low in natural fertility for crops that rapidly lose this fertility upon clearing, exposure, and cultivation. It is thus more practical for these cultivators to shift fields rather than to cope with declining fertility as well as the problem of the invasion of natural plants. Fire provides a simple means of vegetation destruction and removal and has the beneficial value of adding ash to the soil. The return of natural vegetations, if for long enough, serves to regenerate fertility for a cyclic reuse of the land for crops. There is evidence to suggest, however, that when steeper slopes are bared too long, the period of cultivation is too prolonged, or burning is too frequent vegetation modification to grass, soil erosion, and/or watershed deterioration occurs. Obviously in the absence of scientific and technical tools, a delicate harmony with nature is required if these systems are to persist.

SEDENTARY CULTIVATION

Shifting agriculture may typify native activity scattered over the bulk of the area of the Rainy Tropics, but large numbers of people are supported by sedentary or settled agriculture. Many groups remain permanently in one area, cultivating the same fields year after year. Several conditions seem to have induced sedentary farming. (1) Population pressures have forced continuous use of the same land. (2) Settlement has been upon fertile soils whose enduring productiveness encouraged continuous use. (3) Techniques that have made possible the continuous use of the same land have been acquired, developed, or adapted from other cultures.

Sedentary agriculture may be somewhat more advanced and the society practicing it more stable; however, not all groups are on the same level. In general, the crops are similar to those grown by shifting agriculturalists, but more stress is placed upon perennial crops such as coconuts and plantain. In some cases families live on isolated farms—this is especially true along the Amazon River, where they raise subsistence crops, keep livestock, and supplement their living by fishing. For cash income, wood may be cut and sold for fuel to passing river steamers. Products are also gathered from the forest, such as nuts, gums, vines, rubber, and so on. Where people live in clusters the villages are permanent, but the houses, buildings, furnishings, and implements are of simple construction.

The imprint of western contact is often greater upon the sedentary farmer than upon the migratory groups. Many have developed tastes for imported materials such as cloth and various items of manufactured equipment. Sewing machines may be found in many homes throughout the tropics. Cultures in these cases are definitely in transition; some of the traditional ways of life are disappearing.

The prime example of sedentary subsistence agriculture is found in Southeast Asia. Java is distinctive as one of the most densely populated lands of the world (Figure 5.11). Prior to the coming of white man to the area the fertility of the lava soils had been discovered by the natives of Java, who were largely permanently established agriculturists. Dutch colonization brought stability to the government, provided agricultural guidance that developed an increased food supply, and improved sanitation and

FIG. 5.11. Javanese farmers carrying their produce to market in Bandung. This is a typical mode of transporting goods in the Orient. The modern bridge and road are evidences of the former Dutch colonial administration. (Standard Oil Company of New Jersey.)

health, paving the way for great increases in population. Today the Javanese are practicing an intensive subsistence agriculture in which rice, the highest yielding cereal, is the principal food crop. Farming here is similar to other parts of the Orient and this system will be discussed in succeeding chapters.

NATIVE COMMERCIAL FARMING

Native commercial agriculture on a small scale existed in Southeast Asia before the Christian era, and trade was carried on with merchants from India, China, and Arabia. Beginning with the fourteenth century, spices from Malaya and the East Indies became an attraction for European trade. These products, so important in Europe for preserving and flavoring food and for body ointments, were produced entirely in native gardens. In time, demands from Western Europe and the United States encouraged the production of other crops, such as coconuts, palm oil, cacao, and rubber.

Native commercial agriculture in the late nineteenth and early twentieth centuries was overshadowed by the plantation system; however, considerable amounts of commodities continued to be produced on native plots, and the trend in this direction is increasing at a rapid pace with encouragement from domestic governments as well as foreign firms. Most of the native commercial agriculture is found in the more accessible areas, in a number of cases within or near plantation areas.

Along the Guinea Coast of Africa, most of the cacao and palm oil production is in the hands of the natives. In Southeast Asia and adjoining islands, much of the coconut production is in native hands, and the native output of natural rubber supplies about one-half of the world's total. The native holdings are generally small; the average Malayan native rubber grove is approximately $4\frac{1}{2}$ acres. Unscientific and lackadaisical methods often lower the quality of native production, but this is a peasant enterprise with little fixed cost, in contrast to the plantations. It seems unlikely that small-scale native commercial farming will soon entirely supplant the occidental plantation system in supplying tropical foods and raw materials; but methods and products are improving, and it would appear that this phase will be accelerated by growing nationalism and increasing technical knowledge. Many small landholders of Middle America today are growing bananas for export with the encouragement of the plantation marketing firms. This traditional commercial plantation crop is being grown on farms ranging from half an acre to 25 acres. The problems of controlling disease in the larger plantings and obtaining suitable land and labor supply for the large operations have helped speed the trend toward the smaller production pattern.

NONINDIGENOUS OCCUPANCE

The exotic products of the Rainy Tropics have long been the lure attracting white man to these regions. Exploitation began with the Phoenicians, followed much later by the Portuguese, Spanish, Dutch, British, and other European nations, and finally the United States. Much of the area has been under the political domination of these nations, and development has revolved around their economic activities. Early trade was chiefly in sugar, spices, gold, ivory, and slaves. Today's exploitation is concentrated on the products of tropical agriculture, tropical woods, and minerals. Although white men have had long association with parts of the Rainy Tropics, few have elected to make this region their permanent home. Most have gone to the Rainy Tropics under the auspices of a company or government.

With the rapid development of industry and urbanization in Western Europe and the United States in the latter part of the nineteenth century, native agriculture and gathering failed to supply the increasing needs for tropical foods and raw materials. This growing market, along with improvements in transportation, refrigeration, and tropical sanitation, gave impetus to the plantation system. Plantations have brought the products of the forest to cultivated fields to meet the needs of middle-latitude consumers.

PLANTATION AGRICULTURE

The plantation system, one of the earliest forms of market-oriented farming, is commercial tropical agriculture specializing in particular crops to supply areas outside of the tropic regions. The system was developed during the early colonization of the humid tropical Americas. Today, its greatest development is in Malaya, Indonesia, and Middle America.

Plantation locations are usually relatively close to the sea—the islands of Indonesia and the Malay Peninsula have long been favored. Littorals provide the best sites—floodplains have rich alluvial soils; there is accessible water transportation; sea breezes tend to alleviate the heat and humidity. The Southeast Asia area has the further advantage of being located on one of the great world trade routes, near large population centers from which low-cost, industrious labor has been readily available. If plantations are inland, their sites are often contiguous to rivers or railroads for ease of transporting commodities.

Plantations vary from area to area and with different crops, but their general characteristics are similar. They are based on large landholdings, and because of capital risks involved, are commonly corporate organizations. Units range in size from simple developments consisting of only the overseers' dwellings and outbuildings, with a few hundred acres of cropland, to the more elaborate plantations, with homes for workers and technicians, stores, hospitals, schools, and several thousand acres of cropland (Figure 5.12). The average size of a rubber plantation in Malaya is about 1500 acres.

The region supplies the proper environment and usually the labor. Technicians, tools, capital, fertilizer, clothes, building materials, and even canned and refrigerated foods are imported. The work has been largely performed by hand labor (Figure 5.13). In the past, plantation owners found it cheaper to use native laborers instead of mechanical equipment. Now operators are beginning to use some machinery; the greatest advances in this trend are found on banana plantations of Middle America and rubber plantations of Southeast Asia. In some areas small herds of livestock are maintained to keep down the undergrowth.

The chief crops of the plantations in the Rainy Tropics are rubber, coconuts, bananas, palm oil, cacao, and abaca (manila hemp).

The Malayan-Indonesian area provides over 90 percent of the plantation rubber. Banana production centers in Middle America. Cacao is grown on the Bahia Coast of Brazil and in Ecuador. Palm oil plantations are centered in Indonesia and Malaya, but much of the product is obtained from native plots in West Africa. About 50 percent of coconut production is in the Philippine Islands; the balance is in Indonesia and on scattered Pacific islands. Abaca is produced on both plantations and small holdings in the Philippines.

The plantation system is not without problems. Growth in nationalism within the countries of the Rainy Tropics is resulting in policies which are not as favorable to foreign investors as previously. Regulations regarding benefits to native workers are more stringent and taxes are higher, as tropical countries are attempting to increase their own economic returns. Plant disease is a problem in the production of certain crops such as bananas, where control programs, including occasional abandonment of one area and reestablishment in another, leads to increased cost. Marketing is another problem. The huge areas available for the growth of commercial crops in the Rainy Tropics make the production of surpluses easily possible. Inability to maintain production volume in the proper relation to demand

FIG. 5.12. (top) Homes of workers on a rubber plantation in Malaya. Workers live in neat, attractive villages furnished by the management. Families have their own individual garden plots, and grazing areas are provided for livestock. (Natural Rubber Bureau.)

FIG. 5.13. (bottom) Latex collection center. Tappers converge at the loading stage to await the factory truck. Notice the orchard-like appearance of the rubber trees in the background. (British Information Services.)

results in lowered prices and profits—this has been notably true regarding rubber, which has the further limitation of a synthetic competitor. Nevertheless the plantation system will undoubtedly continue and even increase in importance. It may well be that the tropical countries will assume an expanded role in plantation management, through encouragement of local investment. Moreover, new forms of relationships are developing, as exemplified by the United Fruit Company, which has been fostering an associate producers program in recent years—encouraging the participation of local nationals. The company gives technical assistance and provides the marketing arrangements. In some instances it provides the land; in others it encourages small-farm owners to shift production to bananas.

There are many more species of vegetative life that thrive in this environment that could furnish food and raw materials. Commercial production awaits the development of tastes and markets.

FOREST INDUSTRIES

The tropical selva is the most vigorous vegetative growth on earth and covers large areas, yet wood products have been a secondary industry. The basic difficulty is found in the great variety of species and in the fact that many are at present of little economic importance. In contrast to forest stands of the middle latitudes where much of the cover can be used commercially, the valuable trees in the selva are widely scattered. Transportation is complicated by difficulties encountered in the movement of extremely heavy logs. Roads are hard to build and maintain on the rain-soaked terrain (Figure 5.15). Proximity to water is one of the major considerations in forest exploitation; streams are the main means of transport, but often the logs are too heavy to float and must be shipped on rafts. Labor is scarce, workers lack stamina and are adversely affected by disease. Modern logging equipment, especially crawler tractors, is used in a few areas, but the bulk of the labor is still performed by hand.

Despite the numerous limitations upon the forest industry in the Rainy Tropics, certain hardwoods, especially mahoganies, are being exploited and have expanding markets for furniture, veneering, decoration, and other special purposes. The South American selva contributes Brazilwood, mahoganies, rosewood, cedars, and Brazilian teakwood. Valu-

FIG. 5.14. Worker making the delicate diagonal slash into the domesticated rubber tree. The oozing latex may be seen in the cut. (J. Granville Jensen.)

FIG. 5.15. Construction of the Trans-Cameroon Railroad in Africa. Heavy equipment such as that shown here is a boon to road construction in the tropical forests, as well as elsewhere. (Caterpillar.)

able commercial species from Middle America include dyewoods, mahogany, lignum vitae, and balsa. From Africa come rosewood, ebony, and African mahogany. Several hardwoods, but principally *lauan* (Philippine mahogany), are exported from the Philippines. The hardwoods are seldom prepared for consumer use in the region but are exported as logs or rough lumber for further processing. This practice has called for logging in the most accessible locations. As a result overcutting has occurred in some, necessitating movement of operations to less desirable locations.

Undoubtedly lumbering will take on greater significance as new uses are discovered for tropical woods, intermediate zone supplies are further exhausted, and mechanization in operations increases. The selva, furthermore, is a vast storehouse of cellulose raw material for the future. Forest plantations also offer possibilities, since growth is rapid under the favorable climatic conditions. Balsa plantations in Latin America may be indicative of a trend.

The tropical forests provide many other products—fibers, gums, drugs, nuts, and saps—that enter the channels of trade through the native gathering industry. For example, chicle, the base for chewing gum, is the sap of the zapote tree of the Caribbean area; Essentially all of the Brazil nuts of commerce are products of native gathering. Unlike most tropical trees, this species is commonly found in groups, rising well above the surrounding trees. Brazil is the leading source of nuts with Belem and Manaus acting as major assembly points. Venezuela, Bolivia, and the Guianas are lesser producers. The outer shells or pods, three to eight inches in diameter, are removed by the gatherer to obtain 12 to 20 nuts. Wild rubber, and balata are also gathered in Amazonia, and palm nuts for vegetable oil, in Africa. As previously suggested, gathering is practiced by both the "jungle nomads" and the farmers who carry on the activity to either earn cash or have a medium for trade. The greater quantities are supplied by the farmers who, instead of wandering aimlessly through the forest, have somewhat systematized their activity. Sources of supply are well marked, and definite routes are followed through the forests. Although the flow of products is often sporadic, the gathering industry makes valuable contributions of raw materials for which the demand is still too limited to make plantation production profitable.

MINING

The mineral wealth is not fully known. The remoteness of many areas and the heavy vegetative cover have made mineral surveys difficult. With the increasing needs of industrial nations, greater attention is being given to these regions for possible mineral resources.

It is difficult to generalize about the factors related to the localization of mining activities either in the Rainy Tropics or elewhere in the world, because these differ somewhat with mineral types. But it is possible to recognize the more common factors in the total complex of relationships. These include: (1) the physical occurrence of minerals (they can be mined only where they occur); (2) the physical characteristics of deposits including: size, quality, depth, and ease of mining; (3) the situation of the deposits, or accessibility to industrial or other markets; and (4) a complex of development relationships including ownership and min-

FIG. 5.16. Hydraulic tin mining in Malaya. This is one of the many processes employed in tin mining. Powerful water jets cut and disintegrate the ground of the mine. A mixture of sand and clay, ore and water collects in the bottom of the mine whence it is pumped to the head of a flume. There the heavy tin ore is deposited while the lighter sand and clay are carried away with the stream of water. (British Information Services.)

eral rights, available mining technology, capital availability, transportation, and market connectivity.

Not only are minerals scattered in physical occurrence, but they are commonly hidden from surface view. The problems related to the search for minerals, although lessened by advanced technology, along with the cost attending the basic developmental work necessary to begin mining, have tended to make the complex of factors noted in the fourth point above highly important. These, plus the efficiency gained in large operations, also have caused the focus of modern developments to be upon deposits of large size. In fact, these conditions, together with the advances in technology that permit the use of lower-grade ores, have removed the resource values of the small but richer deposit of the sort that received attention in the earlier days of mining.

Thus, although in technically unadvanced nations examples may be found of small-scale mining where the pick, shovel, and handcar comprise the tools, a major share of the mineral output is from large operations utilizing modern machines. The operation scale, however, is measured in terms of capital investment and production, owing to vast increases in labor efficiency (some of the smaller hand mining operations in technically unadvanced areas may employ more manpower than larger-scale machine operations elsewhere). This is also true of many of the more recent developments in less advanced nations, where, in many instances, the capital, technology, direction, and marketing arrangements are provided by foreign firms.

Several areas are already prominent in mineral production. An area on the Isthmus of Tehuantepec in Mexico is one of the world's great native sulfur-producing regions. Much of the world's tin is mined on the Malay Peninsula (Figure 5.16) and the Indonesian islands of Bangka, Belitung, and Singkep. Petroleum is produced on Sumatra, Borneo, and Java. Small deposits of iron are worked on a limited scale in Malaya and Mindanao. Bauxite is mined in large quantities in Surinam and Guyana.

Undoubtedly the Rainy Tropics will become more important sources of minerals as industrial demands multiply and stimulate further search. The control and exploitation of the mineral resources of the Rainy Tropics are largely in the hands of the United States and European interests.

MANUFACTURING

Natural resources, for the most part, gain their utility after they have been refined, shaped, joined, or otherwise processed. Many foods also have to be processed before they can be utilized or to improve their shipping, keeping, or storing qualities. These various activities are encompassed within manufacturing.

As a background for understanding the limited amount of manufacturing in the Rainy Tropics as well as the world pattern, it will be of value to recognize some of the major conditions that localize these activities. For this purpose it is helpful to recognize two classes of manufacturing: primary and secondary.

Primary manufacturing as suggested by the identity, is a first stage—first in time and order in the manufacturing hierarchy. In contrast to secondary manufacturing, it is more directly dependent upon natural resources and agricultural productions and is more oriented to the production of commodities that become the raw materials for further manufacturing. Thus the location of natural resource developments and industrial agricultural crop production is of prime importance in the localization and distribution of primary manufacturing. On the other hand, the products of these activities are dispersed widely as raw materials for secondary manufacturing.

Other influences may localize primary

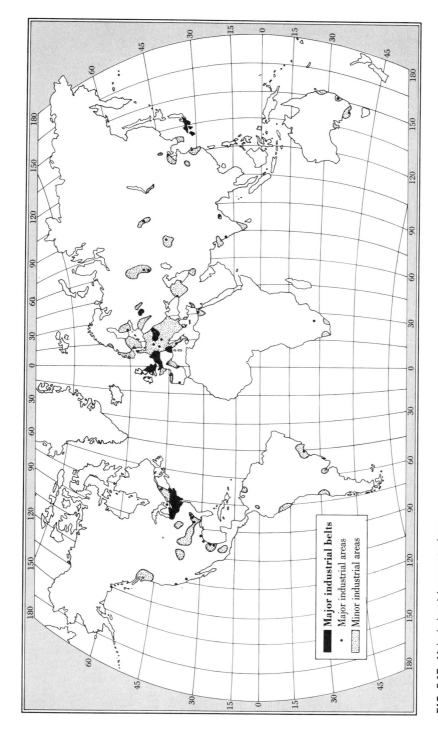

FIG. 5.17. Major industrial concentrations.

FIG. 5.18. The opera house in Manaus. The opera house represents an exotic period in this far up river city of the Amazon when wild rubber was king and the commercial rubber plantations of southeast Asia were only a botanist's dream. (J. Granville Jensen.)

manufacturing besides the primary resource production areas. Availability of low-cost transportation, a requirement for an energy source, the inertia and pull of an established center (with established plant, technology, labor supply, market linkages, and so forth) that may have depleted its original source of raw materials, or even government-promoted rationalization of industrial development are some of these influences.

It holds that on the world level the major localization of primary manufacturing is strongly influenced by and geared to primary raw material production. This results from the benefits gained from the condensability and upgrading of value of commodities shipped to secondary manufacturing centers. There is thus a reduction in shipping costs as well as profit to be gained by those who control the primary raw materials. Therefore we may notice that lumber, plywood, pulp, and other primary wood products tend to be manufactured in the locale of wood supply.

From the foregoing, it can be reasoned that in the Rainy Tropics the bases for manufacturing are largely lacking under existing conditions. Vast areas are either lightly occupied or occupied by preindustrial people whose limited needs for manufactures are met by handicrafts or imports. The commercial activities of the region are largely concerned with the production of raw materials for middle latitude northern hemisphere markets. The normal procedure has been for the direct movement of these materials. However, there is some semiprocessing of certain agricultural products and tropical woods, oil refining, and tin smelting for ease and lower cost of shipment and/or improved keeping qualities. Secondary manufacturing is restricted to a few of the major cities, where combinations of assembly, distribution, financial, and administration functions have both concentrated a market and the bases for tapping it.

On the windward coast of Brazil, Rio de Janeiro and Salvador have expanding industries based on raw materials from hinterlands beyond the Rainy Tropics to which they are connected by a net of railroad lines. Rio de Janeiro is a leading manufacturing city of the nation with products including cotton and wool textiles, leather goods, furniture, and flour made in modern factories.

URBAN CENTERS

The lack of many urban centers in the Rainy Tropics reflects the limited commercial and industrial activities as well as the relatively small populations. Most cities of importance are commercial centers owing their development to favorable water transportation. Only two large urban centers of significance are found in the Amazon Basin. Manaus, 12 miles above the confluence of the Rio Negro and the main stream, is the central collecting point for the upper portion of the Basin (Figure 5.18). Belem, near the mouth, is the gateway city to the entire Amazon Basin. Rio de Janeiro and Salvador, the second and fourth largest cities of Brazil, are on the windward coast (Figure 5.19). Both are important commercial centers with rich agricultural hinterlands. Maceio, capital of the thriving sugar and cotton producing state of Alagoas, is also on this coast. Port of Spain is the chief city of Trinidad, serving as the seat of government and main port. Georgetown, Guyana, and Paramaribo, Surinam, have populations of around 100,000 and are the chief ports of the former British colony, Guyana, and Surinam.

There are few important centers in the Rainy Tropic portion of Africa. Kinshasa on the southwest margin of the Congo Basin functions as the outlet for the entire Basin. Situated 250 miles from the sea at the outlet of Stanley Pool on the Congo River, it serves as the breaking point where river traffic meets the rail line to Matadi, the ocean port on the Congo River. Stanleyville is the collecting and distributing center for the upper part of the Basin.

Singapore is the master city of the Southeast Asia realm with a population in the million class. It owes its significance to several factors of location. One of the world's most important trade routes winds through the Straits of Malacca and bends around the Malay Peninsula. Singapore, a strategic port of call on a small, independent island at the southern tip of the peninsula, has an excellent harbor and has become a notable collecting and distributing center, particularly for rubber and tin, due to its situation in the heart of the most highly developed tropical area. Djakarta, Java, is the main city, capital, and principal port of Indonesia, serving as the *entrepôt* for the entire archipelago. Farther east is Semarang, the fourth largest city and the port for north central Java. Bandung, in the western hills, is the third largest city of Java and has a population that is believed to be approaching one million.

FIG. 5.19. Rio de Janeiro, Brazil. The picture shows the entrance to Guanabara Bay and Sugar Loaf. (J. Granville Jensen.)

six

WET-DRY TROPICS

The Wet-Dry Tropics form zones of transition between regions of great contrast—the hot, wetlands of the Rainy Tropics and the hot, drylands of the Tropical Deserts. From one border to the other there are marked changes in rainfall, vegetation, and agricultural potential. The unifying feature is a climatic rhythm, characterized by a season of rain and a season of drought, as distinguished from the sameness that prevails in the bordering lands. The heart of the Wet-Dry Tropics contains the great tropical grasslands, or savannas. The season of drought and the resource of grass provide the people of this realm with materially different conditions for living than are found in the Rainy Tropics or in the Tropical Deserts. In contrast to the former, drought causes a cessation of plant growth that necessitates storage of seasonal food supplies by indigenous farmers.

Like the Rainy Tropics, these regions are strongholds of indigenous, primitive cultures and rudimentary economies. Similarly, parts of these regions have had a long history of Western influence, especially Latin America where European colonization began in the sixteenth century.

Location

The Wet-Dry Tropics lie on the poleward margins of the Rainy Tropics of Latin America and Africa and include the southern fringes of Indonesia and northern Australia—chiefly between 5° and 20° north and south latitude (see Figure 6.1). A definite belt is formed only in Africa and Australia, where the savanna lands extend across the continent; elsewhere highland barriers and continental configurations produce irregular shapes. These are particularly noteworthy in Latin America, where several portions represent this climate type. The largest area in South America lies south of the equator. The greater portion is known as the Campos, the great grassland of Brazil; also included is the Gran Chaco of Bolivia, Paraguay, and Argentina. A second area comprises the basin of the Orinoco River, known as the Orinoco Llanos, and reaches

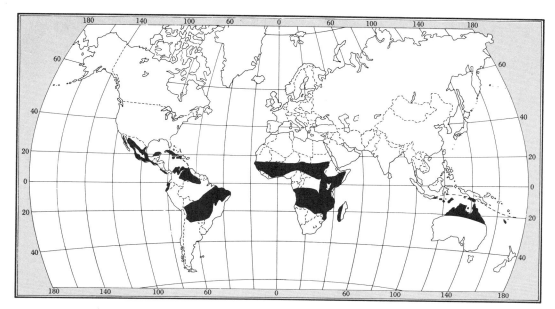

FIG. 6.1. The Wet-Dry Tropics

northwestward to include the Maracaibo Basin of Venezuela and the Bolivar Savanna of northern Colombia. West of the Andes Mountains a smaller area of Wet-Dry Tropics is found on the coast of Ecuador (Figure 6.2a). The Wet-Dry Tropics extend along the southern and western coasts of Middle America, continuing through the Pacific and Caribbean lowlands of Mexico as far as the Tropic of Cancer. The major portion of the Yucatan Peninsula, Cuba, and other western islands of the West Indies are also included (Figure 6.2b).

In Africa, Wet-Dry regions form a crescent partially encircling the Rainy Tropics. North of the equator the *sudan* stretches from the Atlantic Ocean eastward to the highlands of Ethiopia.[1] This region also extends southward to the Gulf of Guinea. South of the equator the Wet-Dry Tropics extend from coast to coast and include the west coast of Malagasy (Figure 6.3a). Many islands in the Central Pacific between 5° and 20° north and south latitude may be classified in the Wet-Dry Tropics. They receive their rainfall principally as a result of shifting of the convectional belt and the trade winds (Figure 6.3b).

Physical Environment

CLIMATE

The shifting of the sun's vertical rays rules the climate of the Wet-Dry Tropics. The equatorial low, or convectional belt, always dominant over the Rainy Tropics, migrates north and south with the movement of the sun's rays. The poleward extremes of this migration mark the outer boundaries of the Wet-Dry Tropics. Following the spring, or vernal equinox in March, as the sun

[1] This region is not to be confused with the Republic of Sudan.

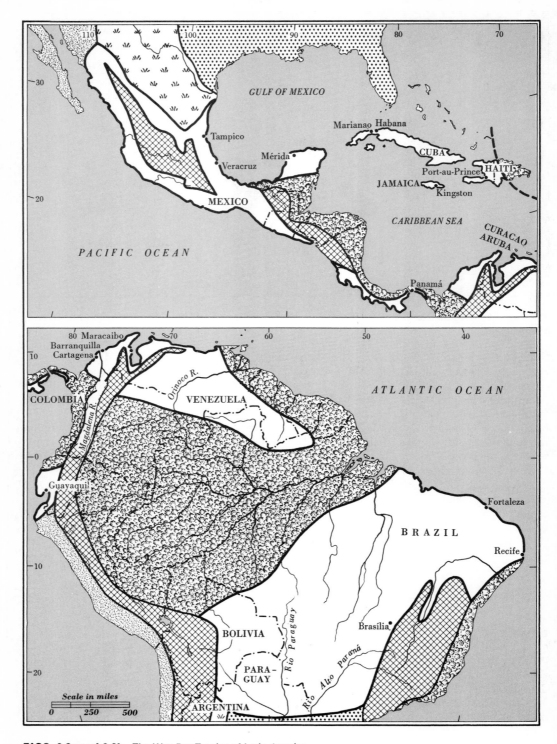

FIGS. 6.2a and 6.2b. The Wet-Dry Tropics of Latin America

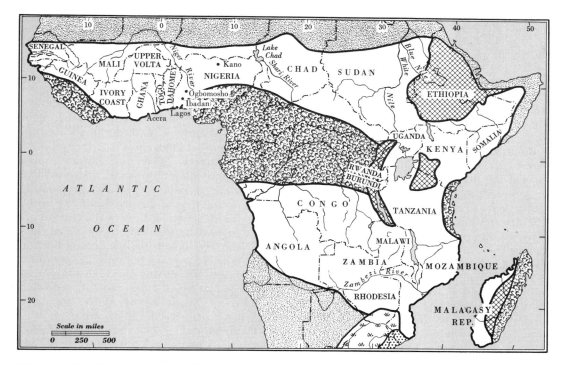

FIG. 6.3a. The Wet-Dry Tropics of Africa

FIG. 6.3b. The Wet-Dry Tropics of northern Australia and neighboring islands.

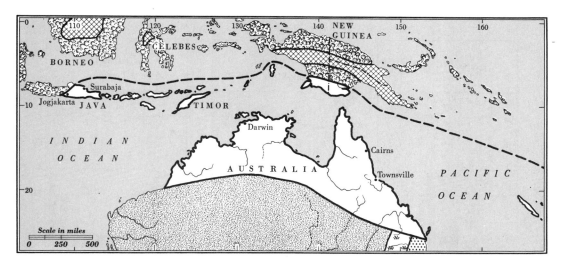

moves northward, areas bordering the Rainy Tropics in the northern hemisphere begin to acquire climatic conditions that are characterized by high humidities and thundershowers; similar conditions are present in the southern hemisphere following the autumnal equinox. When the convectional belt is in control, the climate closely parallels that of the Rainy Tropics (Figure 6.4).

During the low-sun period, the convectional belt is replaced by the trade winds or the dry, settling air of the subtropical high. The low-sun period is the dry season in each hemisphere; a blazing sun from an almost clear sky often raises temperatures to above 100°. Relatively clear night skies allow heat radiation, and diurnal ranges surpass those of the Rainy Tropics. The humidity is low, and the air is hot, dry, and dusty. In some areas strong winds are characteristic; the Harmattan in the western *sudan* blows directly from the Sahara Desert, lowering humidities and desiccating the ground and plant life. During the dry season, many small streams disappear, and nature appears to be in hibernation. All this is changed when the high sun returns and the rains arrive. Rivers are transformed from sluggish streams to raging torrents, green shoots of grass appear, trees send out new leaves, and all nature is rejuvenated. The Wet-Dry Tropics are truly regions of marked seasonal contrast. The seasonal regimen of differential heat of land and water is a marked climatic control along several coastal areas.

Temperatures. There is a constant supply of heat throughout the year, and frost-free periods are continuous. The average monthly temperatures during the dry season are in the 80s, with highest recordings of the year occurring just prior to the rainy season when skies are clear and desert-like conditions prevail. A lowering of average monthly temperatures to 70° and 80° during the high-sun period is due to the heavy rains and the associated cloud cover. Yearly average ranges are above 5° but seldom exceed 15° of variation.

Precipitation. The seasonality and amounts of precipitation are the significant factors setting climate apart from that of the Rainy Tropics. Concentrations occur during the season when the sun is overhead; average amounts range from 60 inches or more down to 10 inches. In general, rainfall decreases in amount progressing toward the desert margins as a result of shorter and

FIG. 6.4. Typical climatic graphs of Wet-Dry Tropic stations.

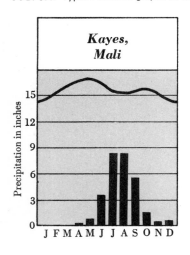
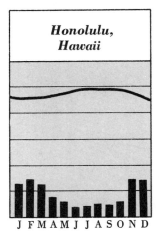
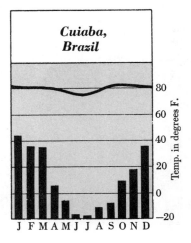

shorter convectional belt control. On the basis of length of the wet season and total precipitation, these regions are frequently subdivided into savannas with long wet seasons (with precipitation of 30 to 60 inches) and savannas with short wet seasons (with precipitation of 10 to 30 inches). Convectional precipitation or thunderstorms are characteristic, but in areas where mountain barriers interrupt moisture-bearing winds, orographic precipitation is prevalent and rainfall is often excessive. Mount Waialeale, on the Hawaiian Island of Kauai, has an annual rainfall of about 460 inches—on the leeside of the island the amount diminishes to about 20 inches annually. Rainfall is not reliable; the beginning of the rainy season is unpredictable, and there are wide variations in amounts from year to year. Generally, the reliability of precipitation decreases as the annual totals decrease.

SURFACE FEATURES

South America. The Campos of Brazil, a west sloping plateau, gradually merges into the lowland of the Parana-Paraguay River system. Some diversity of surface is produced by the many headwater streams of the Amazon, Parana, and Paraguay Rivers, which rise in the tabular surface of the Mato Grosso Plateau and the rolling surface of the Goias Plateau. These streams have carved, broad valleys radiating from the higher land. In the southwest the level plain of the Paraguay River separates the Brazilian Highlands from the Andes Mountains.

The Orinoco Llanos, largely the alluvial plain of the Orinoco River, forms an elongated lowland, generally less than 1,000 feet in elevation, between the plateau-like Guiana Highlands and the Venezuelan Andes. In the west it merges with the Andes through several terraces and in the east juts seaward in a large delta. The Orinoco is the principal river and transportation artery of the Llanos.

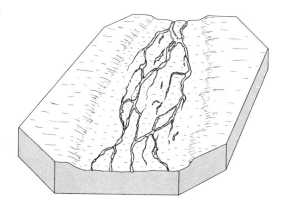

FIG. 6.5. Braided stream pattern.

This great stream and its tributaries can be navigated for a distance of 1,500 miles by boats drawing 12 feet of water. During the wet season such boats can proceed up the main stem to the Colombia border.

The Orinoco and other streams of the Wet-Dry Tropics have a marked yearly rise and fall. During the dry season they slowly lose volume and become unable to transport their load of silt; they then frequently deposit sediments in their beds. Often they split into a maze of channels forming a braided pattern through the alluvium (see Figure 6.5). When the rainy season comes, the clogged channels of the plains often cannot carry the increased volumes, and vast sections of the riverine areas may be inundated. When rivers are incised in plateaus, seasonal flooding is restricted to the valleys.

The Maracaibo Basin is a depression situated between two north-trending branches of the Venezuelan Andes. Lake Maracaibo, 120 miles long and 60 miles wide, occupies its central portion. The shallow waterway connecting this freshwater lake with the Caribbean Sea has been dredged to provide a channel 35 feet deep and 600 feet wide to allow passage of ocean-going vessels. The Magdalena Valley of northern Colombia lies

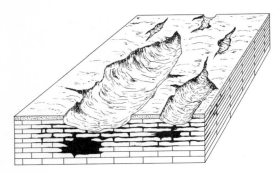

FIG. 6.6. *Karst* topography. Such landforms occur only in limestone areas.

between the central and western ranges of the Northern Andes. Its surface is a low, flat plain mantled with stream-deposited materials.

The west coast Wet-Dry Tropics region is contained largely in the Guayas Lowland of Ecuador, an alluvial surface 40 or 50 miles wide extending north between the coastal hilly belt and the slopes of the Andes.

Middle America. The Wet-Dry Tropics of Middle America occupy the coastal lowlands and the lower slopes of the central mountains. The Yucatan Peninsula is an area of *karst* topography—unique to regions whose bedrock is soluble limestone (see Figure 6.6). The undulating surfaces are usually characterized by the general absence of surface streams and numerous shallow depressions, sinks, caused by subsidence of the roof of solution caverns. Some areas are honeycombed with caverns. The Yucatan Peninsula receives moderate rainfall, but owing to the absorptive quality of limestone the landscape appears arid.

The smaller islands of the West Indies have central mountains and coastal fringing plains. Cuba, the largest island in the group, contains nearly half the land area of the West Indies, but lacks the high central mountain range. Distinctly mountainous country is found only in the eastern portion.

The major share of the island is flat to gently rolling, and the surface is suited to the use of modern farm machinery and the construction of transportation lines.

Africa. The two major areas of Wet-Dry Tropics in Africa are plateaus with broadly similar features. The average elevation of the Sudan is less than 1,500 feet, and the regional slope is northward. The most prominent relief features are isolated residual masses of harder rock that withstood the erosional forces that have worn down the surrounding country.

Three rivers—the Niger, the Shari, and the Nile—provide the drainage of the *sudan*. The Niger River in the west rises in the highlands of Guinea and flows northeast toward the Sahara. It is probable that the Niger once terminated in a vast area of inland drainage—this is evidenced by the remnants of a former great lake on the margins of the desert south of Tombouctou where the sluggish Upper Niger inundates vast areas during the rainy season. East from Tombouctou it bends southward, and here the more powerful Lower Niger has captured the drainage of the Saharan-bound Upper Niger by a process of gradual headward erosion and *river piracy*. The river then flows southeastward, sometimes through gorges, to the Gulf of Guinea.

The major drainage of the central *sudan* is provided by the Shari River, which rises on the northern rim of the Congo Basin and seasonally directs an abundance of tropical rain runoff toward the Sahara. This stream resembles the Upper Niger but terminates in Lake Chad, a great expanse of shallow water frequently under 10 feet deep that covers an area of 10,400 square miles, nearly the size of Belgium. This island-strewn lake has no definite banks, often merging into swamps and varying in size and shape from season to season with fluctuations in precipitation. There is no visible outlet, but underground drainage to the lower basin in the north is probable.

The greatest river in the *sudan* is the Nile. In contrast to the other streams, it succeeds in traversing the entire width of the Sahara Desert to empty into the Mediterranean Sea. Two Niles cross the Wet-Dry Tropics of Africa—the White, or Clear, Nile and the Blue, or Muddy, Nile. The White Nile, which resembles the Upper Niger and Shari rivers, rises in the lakes south of the equator, flows northward and, several hundred miles before reaching the Sahara, spreads out in an alluvial section of marshes around its confluence with the Bahr-el-Ghazal, and then bends toward the east for about 100 miles before continuing its northward course. The Blue Nile has its source in the highlands of Ethiopia, which provide an abundant supply of water. This river has great erosive power; the sediments gathered as it flows down the steep slopes account for its dark color.

The region south of the equator—the second major area under discussion here—has an average elevation approaching 3,000 feet. Only near the narrow coastal plain do elevations decrease below 1,000 feet. The Zambezi River is the main stream of the southern region. Fed by numerous tributaries originating from the drainage divide with the Congo System, it flows in a generally easterly direction and discharges into the Indian Ocean midway down the coast of Mozambique. The Victoria Falls on the Zambezi River—one of the most spectacular natural features in Africa—are over a mile wide and 360 to 400 feet high.

Australia. The Wet-Dry Tropics cut latitudinally through the three major physical provinces of Australia—the western plateau, elevated about 1,250 feet above sea level, the central lowland which here drains into the Gulf of Carpentaria, and the Eastern Highlands which in northeast Queensland are usually under 3,000 feet elevation.

Across the Arafura Sea, the largest lowland of New Guinea is also included within the Wet-Dry Tropics. This is essentially a deltaic plain formed by the Fly, the Digul, and other streams. Much of eastern Java and the Lesser Sunda Islands are comprised of mountains that are continuations of the mainland systems of Asia. Plains are usually associated with streams and border the coasts.

Central Pacific Islands. Based upon manner of formation and present surface, two principal types of islands may be distinguished in the mid-Pacific Ocean. These are the "high" islands of volcanic and diastrophic origin and the "low" islands built of reef limestone. The Hawaiian chain, representative of the former, owes its origin to flow after flow of volcanic material accumulated on the top of a submarine platform or ridge. Other "high" islands, especially in the western

FIG. 6.7. Atolls. This diagram shows various stages of atoll development.

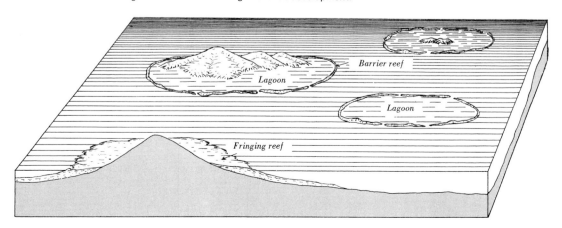

Pacific, are the summits of submarine mountain ranges. The "low," or coral, islands have been constructed by lime-secreting organisms that thrive in the warm, shallow, tropical waters. Coral reefs frequently form offshore and with emergence may be added to the land. In other cases small reef-encircled islands, perhaps of volcanic origin, appear to have slowly submerged while the coral reef continued to grow. Such reefs now appear as more or less complete coral rings, called atolls, which encircle a central lagoon (see Figure 6.7).

Breaks in reefs, along with deep water lagoons in some places, form excellent harbors. The most spectacular example of coral is the Great Barrier Reef, nearly 1,300 miles in length. This vast reef is a maze of coral reefs, islands, and shoals, and marks the submerged margin of the continental shelf along the northeastern coast of Australia. The barrier reef of New Caledonia, another major example, extends beyond the limits of the island and is almost 400 miles in length.

NATURAL VEGETATION

The transitional position of the Wet-Dry Tropics between two areas of climatic extremes results in a vegetative zone of transition from tropical selva through woodlands, then from grasslands to almost barren desert (see Figure 6.8). There are no sharp boundaries between grass and forest. The forest continues from the equator side of the region and slowly loses its identity poleward. The merging of forest and grass is imperceptible; the forest continuity is gradually broken by low bushes and patches of grass. The savanna soon governs the vegetation, with occasional trees dotting the landscape to give it a parkland appearance (Figure 6.9). The forest remnants are low, with umbrella-shaped crowns. In the African savanna an occasional baobab tree appears with its scraggly branches drooped over a huge bottle-like trunk. The hardy, gray-green eucalypti are distinctive, being indigenous to Australia, including Tasmania. There are numerous species—about 600 in all. Some closely rival the redwoods of California in size. Bordering the banks of streams entering or emerging from the Rainy Tropic regions are *galeria* forests, long tongues of the jungle or selva that penetrate into the savanna. Approaching the desert margins the grass becomes shorter and the cover less continuous, finally giving way to scattered scrub growth.

The heart of the Wet-Dry Tropics is dominated by seas of grass broken only by isolated trees. The grasslands often present a nearly impenetrable wall of rank vegetation, making travel on foot difficult unless wild-game, cattle, or native trails are available. The seasonality of the rainfall is reflected in the colors of the landscape; a monotonous vista of brown during the dry season is replaced by pleasing successions of greens following the rains. In some of the grazing areas, the end

FIG. 6.8. A simplified cross-section of the natural vegetation of the Wet-Dry Tropics. The graduation from trees to grass parallels the decreasing rainfall.

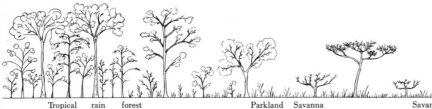

Tropical rain forest   Parkland Savanna   Savanna   Steppe

FIG. 6.9. Aerial view of a park savanna. Notice the spacing and the flat crowns of the trees. (Ray Bassett.)

of the drought is marked by dense clouds of smoke filling the skies when the parched grass is burned to make way for the new growth. The savanna grass usually grows as individual stalks and attains heights ranging from 4 to occasionally 20 feet. The first early shoots are tender and palatable to the grazing herds, but the grasses later turn coarse and stiff. Burning—purposeful for clearing croplands or to remove the coarse, dead grass to expose new shoots for grazing animals, as well as uncontrolled fires caused by both man and nature—has modified the vegetation of many areas. Some biogeographers suggest that the forest margin has been pushed back significantly and that the grass area has expanded accordingly as the result of fire.

NATIVE ANIMAL LIFE

The distribution of animals in the Wet-Dry Tropical regions is uneven. The savannas of Africa, in general, have greater numbers and varieties of species than their counterparts in the Americas and Australia. An abundance of insect life is common to all, but few large animals are present in the tropical savannas of the Americas. In Africa the great grasslands support vast numbers of herbivorous animals. Herds contain thousands of antelopes of various species such as the kudu, hartebeest, and gnu. Striped zebras blend with the grassy background. Towering giraffes browse near isolated tree groups (Figure 6.10). Carnivores such as lions and the scavenging jackals and hyenas prey on the grass-eaters. The savanna is also the habitat of the fierce African buffalo, one of the most dangerous of all game animals, and the home of the rhinoceros and the African elephant. Hippopotamuses sport in the streams, and crocodiles bask on the muddy banks. Monkeys and baboons inhabit the tree islands and surrounding areas. Snakes and lizards are common. Great ant nests serve as landmarks on the plain; some even provide shade for solitary animals. Many types of birds, including the swift-footed ostrich,

FIG. 6.10. Rhino in the savanna. Africa's unique wildlife constitutes an important resource, attracting large numbers of foreign tourists each year. (William C. Rense.)

live on the grasshoppers, locusts, and enormous numbers of other insects swarming through the air. In no other region does the animal life form such a colorful pattern in the panorama of the landscape.

The advent of the dry season places restrictions upon the animal activity of the African savanna. Large herds of grass-eaters, followed by the lions, leopards, and hyenas, must leave the dry areas and migrate to places where water is available. Elephants seek the galeria forests or climb the slopes of mountains to the higher forests. Insects and many animals become inactive during the dry season. When the rains begin, the savanna comes to life. Herds disperse from the water holes and divide into smaller groups. The air is soon filled with winged ants, termites, and newly emerged butterflies. All animal life responds to the termination of the drought.

The native animal life of the African savanna has been significantly modified by human interference. The advance of agriculture has reduced the area of habitat, overgrazing by domestic livestock and uncontrolled burning of vegetation has reduced the quality of the habitat, and these together with overhunting have reduced the numbers of animals. Some species have been diminished to near extinction and the dynamic equilibrium of the life forms of the natural environment has been upset. Control measures are starting to be applied based upon ecological principles. These measures include the development of reserves, control zones around reserves where excess game population can be reduced by hunting, hunting seasons and game limits, etc. In the face of increasing human population, it is doubtful, however, that animal life over large areas can be returned to the diversity and numbers of pretwentieth-century times.

In Australia animal life is less varied than in Africa. Insects, birds, crocodiles, and reptiles are common, but the unique feature of this continent, which has long been isolated from other landmasses, is its archaic types; here live the greater part of the world's surviving marsupials, the most famous of which is the kangaroo. Australia is known as "the land of living fossils."

SOILS

Owing to the transitional nature of climate and vegetation and to the diversity of parent material, it is difficult to make generalizations concerning the soils of the Wet-Dry Tropics. The soils have greater general fertility and better structure than those of the Rainy Tropics because of less rainfall and the dry season. They do show transitional characteristics paralleling the climatic change. On the wetter margins, soils are leached and are classed as lateritic. Toward the drier margins, the soils have developed under conditions of high evaporation, low rainfall, and grass vegetation—they thus contain more fertility elements. The primary agricultural problem on the dry margins results from the limited and unreliable rainfall. Throughout the areas where rainfall is sufficient for crop production, natural soil-degrading processes operate on the cultivated land, and permanent agriculture requires fertilizers and care in management.

Occupance in the Wet-Dry Tropics

Immense spaces in the Wet-Dry Tropics remain virtually unoccupied. These regions, up to the present time, have discouraged widespread settlement for many of the same reasons as the Rainy Tropics. Only the most favorable areas have been developed, and for the most part population densities are below 15 persons per square mile. Notable concentrations are found in the West Indies, where Jamaica has an average of 480 persons and Cuba 188 persons per square mile. In Mexico and along the Pacific coast of Middle America several prominent areas of concentration are found; in South America the major distribution is in eastern Brazil. The greatest densities in the African regions are in the *sudan* and especially in Nigeria. As in the Rainy Tropics, occupance is based mainly upon primary activities.

In recent years several nations have been placing considerable emphasis upon development in these regions. Many of the problems that have so long deterred progress and settlement are being met with new technology. Although advances are yet small, there appears to be little doubt that the future will bring greater utilization of the Wet-Dry Tropics.

INDIGENOUS OCCUPANCE

Some of the native peoples are in a primitive stage of development, whereas others, by comparison, are rather advanced. As a general rule, all are more advanced than is average for the Rainy Tropics. The intermittence of plant growth resulting from the seasonality of rainfall eliminates the possibility of obtaining a yearly food supply by simple gathering and continuous crop production; thus the people of the Wet-Dry Tropics must plan and provide for the dry season.

Few peoples are exclusively hunters, although game has been abundant in Africa. Hunting is carried on in a primitive manner; weapons usually consist of bows and arrows, spears, and occasionally blowguns. Some tribes smear poison on the tips of their weapons—this may either paralyze or kill the game in a few minutes. Pits, excavated and carefully concealed along game trails, are used to trap animals. The rifle has been introduced as the hunting weapon for some groups in recent years, greatly increasing their efficiency. In turn, this more effective weapon is partially responsible for a reduction in game supply. This in turn has resulted in the institution of control measures in some countries. In the long run it would appear that this form of livelihood will decline markedly.

Where lakes, streams, and the sea are available, fishing may become an important means of livelihood. This activity is especially noteworthy in the Bangweulu Swamps of Zambia. In a few areas, especially the West Indies, some people are commercial fishermen. Hunting and fishing are usually engaged in for the purpose of augmenting food supplies; native man is dependent either upon livestock grazing or upon crop farming to provide him his principal means of earning a living.

The Wet-Dry Tropics of Australia support only a few people. Most of the remaining 45,000 aborigines are scattered through the area; their sparse density is indicated by the fact that only about 4,000 occupy the 31,200 square mile Arnhem Aboriginal Reserve in the Northern Territory. These people, the vanishing remnant of an ancient stock, hunt, fish, and gather nuts, roots, and seeds for their sustenance, using largely the primitive weapons and implements of their ancestors. Only about 200,000 white people live in the entire northern third of the continent. They are concentrated mainly on the Queensland Coast, where they engage in commercial agriculture and associated trade and service activities. There are scattered stock stations on the savanna and a few small towns along the north coast; the inhabitants of some are employed in pearl shell fisheries.

NATIVE AGRICULTURE

On the equatorial side of the Wet-Dry Tropics, where the forests occur, the natives practice shifting agriculture. The tribes use similar methods and are faced with the same problems common to primitive farmers in the Rainy Tropics. Here, too, is the need for crudely-cleared patches in the forest and the abandonment and rotation of fields owing to short duration of soil fertility. The old plots revert under favorable conditions to a second-growth forest or to brush and are not cleared again for a period that may vary from five to fifteen years or more, depending upon population density and need. Soil fertility and brush or trees (for ash) have an opportunity to regenerate. Unfortunately, conditions sometimes are not sufficiently favorable to allow the development of a good second growth. Over large sections of the tropical forests there are extensive open grasslands, resulting from too frequent fire-clearing of the fields. In some areas pressure on the land has led to such a shortening of the period of rest from the traditional system of cultivation that serious soil erosion has resulted.

In recent years the destructive qualities of this form of agriculture have come under attack. It has been also labeled a backward stage of culture, with concern expressed for the increasing gap in the social status between its practitioners and those in the advancing economic sectors in the humid tropical countries. Increased scientific effort to lift the main ecological limitation of sedentary farming—rapid loss of soil fertility —along with education and restricting legislation can be expected as these countries press for economic and cultural advance.

Progressing away from the equatorial margins of the Wet-Dry Tropics, the dry season lengthens, grass replaces forest, and soils undergo less leaching. The farming possibilities are materially improved with the land easier to clear, the soil fertility more enduring, and the dry season favorable for land preparation, planting, and the complete maturing of crops such as grains. Under these conditions, it is not surprising that some have developed a rudimentary agriculture of the sedentary type in the savanna lands. The fields are cleared by burning the grass during the dry season. The degree and manner of tillage vary from group to group. Some simply use sharpened sticks; others use hoes or mattocks to work the soil. The rainy season is the growing period and the dry season, the ripening and harvesting period. Principal crops include millet, corn, wheat, peas, beans, sweet potatoes, and pumpkins, and in some African regions groundnuts (peanuts) and cotton are also important.

Regular trading with the grazing tribes is frequently carried on by the farming groups. Even though these cultivators are more advanced than the shifting farmers, individual land ownership has not been common. The farmers live mainly in communal villages. Houses are simple and often constructed with mud walls and grass-thatched roofs. Toward the drier margins farming becomes precarious as the result of less rainfall and the dangers of drought—here agriculture gives way almost entirely to grazing.

In the areas where there is contact with foreign settlement, the natives frequently produce agricultural products for sale. Developmental programs are under way in a number of the new independent countries. These programs are planned to lift the level of living of the natives as well as to increase exports and open new markets. Assistance is in the form of technical advice, machinery, capital, plans, and guidance. Complete sub-

FIG. 6.11. Sennar Dam on the Blue Nile. This dam stores floodwaters for use in Sudan's long dry season. The canal on the right carries water to the irrigated fields of the smooth, featureless plains of Gezira. (British Information Services.)

sistence farming is gradually giving way to farming combining subsistence crops with one or more cash crops. One of the most significant developments in agriculture is in Sudan, where the British sponsored several irrigation projects. The largest takes water from the Sennar Dam constructed across the Blue Nile to irrigate a portion of the Plain of Gezira between the two Niles (Figure 6.11). The dam was completed in 1925 and the canal system, in 1929. The project, totaled close to a million acres by 1952. Since Sudanese independence in 1956, a number of extensions to the Gezira have been planned to increase agricultural output and relieve overpopulation in the north. The main aspects are shown on Figure 6.12. In addition to the government-sponsored schemes there has also been some private investment in irrigation facilities.

With irrigation, the Gezira area has changed from a land of marginal agriculture limited by uncertain rain to one of a stable agrarian economy based upon a certain water supply and improved farming techniques. Today, it is a major producer of cotton, with nearly a quarter-million acres devoted to the crop. Cotton is systematically rotated with *dura* (millet) for food and *lubia*, a leguminous fodder for workstock and other animals.

In Latin America considerable portions of the export products are grown by nonwhite tenants[2] and small landowners. Their economy and way of life have long been influenced by association with white men and their techniques and by the nearness of American markets.

NATIVE GRAZING

Many of the native African tribes of the savannas depend chiefly on cattle for subsistence. The native breeds are able to withstand the heat, penetrate through the high grass, and defend themselves against wild

[2] The Negroes are native to the tropics but not to the Latin American region.

animals. They are used chiefly for milk and blood and are seldom butchered other than for festive or ceremonial purposes.

In addition to their use as a source of food or as beasts of burden, ownership of many animals elevates a man's social position —they are often traded for wives. Among these tribes, animals are the chief source of wealth and the medium for barter; a man's social position often depends upon the size of his herd. Around settlements located near a source of water, natives sow millet at the beginning of the rainy season. When the crop is in, the younger and more able-bodied members of the tribe move their herds in search of grass, while those who are unable to follow remain at home to care for the fields. The tribe returns during the dry season to harvest and store the grain.

NONINDIGENOUS OCCUPANCE

The presence of the white man in the Wet-Dry Tropics is explained largely in commercial terms. Until recently, at least, he has been in these regions to manage the exploitation of climatic advantages for agriculture or to exploit the grasslands, forests, and mineral resources. Much of the African savannas was held on a colonial basis, and the white man was there as an administrator or technician. Between the end of World War II and 1970 most of these ties were broken as independent nations

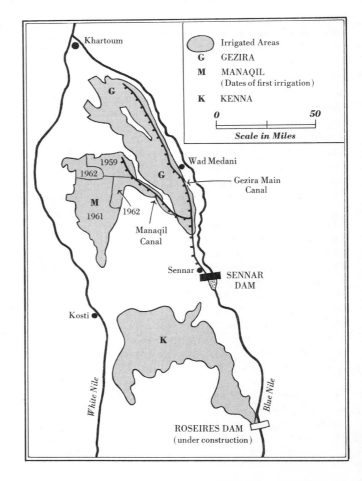

FIG. 6.12. Irrigated areas between the White and Blue Niles, Sudan.

emerged. In Latin America, where the white man came from Western Europe as a conqueror and exploiter, foreign political ties were, for the most part, broken earlier, although several of the Caribbean Islands and the former British and Dutch Guianas did not achieve independence until recently. Descendants of the Spanish and Portuguese are the dominating elements in many Latin American countries, but they have preferred to settle in the higher altitudes where climate is more moderate. The long period of their occupance has had profound effects on the population make-up, owing to the mixing of Europeans with the indigenous Indians and the African Negroes brought in during the colonial period as slave labor. Pure whites are in the minority.

COMMERCIAL AGRICULTURE

The establishment of commercial agriculture followed almost on the wake of the first European conquests in Latin America. The Portuguese developed important sugar colonies in northeast Brazil as early as the mid-sixteenth century, and that area continued as the world's chief source of sugar during most of the seventeenth century.

Late in the seventeenth century the West Indies' advantages for sugar cane production were recognized. These favorable elements included climate (tropical temperatures and long wet and dry season) an abundance of suitable land, accessibility to water transportation, and closeness to European markets. Slave labor was brought in from Africa, and soon a flourishing sugar production was established under European management. The main sugar islands have gained their independence, but commercial agriculture remains a major component of the economy. Although other crops are produced, sugar remains the leading enterprise with Cuba the principal world cane sugar exporter.

Important quantities of sugar cane are also produced in other portions of the Wet-Dry Tropics, including Brazil, the Middle American countries, and the Hawaiian Islands, where irrigation is necessary to compensate for the short wet season and dry shadow locations. The Hawaiian Islands are also the leading world commercial producers of pineapples. Improved varieties, fertilizers, mechanization, and chemical control of pests have made commercial agriculture here more advanced than elsewhere in the humid tropics. Cotton is a significant crop under lower rainfall inland from the sugar-producing area of northeast Brazil as well as in northern Argentina. The northern portion of the Yucatan Peninsula is the major producer of a hard fiber plant, henequen, which is a xerophyte structurally adapted to growing with a scarce water supply. Tobacco is grown in many places; commercial production is especially important in the West Indies and Rhodesia.

Commercial agriculture in the African realm, less developed than in Latin America, is found mainly along the rivers of the Sudan. As noted earlier, the French and British left a legacy of developmental effort, especially in irrigation. In several sections cotton, sugar cane, the hard fiber plant sisal, and rice are produced for commerce (Figure 6.13).

Much of the more advanced development in the tropical portion of Africa occurs in the highlands. The elevation-moderated temperatures have been especially attractive to Europeans, and hence their greater developmental effort was commonly in these areas. This will be discussed in a later chapter.

The Queensland Coast of Australia is notable among the developed areas of the tropics; sugar cane is produced by white farmers on family-size units. This activity was first organized as a plantation enterprise based on labor recruited from the southwest Pacific islands. When this *Kanaka* labor was prohibited early in the twentieth century, sugar production was reorganized on a family-farm basis with white ownership and labor, largely Italian.

FIG. 6.13. Sisal plants in the African Wet-Dry's. Sisal is an important source of fiber. (Ray Bassett.)

COMMERCIAL GRAZING

Despite the tremendous expanses of grasslands, the Wet-Dry Tropics produce relatively few livestock for commercial purposes (Figure 6.14). The grazing industry is confronted by a number of handicaps and problems: (1) During the dry season potable water is scarce, grass becomes harsh and dry and is poor forage for fattening cattle. (2) Floods, in the rainy season, inundate many acres of grazing land, especially in the Orinoco Llanos. (3) Cattle are subject to many diseases due to the prevalence of ticks and other insects. In Africa snakes and predatory animals take a toll of livestock, but the most serious problem is the tsetse fly. This pest, found throughout the Wet-Dry Tropical regions of Africa, accounts for a high death rate among the cattle population. Cattle raising is practically impossible in areas of heavy tsetse-fly infestation. (4) European beef breeds have not proved successful under existing conditions. (5) Marketing is difficult because of inadequate transportation and refrigeration—cattle are usually driven to the slaughterhouses and arrive in poor condition. In addition, higher quality beef meets sharp competition from more favorable producing areas.

The American savannas, due to early development, better transportation, and fewer wild animals, are utilized more for commerical grazing than the larger African grasslands. Much of the grazing industry in the Latin American savannas stems from hardy stock, descendants of cattle brought from Spain over 400 years ago. The industry is not based entirely on meat and hides, since milk for cheese is also significant in the local economy. In the Orinoco Basin, there are some large cattle ranches on which the herds are tended by *llaneros* or Venezuelan cowboys. The llanero lives on the plain in a simple mud hut with a thatched roof. His costume consists of a short-sleeved shirt and short breeches, and his equipment is the machete and a 60- to 75-foot rawhide rope. His chief activities are herding and, unlike the North American cowboy, in some instances milking and cheese making. The llanero is a man on horseback—he scorns any activity associated with crop production.

Several of the Latin American countries are giving attention to the improvement of the livestock industries. They are giving increasing care to disease control, and some have brought improved breed stock from the United States, including crosses between the tropic-tolerant zebu cattle (native of India) and European breeds, such as the Hereford and Shorthorn.

The northern fringes of Australia contain an advancing commercial cattle grazing activity. Stations (or ranches) on the Australian savanna are large; some in the Northern Territory measure their herds in thousands of head and land leases in thousands of acres. In Queensland, particularly on the coast, the operations are usually on a smaller scale. A better grade of beef is raised here than that produced in the Llanos or Campos of South America. Australian stockmen, however, are faced with problems similar to those encountered by the South Americans—seasonal drought, lack of supplemental feed, pests, and long distances to markets. Some advancements, yet minor in extent, are being made; these include the introduction of improved stock, the development of water sites, and the production of dry-season feed by irrigation in some of the coastal stream valleys. Experimentation with aircraft in transporting meat products to markets is showing promise.

FOREST INDUSTRIES

Lumbering is not a widespread activity since climatic conditions impose numerical limits on trees of commercial significance. Only on the equatorward margins, the hill slopes which receive heavy orographic precipitation, and in the galerias do we find forests comparable to the selva. One of the outstanding forest products comes from the quebracho tree, which grows in the open forest of the Gran Chaco in South America and forms the greatest single source of natural tannin in the world. Quebracho logs are assembled along river banks where processing plants chip the extremely hard wood and extract the tannin. In Africa, particularly in Zambia and Malawi, the governments are aware of the potential value of the trees of the woodland savannas. Teak and mahogany are presently being logged. Hardwoods have long been commercial products of the West Indies.

MINING

Minerals were the first lure that attracted white man to the Wet-Dry Tropics. The European conquistadors were spurred by the legends of hidden Indian accumulations of gold in Latin America. Extensive searches for natural deposits were made later, but in both cases the sources discovered were soon exhausted.

FIG. 6.14. Two young Masai herdsmen. Numbers of cattle constitute the sole measure of personal wealth to these people.

In the modern era, petroleum and rich deposits of iron, copper, sulfur, and bauxite have contributed to the importance of these regions. Development has been largely by foreign interests, and, until recently, a relatively small segment of the benefits of exploitation has accrued to native inhabitants. Venezuela is a leader in petroleum production, with major fields around the east shore of Lake Maracaibo, in the north central area, and west of the Lower Orinoco River. Across the Andes, Colombia has an important field in the central portion of the Magdalena Valley. Mexico also has a significant oil production on the east coast, with the principal field near Tampico. A major portion of the petroleum from the Latin American countries moves into world commerce—for example, about 70 percent of Venezuelan production is exported, nearly half coming to the United States and Western Europe. Latin America today is exercising more rigid control over development than was formerly true. Colombia and Venezuela have divided their oil areas into reserves; they are controlling concessions and requiring that foreign developers carry on the refining within the country. Greater benefits are thus resulting, with larger numbers employed and the life of the reserves lengthened. Concentration

FIG. 6.15. Cerro Bolivar, Venezuela iron mining development. This major tropical producer is located west of the Caroni River and south of Ciudad Bolivar. A 91-mile railroad has been constructed to the Orinoco River. The channel of this stream had been dredged to permit ore boats access to the loading terminal 195 miles inland. The ore is shipped to steel centers on the eastern seaboard of the United States. (United States Steel Corporation.)

of workers in the oil fields and refining areas is encouraging the establishment of agriculture to provide the necessary food supplies. Development of rich iron reserves in the Guiana Highlands south of the Orinoco River by United States interests is bringing added importance to Venezuela, now a major world producer and source of supply for the eastern seaboard plants of the United States (Figure 6.15). Jamaica has become an important producer and processor of bauxite in recent years. It also ships ore to the United States and Canada.

One of the major copper-producing areas of the world is in Africa, extending from the Katanga District south of the Democratic Republic of the Congo into the adjacent portion of Zambia. This is possibly the largest source area for copper—it is over 250 miles long and is in places 50 miles wide. These rich deposits, the ores of which have a very high percentage of copper, are being exploited by Belgian and British interests. Copper mining, smelting, transportation, and the production of food for workers have stimulated a concentration of population. This same region is one of the world's principal uranium sources, with the mine at Shinkolobwe the leading producer. About three-quarters of the world's supply of cobalt is also mined in Katanga. Tin and coal, mined in Nigeria, are the most notable mineral products of the Wet-Dry Tropics north of the equator in Africa.

Considerable mining activity has occurred in the Australian region in recent years. Large bauxite reserves are being developed on the northwest side of Cape York peninsula, and Yampi Sound is a major domestic source of iron ore. Also, petroleum is being exploited at Moomie in Queensland; a pipeline connects this field with Brisbane, 190 miles to the south.

New Caledonia, a long time producer of nickel, has been expanding its production of this metal recently. A major stimulus has been the decision of the French Government to permit competition in the nickel mining industry. As a result, production doubled between 1965 and 1970.

MANUFACTURING

Manufacturing in the Wet-Dry Tropics is concerned with primary stages of production, mainly of agricultural, mineral, and forest items. Plants are localized near the raw materials. Sugar is processed in numerous mills (called *centrals*) with cane by-products of molasses, rum, and bagasse, the residue fiber. Tobacco factories in Cuba produce cigars for export, pineapple is canned and sugar refined in Hawaii, and a few tannin extraction mills operate in the Chaco. Much of our hard fiber for twine comes from Yucatan; and Recife, Brazil has significant textile mills. A few South American slaughterhouses process the beef of the savannas. Venezuela has a number of oil refineries localized in the principal producing fields and the coastal termini of the pipelines; the offshore Dutch islands of Curacao and Aruba also have refineries.

An early concern of developing nations is the establishment of portland cement manufacturing facilities. This essential binder of concrete is a relatively bulky commodity that can not stand the cost of distant transport. Hence it comes to be produced as close as possible to the centers of major use. The raw materials for portland cement are limestone and a shale or clay (sometimes limestone has impurities of clays in desired composition). Also required is a source of heat. The raw materials are relatively widespread and in response to increasing development in a number of the Wet-Dry Tropic countries, cement manufacturing is now widespread.

A new industrial development area (about 60 square miles), financed by the government and directed by a government corporation, is taking shape at Ciudad Guayana, at the confluence of Rio Caroni and Rio Orinoco in eastern Venezuela. The area is on high ground above the high water potential of the flood prone rivers. Near here the railroad from the iron mining district about 100 miles to the south terminates at the ore dock on the navigable Orinoco. Hydroelectricity is available from one of the world's largest installations on the Caroni, and oil from the field to the north; alumina is imported. An aluminum plant and a steel mill are presently in operation. The steel mill, designed and built by an Italian firm in 1962, had a 1970 annual capacity of about three-quarters of a million tons. An important produce of the mill is steel pipes for the petroleum industry.

TOURISM

The Wet-Dry Tropics contain two groups of islands, the Hawaiians and the West Indies, in which the tourist trade is becoming increasingly important. The West Indies, favored by their proximity to the populated eastern section of the United States, are invaded by tourists during the winter season. The remoteness of the Hawaiian group has been broken by regularly-scheduled passenger service by sea and air. "Packaged tours" and installment payments are making island vacations possible for people in low and medium income brackets. These vacations include standard transportation, food, lodging, and directed scenic tours at one price. Tourist attractions are climate, sea and mountain scenery, exotic vegetation and animal life, fishing, and sandy ocean beaches with warm sea water suitable for bathing.

The "South Sea" islands of the Pacific are also attracting tourists. The newly independent Fiji had upwards of 85,000 tourists in 1971 and is planning for steady expansion. The West Coast of Mexico draws increasing numbers from the United States, especially during winter months, and tropical sport fishing is carried on at all seasons. As a result a number of tourist centers are evolving. These include Acapulco, an older attraction, Mazatlan, Guaymas, and Puerto Vallerte.

The big game of the African Savannas stimulates a different clientele—the hunter-sportsman—particularly in former European colonies of East Africa. Economic benefits to the locale accrue from guide services, provision of food and lodging, and sale of supplies. The safari, a big game hunt, once limited to sportsmen of substantial means, is now possible for those of the middle class.

URBAN CENTERS

There are more community and village settlements in the Wet-Dry Tropics than in the Rainy Tropics. This is particularly true in Caribbean America and the Sudan of Africa. The existence of only a few large urban centers again reflects the general absence of the city-forming factors of commerce, industry, services, and large populations. Many of the large cities that do serve these areas are found in the highlands and will be discussed in the chapter on the Tropical Highlands.

In the Wet-Dry Tropics proper, most of the major centers are located near the coast at favorable assembly and shipping points. The greatest concentration is found in the Caribbean area. Havana, Cuba's capital and chief port, has a population of over 1.5 million. Port au Prince, Haiti, and Kingston (including Port Royal), Jamaica, both capitals and commercial centers, have populations of 250,000 and 460,000 respectively. Veracruz, Tampico, and Merida are the main east coast centers of Mexico. Panama City, with about 375,000 people, located on the Pacific side of the canal, is the major city and capital of the Republic of Panama (Figures 6.16 and 6.17).

Barranquilla (625,000) and Cartagena (310,000) are large lowland cities, and are major ports of Colombia; the former is

FIG. 6.16. Panama City. This city was founded in 1519 by the Spanish as a concentration and shipping center. Ravaged by the pirate Henry Morgan, the city was rebuilt a few miles from the original site. Modern growth is associated with its location near Balboa, the terminus of the Panama Canal. Here is found a blending of old and new buildings, with sharp contrasts such as that displayed between the modern Hilton Hotel and the antiquated buildings in the tenement district. Note the many buses, the major means of transport for modest-income people. (Oliver H. Heintzelman.)

FIG. 6.17. The University of Panama, on the heights above Panama City. Centers of learning such as this are major keys to the future development of the tropical countries where science, technology, and education in general are at a low level. Note the covered walks, an adjustment to a feature of the humid tropical climate—the downpour! (Oliver H. Heintzelman.)

FIG. 6.18. Barranquilla, major city and port of Colombia. The harbor is ten miles up the Magdalena River, which has been dredged to permit access by ocean-going vessels. The Caribbean is visible in the background. (Oliver H. Heintzelman.)

situated at the main mouth of the Magdalena River (Figure 6.18). In Venezuela, oil development is responsible for the growth of Maracaibo, the only large lowland city in that country. The northeast coast supports the significant cities of Wet-Dry Tropical Brazil. Recife, railroad terminal and chief outlet for an important agricultural area, has a population of around one million. Fortaleza, with over 350,000, serves similar functions for an area farther north. In 1960 Brazil moved her capital to Brasilia, in the Campos. The planned city of unique design has a population approximating 600,000. The

FIG. 6.19. Santa Marta, Colombia. Less than a decade ago, Santa Marta's economy was chiefly associated with banana plantations in the hinterland. Today the splendid beach and accommodations attract numerous tourists. (Anne Hollingshead.)

move was instigated by a desire to obtain a more central position for the seat of government as well as to stimulate interest in the development of the interior. On the west side of the continent, Guayaquil (650,000) is the port and leading city of Ecuador.

The major city of the mid-Pacific Ocean is Honolulu, in Hawaii. Chief port, military and naval base, and commercial and tourist center, it contains about half the population of the island group.

In Wet-Dry Tropical Africa there are five centers with populations above 200,000: Ibadan and Lagos in Nigeria, Accra and Kumasi in Ghana, Abidjan in Ivory Coast, and Omdurman in Sudan.

The underdeveloped character of northern Australia is indicated by the fact that Darwin, the largest center on the north coast, has a population of about 10,000. Somewhat greater activity on the east coast of Queensland has resulted in several centers. Townsville and Rockhampton, with less than 100,000 each, are the largest.

The major cities of Surabaja, Surakarta, and Djakarta are situated in the eastern, Wet-Dry Tropical portion of Java.

seven

MONSOON TROPICS

The Monsoon Tropics of Asia are special wet-dry climatic regions. The seasonal pattern of rain and drought is the same as that of the Wet-Dry Tropics but results from the climatic influences of the great Asiatic landmass, and the wide expanses of tropical ocean waters. This region of Asia is *the* populated land of the tropics—the good areas teem with millions of people. The press of an increasing population stimulates a continual demand for more food, and the imprint of intensive agriculture is deep in the land; the soil has long been the intimate of generations of toilers. Much of the cropland receives little rest, for the threat of famine makes farming an essential activity in dry as well as wet seasons. Man's history in the region had its beginnings thousands of years ago, and culture today is the product of a rich and varied past.

Location

The Monsoon Tropics are found in the same latitudes as the Wet-Dry Tropics in Latin America and Africa. Due to the influence of the monsoons, however, this climatic type extends further poleward in Asia, reaching 30°N in Pakistan and India (see Figure 7.1). The region stretches from the margins of the Thar Desert and the mountains in the northwest through West Pakistan, India, Bangla Desh, Burma, Thailand, Cambodia, Laos, North and South Vietnam, and into south China. Hainan, Taiwan, and the western Philippines are also included (Figure 7.2).

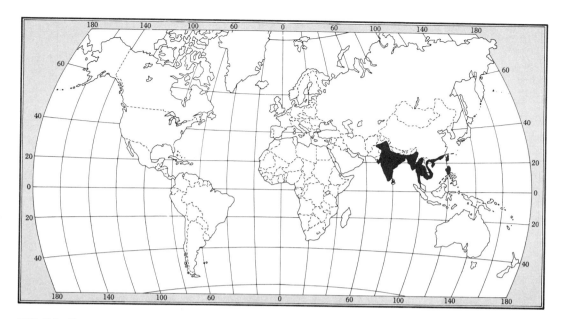

FIG. 7.1. The Monsoon Tropics

FIG. 7.2. The Monsoon Tropics of South and Southeast Asia

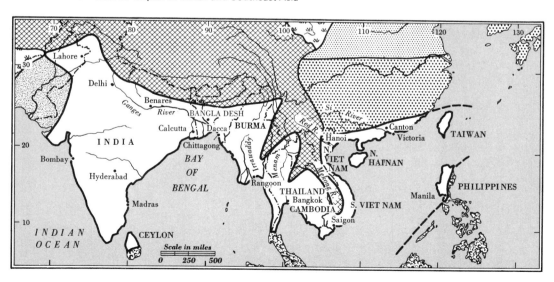

Physical Environment

CLIMATE

Monsoon comes from an Arabian word meaning "season of winds," but to the millions of people living in India and Southeast Asia it is synonymous with life-giving rain. Nowhere in the world do so many people anticipate and depend so much upon a phenomenon of nature.[1]

[1] The phenomenon of the monsoon is found in other continents, especially North America and northern Australia, but the greatest development is in Asia.

The monsoon may be thought of as a seasonal land and sea breeze on a continental scale (see Figure 7.3a and 7.3b). During late May and early June, the low pressures that have been created over the Asiatic landmass induce a flow of warm tropical air from the sea toward the continent. The air has passed over warm ocean water and is laden with water vapor—this becomes the source of precipitation. In the winter months the winds are reversed, since the air from the cooled continent and its high pressures streams toward the lower pressures of the warm ocean.

FIG. 7.3a. The Asian Monsoon—in summer, air moves in a generalized direction from the tropical seas toward the lower pressures.

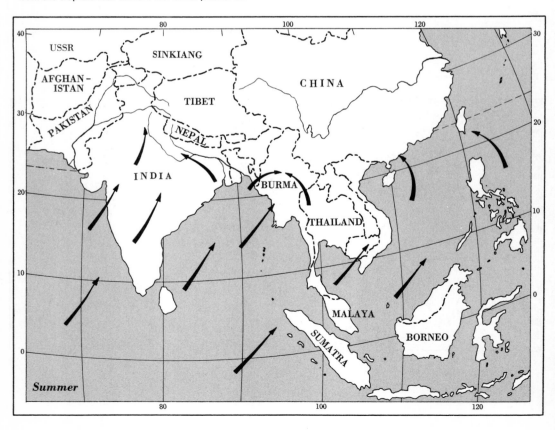

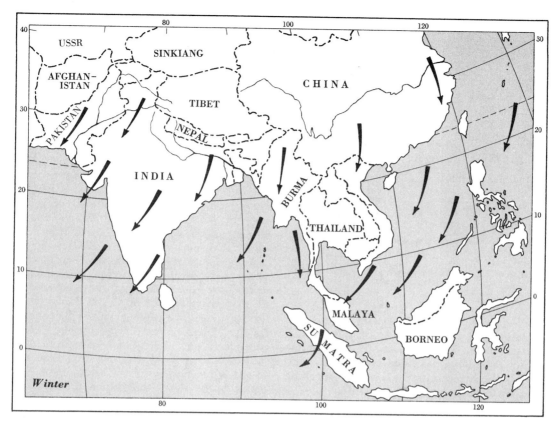

FIG. 7.3b. The Asian Monsoon—in winter, air moves from the higher pressures of the continent toward the lower pressures.

The dry season occurs in the Monsoon Tropics when the winds blow from the land. If the winter winds pass over water, coastal areas receive a second season of rain in winter.

The Monsoon Tropics, demonstrated by the climate of India, have three seasonal divisions: the cool season from October through February, the hot season from March through May, and the season of the rains from June through September. During the cool season, the sky is nearly free of clouds and there is little rain. Temperatures hover in the mid 60s on the northern margins of the country and increase to about 75° in the south. Humidity is low, but increases southward with higher temperatures. There are steady, though relatively weak, winds blowing from different quarters of the north. The cool season is generally the most pleasant time of the year, although lower temperatures occasionally cause discomfort. In March, a rise in temperature begins, skies remain clear, and there is little rain. The heat of the day becomes intense, with thermometer readings rising well over the 100° mark. Hot winds desiccate the vegetation and often fill the air with dust. The heat continues to increase and reaches its maximum just before the "bursting" of the onshore monsoon. June heralds the beginning of the wet season—thick clouds cover the sky, humidities rise, and daily convectional storms send down heavy showers. Plants send forth their foliage, and everywhere

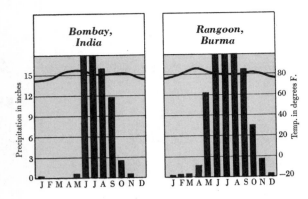

FIG. 7.4. Typical climatic graphs of Monsoon Tropic stations.

the earth is mantled in green. There is a marching effect to the sweep of the monsoon across India. The monsoon first strikes the tip of the Indian peninsula in early June and gradually sweeps up the country, reaching the north areas later in the month.

Temperature. Monsoon Tropic temperatures follow almost the same patterns as those of the Wet-Dry Tropics. The highest readings are recorded prior to the beginning of the rainy season. A majority of Monsoon Tropic stations have their highest temperatures in May and early June, with hottest month temperatures averaging 85° to 90°. January is the coolest month, with a 65° to 70° average.

Precipitation. Approximately 60 percent of the rain comes during the summer months of June, July, and August. Stations, unless subject to heavy orographic rainfall, receive amounts ranging from 30 to 80 inches depending upon distance inland and exposure (see Figure 7.4). Convectional storms account for much of the rainfall, but the rising of moist air along various mountain barriers also provides orographic precipitation. Cherrapunji, for example, located on the south slopes of the Khasi Hills in northeast India, is one of the wettest spots in the world, with a yearly average of about 450 inches. One year the station received a high of 1041 inches; 41 inches have been recorded in one day, and 366 inches in a single month! Cherrapunji exemplifies the wet-dry regime at its extreme—the July average is 109 inches, whereas December's is but 0.2 inches.

There are a few exceptions to the general rule of summer rainfall. Most notable are the southeast portion of the peninsula of India and the Annam Coast of Vietnam, which receive winter maxima as the result of the winds blowing out from the continent, crossing the water, and again moving onshore.

Monsoon precipitation typically fluctuates in amounts from year to year, especially along the drier margins. The impact of drought and flood is quite far-reaching, since either disaster may bring famine to the crowded agricultural villages. The Indian subcontinent is subjected to tropical cyclones, the most destructive disturbances in the atmosphere. For example, the storm that struck the coast of East Pakistan in 1970 caused millions of dollars worth of damage and the loss of many hundreds of lives.

SURFACE FEATURES

Asia. Landforms in the Monsoon Tropics are a significant factor in the production of climate differences and in land-use possibilities. Since this is a region of wet-crop farming, it can be readily understood that landforms are also a main control of population distribution. The plains are densely peopled, but the highlands (which restrict agriculture) have, by comparison, few permanent inhabitants. Although there are no large structural plains, several sizeable river valleys are distinctive for their intensive settlement and are equally easy to locate on physical or population maps.

Two great peninsulas comprise the bulk of the realm: India-Pakistan and Southeast

MONSOON TROPICS   139

Asia. Three major physical divisions are easily recognized in the India-Pakistan peninsula (see Figure 7.5). The Himalayas in the north, the greatest mountains in the world, together with associated ranges in the west and east, effectively isolate India and Pakistan from central Asia. Inside this mountain wall the Ganges Plain, the depositional product of the Ganges River and its tributaries, sweeps 1500 miles in a broad arc from the northwest mountains to the Bay of Bengal, forming the largest lowland in the Monsoon Tropics and largest alluvial plain in the world, comprising some 300,000 square miles. The Brahmaputra, an important river in the northeast, rises in Tibet and flows parallel to the Himalaya Mountains for many miles; it breaks around the eastern end of these highlands, creating an important valley in Assam before turning southward to join the Ganges. Together these

FIG. 7.5. Physiographic diagram of India-Pakistan.

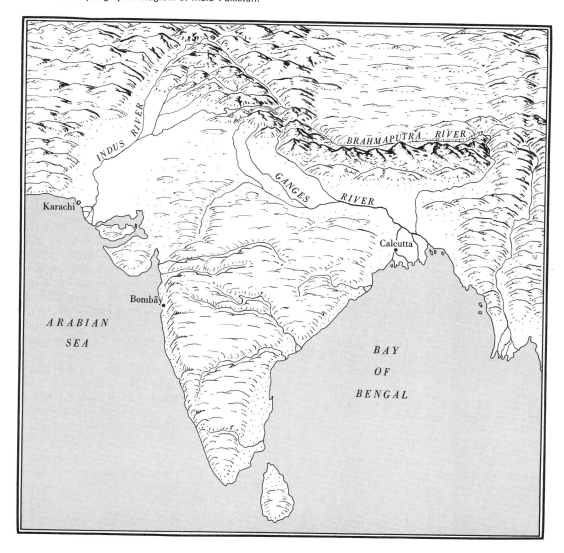

streams have built a large and complex delta at the head of the Bay of Bengal.

South India, the third major division, is primarily a plateau rising abruptly on the west coast. This wall-like edge of the plateau is called the Western Ghats; the land slopes rather gently eastward from it, terminating in the Eastern Ghats, a lower and less continuous line of hills. The plateau, sometimes called the Deccan,[2] an undulating surface with an elevation of about 2000 feet. It has had a complicated geological history. Much of the plateau is composed of very ancient crystalline rocks, but in the hinterland of Bombay enormous lava flows have buried these, and today an area of 200,000 square miles is still covered by volcanic deposits. The narrow plains bordering the peninsula are known as the Malabar Coast on the west and the Coromandel Coast on the east.

Forested mountains dominate the landscapes of Southeast Asia. From the lofty heights of eastern Tibet, many ranges spread fan-wise toward the sea. Several extend through Burma, one forms the backbone of the Malay Peninsula, and the Annam Mountains fill most of the northwestern and eastern portions of the countries of former Indochina. The mountainous interior, characterized by sharp ridges and deep gorges, is largely inaccessible except from the south. Toward the coast several aggrading rivers and streams that deposit sediments have built floodplains by deposition and are projecting their deltas seaward. Five river valleys are the major lands of settlement: the Irrawaddy, including the Mandalay Basin in Burma, the Menam[3] in Thailand, the Mekong in South Vietnam and Cambodia, the Red in North Vietnam, and the Si in China all have densely populated valleys.

NATURAL VEGETATION

Natural vegetation reflects the pattern of wet and dry seasons. Forests dominate on the wetter margins and are replaced by scrub thorn forest and savanna where drier conditions prevail. Vegetation patterns are quite similar to those found in the Wet-Dry Tropics; several factors, however, account for some differences. Human modifications are much greater in the Monsoon Tropics. Any land suitable for crop agriculture in the populated monsoon regions was cleared of its natural cover long ago. The need for wood, moreover, has resulted in repeated cuttings of neighboring forest through the centuries. Only in the more inaccessible and nonarable areas are typical vegetative communities represented. The Ganges Plain of India is devoid of trees in many places due to extensive cultivation; the resulting lack of wood for fuel leads millions of people to burn dried cow dung, thus preventing its use as a fertilizer. The Ghats, Himalayas, and the numerous highlands that fan out into Southeast Asia receive heavy precipitation on their windward slopes and support monsoon rain forests. In contrast to the selvas of the Rainy Tropics, these forests have fewer species, and the trees are smaller in girth. Broadleaf deciduous trees, which lose their leaves during the dry season, are prevalent, and pure stands are common. Several species, such as the teak tree and multiple-use bamboo,[4] are typical of the Asian realm. Monsoon forests have wide tree spacing, resulting in little interlocking of crowns, as is characteristic of the selva, and therefore the penetration of light to the forest floor stimulates the growth of dense jungle understories. Thick tidal forests of mangroves flourish on the low deltaic plains of the rivers

---

[2] The term *Deccan* is variously used; some restrict its usage to the area of lava flows; some apply it to the area south of the Narbada River; others use it to embrace all areas south of the Indo-Gangetic Plain.

[3] The full name of this river is Menam Chao Bhraya, which westerners usually shorten to Menam, which means the river.

[4] Bamboo is botanically a grass, although it often grows to heights of 60 or 70 feet.

along the coasts. An example is the Sunderbans, a thick wall of stilted mangroves along the complex system of channels in the mouths of the Ganges-Brahmaputra.

NATIVE ANIMAL LIFE

The Monsoon lands of Asia have a surprising number of wild animals despite the heavy concentrations of population. The forest and savanna are the habitat of a great variety of species, and animal life is comparable to that of the African Wet-Dry Regions, except for the absence of the vast herds of grass-eaters. Insects, birds, and reptiles are common. Poisonous snakes are especially prevalent, and each year in India thousands of deaths are attributed to the king cobra and to small, deadly vipers. There are several species of monkeys; colonies inhabit the fringes of cultivated land and often make raids on fields and orchards, where they are usually left unmolested by the pious Hindu farmers. The rhinoceros and elephant represent the large vertebrates. Although elephants have been domesticated and are the beasts of all work, many wild herds still exist. Asia is the only habitat of the tiger, king of the cat family. In their old age they sometimes become man-eaters, preying upon the inhabitants of the outlying villages. In India professional hunters may be hired to kill these animals. Excessive hunting for pelts or sport has depleted the tiger to the point where protection may be necessary.

SOILS

The mature soils of the Monsoon Tropics, having developed under similar conditions of temperature, rainfall, and vegetation, display characteristics much like those discussed for the Wet-Dry Tropics. Paralleling the diminishing precipitation and the change from forest to grassland vegetation, there is gradation from red and yellow lateritic soils of low fertility to the better quality chestnut and brown groups of the pedocals. Black soils, known as *regur,* occur over a wide area in the western portion of the Deccan. They have developed, largely upon lava, under modest rainfall and grassy vegetation. The regur have an excellent water-holding capacity which partially compensates for the relatively low rainfall of their locale and, in contrast to other plateau soils, are high in fertility. Much of the agriculture is found on alluvial soils of river floodplains and deltas which, owing to relative youth and diversity of source, are usually productive.

Occupance in the Monsoon Tropics

In contrast to the other regions of the tropics, which for the most part are thinly populated, the Monsoon lands stand out as one of the truly dense population blocks of the world. Over one-quarter of mankind is concentrated in South and Southeast Asia. India, alone, had an estimated population of approximately 555 million in 1970; this gives an average 1970 density of 460 per square mile and an average of 830 per square mile of crop land. A high population growth rate is an overriding problem of the region. The current rate in India is adding about 14 million a year; if sustained, this will give India a population of a billion before this century ends. Recent improvements in productive outputs have not yet resulted in per capita gains because of this problem.

There is unequal distribution of population. The rural densities are so great in the lowlands of the Indian Peninsula and the river valleys of Southeast Asia that 600 to 900 persons per square mile are usual, and concentrations of over 1,500 are not uncommon. The majority live in farm hamlets and small villages, over 600,000 of which are found in India and Pakistan.

The clustering of farm homes and farmsteads, commonly on lands least satisfactory for crops, along with minimal roads, trails, and dispersed buildings, are, in part, associated with the traditional philosophy of productive occupance of good lands. The land capable of supporting people under existing conditions is almost fully occupied in India and Pakistan, but in Southeast Asia there is an incomplete development of cropland. There is extreme crowding in the best lands of the river plains, but large upland areas that could be brought into use with clearing, irrigation, or by transportation development are now only sparsely populated with primitive tribes.

Southeast Asia, along with adjoining mainland areas and the islands to the south and east, is one of the most fascinating centers of cultural contact and movement on earth. A distinguished geographer has claimed Southeast Asia as the probable "culture hearth" of agriculture where man first practiced plant cultivation. Over thousands of years peoples have mingled here, drifting south through the passes and down the great valley trenches, mixing with or driving older groups before them.

The invasions of northwest India by Aryans around 2000 B.C. gradually led to Hinduism, which overflowed India to the coasts of Southeast Asia about the time of Christ. This colonization was matched by migrations of Chinese merchants and the establishment of Chinese rule on the east coasts of the region. From the ninth to the fourteenth centuries, Moslems (Moguls) spread over India and beyond and were reinforced in the fifteenth century by Moslem traders from Arabia who controlled and converted large groups in Malaya and the East Indies. Beginning in the fifteenth and sixteenth centuries, Europeans carried western political and economic systems and Christianity into India and Southeast Asia. More recently, a strong Chinese cultural overlay has been added to many areas in Southeast Asia.

These waves of penetration have influenced every aspect of life in the Asian Monsoon Tropics. Cultural clashes between Hindus and Moslems were the principal reason for the division of India and Pakistan. In addition to the two great religions, there are seven others, plus more than two hundred languages in the Indo-Pakistan realm. Language, religion, and other cultural differences have tended to separate groups into communities—a condition that has been detrimental to economic development and political unification. Racial and cultural diversity also characterize Southeast Asia; there are five major races and at least a score of lesser races in Burma and four major and several lesser races in Indochina. This diversity has led Professor Broek to characterize the region as a "social crazy quilt."[5]

Today the region is one of the world's centers of change. Differences are beginning to be compromised. Following World War II Burma, Indonesia, India, Pakistan, Ceylon, and the Philippines emerged as independent states. All of these are now engaged in programs to elevate the level of living of their people in terms of education as well as material well-being through fuller resource development and improved technology. At the same time, their voices in world affairs are rapidly gaining importance. The rise of nationalism among the peoples of the former French colony of Indochina resulted in the forming of four independent countries in the mid 1950s: North and South Vietnam, Cambodia, and Laos. As yet, however, they have been unable to meet the challenges of independence and build viable national entities owing to the continuing strife of war.

[5] Jan O. M. Broek, "Diversity and Unity in Southeast Asia," *Geographical Review,* 34 (1944), 175–95.

FIG. 7.6. Rural dwelling near Madras. (J. Granville Jensen.)

SUBSISTENCE AGRICULTURE

The people of the Asian region are dominantly subsistence farmers. Farming here is the product of a very long history. Two hundred years ago the practitioners were as advanced in techniques as the farmers of Europe or elsewhere. But following the industrial revolution the Asian farmers fell behind. They gained little from the Western advances in resource-converting and space-adjusting techniques that increasingly brought other elements of nature into the service of man; agriculture based upon traditional techniques remained as the pervading influence in the way of life with the self-sufficient family the basic unit of the economy and society. The farming system has promoted the large family; yet the land base has not been sufficiently large to provide adequate support for increasing numbers of people.

Intensification of the traditional form of occupance has been the principal approach to supporting the larger and larger population; crowding of farmers into the areas suitable for crop production (especially the river valleys and plains), decreasing per capita land base, fragmentation of holdings, and the "gardening" nature of farming are manifestations.

The overriding characteristics of farming in this realm are: subsistence orientation, although considerable amounts of products do move to urban centers and even to world markets (not from individual farms, but from the aggregate); major attention given to food crops, especially grains; the small size of individual farms, often no more than two acres, generally less than five acres, and commonly fragmented into a half dozen or more scattered plots; private ownership of land and family-unit organization, but with considerable landlordism and tenancy; evidences of infinite care of the land—in surface shaping, terraces, dikes for paddies, control and use of weeds, and so on; intense farm practices featuring multiple cropping, intertillage, collection and use of natural fertilizing materials, and so on; dominance of the employment of human power and simple and traditional methodology.

The complete food supply must be produced on the farm. Crops grown are selected with a view to giving balance as well as greatest possible yields. Rice, the highest yielding member of the small grains, is the favorite. The crop probably originated in India and spread to China by 3000 B.C. In the interval the improvement of the conditions for its production has significantly modified the physical system of much of the farmland. Although the rice plant may be considered a facultive hydrophyte, the bulk of the crop is grown on wet fields with three to eight inches of water. Important requirements are smooth surface, impervious subsoil, and supplemental fresh water supply (as well as warm temperatures; 52° to 54°F for germination, 72° to 73° for flowering, and 69° to 70°F for grain formation and seasons ranging from 120 to 160 days depending upon variety). Much effort has been exerted through the centuries to adapt cropland to rice. In areas receiving over 80 inches of rainfall, it is the main crop, but where rainfall is below 40 inches, water supplies become limiting in the distribution pattern; it remains an important crop in the intermediate areas. On millions of acres of the plateau of India, in northwest Pakistan, and the Mandalay Basin of Burma, farmers must grow wheat, sorghum, millets, or barley instead of rice in a single-cropping system. Grains dominate cropland use, but each farmer produces various other items such as sugar cane, sweet potatoes, beans, peas, a variety of vegetables, tropical fruits, and oil seeds, notably sesame, rape, and peanuts.

Irrigation is extremely important in the food production system of the region. For most rice production the practice may be considered supplemental; the purpose is to increase the amount of water provided by natural rainfall to the levels needed for rice. Accordingly through the long history of production a great many irrigation techniques have developed. Some depend upon shallow wells with water drawn by various fulcrum and windlass arrangements, but most are cooperative efforts of villages which depend upon small stream diversion.

In India and Pakistan, irrigation development has increased the cropping possibilities and added stability to agriculture in many sections where uncertainty and irregularity of rainfall restrict normal production. Over most of the area irrigation is necessary for crops during the dry season of offshore winds. The type of irrigation practiced varies with the source and supply of water. Canals are the most important, providing water for at least half the irrigated acreage. The major development is in the middle Ganges Valley and the Punjab, Land of Five Rivers, where rainfall is low and undependable but perennial streams are available. Wells provide the source of water in many districts, but especially in the lower Ganges Valley where the water table is high. Tanks are the principal means of providing water for irrigation on the plateau; by this system the rain runoff of

FIG. 7.7. Bullocks are typical Indian work animal. (Anne Hollingshead.)

FIG. 7.8a. Clumps of rice seedlings are pulled from the seed bed for transplanting in the paddy field. (J. Granville Jensen.)

the wet monsoon is impounded in drainage channels by simple dams for diversion to the fields at the end of the rainy season.

The integration of livestock with crop farming is minimal. Most farmers are direct consumers, not being able to afford the conversion loss that would occur in feeding animals and consuming meat and other animal products. None the less, there are many animals in the region, especially swine, goats, and poultry. These can be produced in confinement and fed waste parts of food plants or they can spread utilization to nonarable lands within the farming areas. Farmers with the larger holdings also may keep some workstock. Bullocks in India and water buffalo in Southeast Asia are used for draft power.

There are a number of significant variations from the general characterizations made above. The opportunity for and practices of multiple cropping varies with climate, population density, and desire—the multiple-cropping index ranges from about 2.0 in Taiwan to 1.14 in India and Pakistan, and the practice is negligible in Burma and Thailand. There are differences in the level of techniques; the farmers of Taiwan (Japan and South Korea) are, in general, more advanced than elsewhere. There are also differences stemming from political organization; espe-

FIG. 7.8b. Rice transplanting in the paddy field. (J. Granville Jensen.)

cially significant in this respect has been the growth of communistic forms of government in parts of the region. The impact thus far has been upon land and labor organization and the allocation of inputs and control of agricultural output. Particularly important in the communist areas has been the disappearance of the family farm unit. Actually, government-level planning and development stimulation have become notable influences on change in most of the countries of the region. Also, there are differences in the relative role of agriculture in the total economies.

COMMERCIAL AGRICULTURE

Although it is primarily a land of subsistence agriculture, many Asian products find their way to city markets and some into channels of world trade. The aggregate is large. Many farmers are tenants and must share the harvest with their landlords, who in turn sell surpluses. Even the poorest farmers require certain essentials such as clothing, salt, and matches—they must sell produce to provide the cash for these purchases. In the area of rich black soils, located inland from Bombay, many farmers grow a crop of cotton for market. Most of the world's jute, a coarse fiber use for making burlap and gunny sacks, is produced on the wet delta lands of the Ganges and Brahmaputra Rivers.

Some commercial agriculture developed under European management and capital. The British were responsible for the tea industry of India, establishing plantations in the hills of central Ceylon (now independent of India), the Assam Province, the Himalayan slopes, and the hills at the southern end of the peninsula. The British also sponsored commercial rice production in the Irrawaddy Delta of Burma, the French developed some plantation agriculture in Indochina, and native landlords spread into the rich delta lands of Thailand. Burma, Thailand, and Cambodia are the principal surplus-rice producers and exporters of Monsoon Asia.

GRAZING

There has been little development of a commercial grazing economy in the Monsoon lands. This statement appears as a paradox, since the Asiatic region raises two-fifths of the world's goats and India alone has 175 million cattle and approximately 100 million goats and sheep. Goats, hardy animals needing little care, are reared principally for milk and hides as part of the subsistence agricultural pattern. The Zebu cattle of India have a religious significance and thus play a very limited role as a source of food supply; they are used, however, for pulling carts and plows and to supply hides and meager amounts of milk. Sheep are raised principally for wool. The Indian Government has recognized the seriousness of the livestock problem and is attempting to improve the situation through such measures as the introduction of better breeding stock, sterilization of low-quality male animals, and encouragement of the use of dairy products.

FOREST INDUSTRIES

Teak in the forests of Burma, Thailand, and Indochina is the leading commercial tree species of the region. Other important forest products include sal, bamboo, and rattan. Teak is used chiefly in ship construction (Figure 7.9); it resists salt water and contains an oil preservative which prevents iron from corroding when in contact with the wood. Resistance to termites and fire also favors its use for cabinets and for construction wood. Teak logging is complicated by the rugged terrain, dense vegetation, and heavy rains. The tree, too heavy to transport when green, must be girdled and allowed to dry for three years before felling. Cutting is done in the dry season, and the logs are hauled by elephants to be floated down the swollen rivers during the wet season to Rangoon, Saigon, and Bangkok, the concentration ports.

MINING

The most significant mineral-producing area in the Monsoon Tropics is situated in the uplands of India 200 miles west of Calcutta. The country's most important reserves of iron and coal, along with deposits of manganese, mica, and limestone, are concentrated here. The iron belt contains one of the largest and best stores of iron in Asia. Manganese is widely scattered, with the largest deposits in the central portion of the plateau. India is one of the major world producers of manganese and is the leading supplier of sheet mica.

The mineral wealth of Southeast Asia is varied and large, but inadequate transportation has limited development to only the most accessible and valuable deposits. Tin, tungsten, lead, zinc, copper, coal, petroleum, and gem minerals are mined in Burma; tin, tungsten, lead, zinc, iron, and phosphate are mined in Indochina; tin is the only mineral produced in noteworthy quantities in Thailand which has recently ranked third, after Malaysia and Bolivia, in output. The major portion of these minerals move out to European and American markets, much in semiprocessed form or as ores. Owing to the in-

FIG. 7.9. Indian sawmill. Hand sawing of lumber is common in the Monsoon Tropics. Slow and tedious, this method results in low output per man. This typifies the tremendous expenditure of human labor and low unit production of the underdeveloped countries. (United Nations.)

creasing world demand, however, the nations of Southeast Asia can be expected to increase their incomes from minerals by further processing as well as through increased mining development.

MANUFACTURING

The Monsoon Tropics—with greater and more diversified amounts of raw material as well as millions of consumers—lead the tropical areas in manufacturing activities. India, as a nation, leads in industrial development. Handicrafts practiced for generations still typify a large portion of manufacturing, with goods produced in small home workshops for local consumption. The country is making rapid strides toward modern industrialization (Figure 7.10), and a number of factories are processing the products of field, forest, and mine. There are three distinct manufacturing regions in India: (1) the Bombay district on the margin of the cotton-growing area is the leader in cotton textile manufacturing; (2) the Calcutta region is more diversified with jute, rice and flour mills, wool and silk weaving, wood products industries, and sugar and oil refineries; and (3) Jamshedpur, in the northeast part of the plateau where coal and iron occur close together, is the iron and steel center. Although far behind India in industrial progress, Pakistan, in the few years since the partition, has been rapidly constructing textile mills, machine shops, and other manufacturing facilities to improve the standard of living. The development of a modern manufacturing sector, although increasing output and total value of manufactures, has not necessarily meant more employment. Modern machines are exceedingly more efficient than human hands.

Under the colonial systems, Southeast Asia functioned as a supplier of raw material, in much the same way as did India, but manufacturing was even more retarded. Industries in Burma are concerned almost entirely with rice, petroleum, and timber; in Thailand, with rice and timber; and in Indochina, with rice, other agricultural products, and timber. The people of these areas are skilled in a variety of native crafts. Some of their products, such as the lacquer wares and ivory and wood carvings of Burma, have found exclusive markets abroad.

More industrial development can be expected in Monsoon Asia with growth in technology and capital. The raw materials are

FIG. 7.10. Indian industry: a worker operating a modern machine tool—a sharp contrast to the traditional village handicrafts. (Government of India Information Services.)

FIG. 7.11. Street scene in Old Delhi.

available, and a market and a need for expanded production exist.

URBAN CENTERS

There are more cities in this climatic region than in any other within the tropics. Although they represent a small portion of the population, there are over fifty cities of the 100,000 class in India. Calcutta is the largest city of the northeast and of India with a population of nearly 8 million. With a port stretching 20 miles along the Hooghly River, a Ganges distributary, it occupies a strategic position for serving the commercial needs of the rich northern plain. Bombay, the second largest city of the nation, with close to 6 million people, is located on an island and serves as the gateway and industrial center of the West. Madras, the principal city and port on the east coast, Hyderabad, the main city on the Plateau, and Delhi, the chief center in the upper Ganges Plain, have populations ranging from 1 to 4 million (Figure 7.11). Most of the large cities of India are commercial centers, some derive additional importance from being provincial capitals, others have developed long standing handicrafts, several have become industrial, and a number, such as Benares in the middle Ganges Plain, are religious centers. Growing cities of Pakistan are Lahore, the great educational and agricultural center of the Punjab, Dacca, and Chittagong, the expanding port in Bangla Desh.

Five notable cities are delta gateways to the river valleys of Southeast Asia, and each tends to dominate the trade of its respective valley. Rangoon on the Irrawaddy delta is the trade center of Burma; Bangkok, near the sea on the Menam, serves Thailand; Saigon, on the Mekong River delta, and Hanoi with its port, Haiphong, on the Red River, are the chief commercial centers of the Vietnams. Canton, the leading city of south China, occupies a delta position with the Si

Valley as its hinterland. Victoria on the island of Hong Kong also lives largely by commercial activity with south China. Manila, chief port and city of the Philippines, is located on Manila Bay on the west coast of Luzon. This center is the hub of transportation for the island group and, with a population nearing 1.7 million, ranks as one of the great cities of the world.

PROBLEMS AND PROSPECTS

If the success of an occupance system is measured in terms of its capacity to provide for the perpetual well-being of the dependent society, the subsistence agrarian system of Monsoon Asia, despite an early flourish, would seem doomed to problems. The area has been almost wholly dependent upon croplands, the expansion of which was bound to be limited in time by physical and technical barriers; it has fostered large families for labor supply; and it has employed traditional techniques that improved little with the passage of time. The farmers of the realm had undoubtedly been among the world's best land managers two centuries ago, but after the industrial revolution, they steadily fell behind. The system made few concessions to foreign advances in resource-converting techniques, continuing to depend upon croplands and traditional methods. Today the results (judged in terms of Western culture and value systems) are seen in the underemployment of natural resources other than cropland, in an unfavorable mancropland ratio, an unfavorable work-effort–output ratio, low per acre yields, and a huge dependent population living at a generally minimal level.

In a society with such a long culture history and with traditions so firmly ingrained, change comes slowly. Yet the modern leadership has recognized the social and economic problems, and it is hoped that through the application of advanced resource-converting and space-adjusting techniques many of these ills can be solved. Technical-aid programs, of both a private and governmental nature, have been accepted from a number of advanced nations. Most of the countries have initiated government-sponsored planning schemes, mustering all resources—natural, economic, social, and technical—and bringing them to bear upon designed programs of development. Programs striking at the problems of low literacy level and burgeoning population are also underway.

In the last three years of the 1960s the fairly rapid spread of new varieties of rice and wheat brought substantial increases in output in several of the Asian countries; this is continuing. The new rice varieties were developed at the International Rice Research Institute in the Philippines in the early 1960s, and the new wheat in Mexico in the late

TABLE 7.1

Change in Agricultural Production, 1969*

Country	Total Output	Per Capita Output
	1957–59=100	1957–59=100
Ceylon	154	118
India	136	105
Pakistan	127	115
Burma	119	95
Cambodia	145	111
Indonesia	127	98
Japan	131	117
Korea	161	120
Malaysia	181	130
Philippines	150	104
Taiwan	150	109
Thailand	190	133
South Vietnam	112	83

* Foreign Agricultural Service, United States Department of Agriculture.

1950s. Both types of grain have short, stiff straw and produce much higher yields than traditional varieties. They do not lodge (beat down by rain), have wide adaptability, produce crops in shorter periods than most traditional varieties, and can make better use of larger amounts of fertilizers. They do, however, have high water requirements and, moreover, there is some consumer resistance to the quality of the new rice; this is resulting in research to overcome objections. The increased output that has resulted has been hailed as the Green Revolution. It does appear possible that several of the Asian countries may become nearly self-sufficient in a few years; some already have surpluses (although not necessarily real, but above what can be sold in the domestic market).

Both the recent accomplishment and the prospect, however, must be kept in perspective. In addition to better seeds, the widespread alleviation of hunger will require consideration and viable improvements in many other facets of the system: fertilizers, pest controls, cultivation and planting techniques, land reform, storage facilities, transportation, distribution, and, most certainly, capital. Farmer education and population curbs are basic. Although progress has been made in overall production, the failure of per capita production to increase significantly more rapidly than the population has perpetuated the food problem (Table 7.1).

As noted earlier, there have been significant advances in developing other sectors of the economies—in resource development other than cropland and in manufacturing; however, these advances are relative to the low level of the past. Six to eight persons out of every ten are still directly dependent upon agriculture. At the village level, where the problems truly exist, national development programs have as yet had very limited impact and few improvement results.

Although theoretically a technological upswing should be able to solve many of the social and economic problems, the course of implementation is most difficult. The limited available capital, the deeply intrenched human and institutional barriers, and the low technological level of present production retard advancement. Moreover, discord in political and social ideology exists in several countries. Forward movement in these countries awaits favorable settlement of those conflicts.

eight

TROPICAL DESERTS

The Tropical Deserts are lands so hostile that life scarcely exists over vast areas. Sunshine, heat, and wind comprise the climate with rain only a whim of nature, having little annual or seasonal pattern. In no other regions are life and water so intimately related. The meager vegetation and the limited animal life must make special adjustments to the heat, aridity, and excessive evaporation in order to live. Homes are constructed with thick walls for insulation, and light-colored clothing is often worn to reflect the daily stream of the sun's rays. There is no semblance of evenness in the distribution of the scanty population. The location of agriculture is strictly determined by the presence of water. Subsistence grazing exists only through a constant nomadic quest for forage and water. Dependence on animals and crops from oases has dominated desert livelihood for centuries. Aside from a modest amount of trade activity, other economies were lacking until mineral discoveries, especially that of petroleum in the Middle East, stimulated mining activities. Oil development is having a significant impact upon the Arab way of life, bringing the capital and technology for change and improvement. The climate of the American Southwest has also been "discovered" in recent years—the abundant sunshine, warm to hot temperatures, and dryness, which in combination are unattractive to farmers unless irrigation is possible, but are attractive to many people in retirement, seasonal vacationers, and some in footloose activities. Increased mobility, high income, increased number of retired people, the development of air conditioning, and the increasing popularity of outdoor living are among the features which have underlain population growth in the United States portion.

Desert Causes

Deserts occur as the result of one or several of the following: (1) location in areas dominated by the subtropical high pressure belt of settling, therefore dry, air; (2) location in the belt of the persistent trades, whose drying action gives rise to the term "Trade Wind

Deserts"; (3) a dry-shadow location on the leeward side of high mountains where the barriers effectively block out moisture-laden air; (4) location in the interior of continents far removed from sources of oceanic moisture; and (5) a land location leeward of cold water due to cold ocean currents, such as the Benguela Current paralleling the coast of the Kalahari (Namib portion) and the Humboldt Current offshore of the Atacama, and the upwelling of cold water from ocean depths. Winds that have passed over these cold waters reach the land with lowered temperatures but become warmed as they pass over the land and act as drying agents inland .

General Location

The Tropical Deserts are usually situated between 15° and 30° north and south latitude (see Figure 8.1); however, the latitudinal extent varies from one desert to another as the result of differences in surfaces, continental size, and configuration. They seldom extend to east coasts because of rain resulting from onshore winds, but are deep in the continental interiors or in the lee of mountains. Only the Sahara of Africa spans the width of a continent, and in this case the Arabian peninsula and the landmass of Asia prohibit the trades from crossing an ocean. The western margins of deserts usually reach the sea; in South America and South Africa they are extended equatorward by the effects of bordering cold currents.

MAJOR DESERTS

There are seven major tropical deserts. The Sahara (1) is the "Great Desert" filling the northern portion of Africa and stretching 3,500 miles from the Atlantic Ocean to the Red Sea and 1,200 miles from the Sudan to the Mediterranean Sea and the Atlas Moun-

FIG. 8.1. The Tropical Deserts.

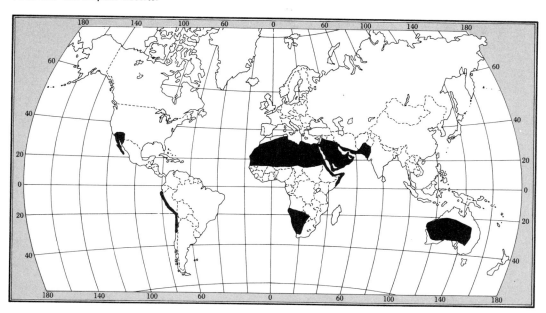

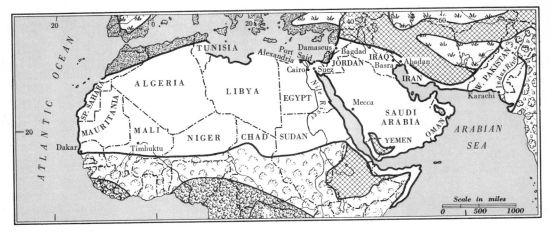

FIG. 8.2. The Sahara and the deserts of southwest Asia.

tains. This vast and often sterile expanse exceeds the size of the combined United States (Figure 8.2).

The desert of southwest Asia (2) contains an area at least half the size of the Sahara. All of the Arabian Peninsula is included, except for the highland corners (Yemen and Oman), as well as southern Israel, Jordan, and most of central and southern Syria, Iraq, and Iran.[1] This broad, continuous tropical dryland, which comprises most of the Middle East, joins the Thar Desert (3),

[1] The northern fringes of the desert of southwest Asia could be classed as subtropical, but they are included in this classification because the physical character (aside from a short cool season) and human development are so closely related to tropical areas to the south.

FIG. 8.3. The Australian Desert.

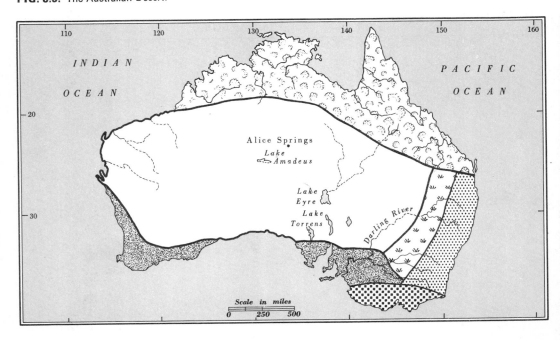

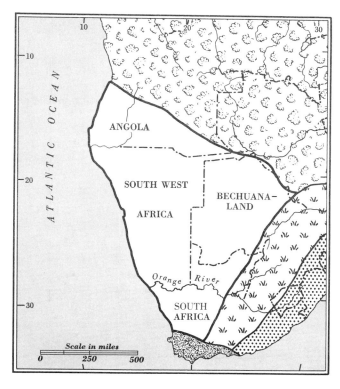

FIG. 8.4. The Kalahari.

FIG. 8.5. The Atacama Desert.

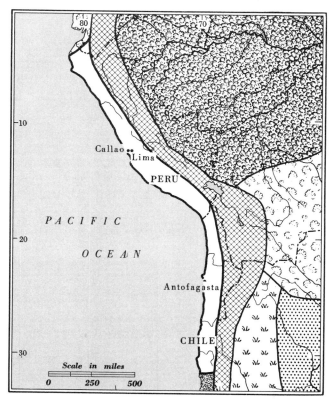

which centers upon the middle and lower parts of the Indus Valley of West Pakistan and the adjoining area of India.

The Australian Desert (4) is the largest in the southern hemisphere, comprising some 40 percent of the continent (Figure 8.3). The Kalahari-Nabim (5) of southwest Africa consists of a half-million square mile expanse (Figure 8.4). In South America, the Atacama Desert (6) is restricted by the Andes Mountains to a long, narrow belt fringing the continent from 4°S to 31°S latitude. It includes all of coastal Peru and the northern third of Chile (Figure 8.5).

North America has only limited land area in the latitudes of the Tropical Deserts, but this climatic type is represented by the Sonoran Desert (7), which includes Lower California, the northwest coast of the Mexican mainland, and a portion of California and Arizona within its boundaries (Figure 8.6).[2]

[2] An examination of an atlas would indicate many localized desert names within the broad name framework used by the authors. The Gibson and Simpson deserts in the Great Australian desert, the Desert El Djouf in the Sahara, and the Namib which borders the west edge of the Kalahari are examples.

FIG. 8.6. The Sonoran Desert.

Physical Environment

CLIMATE

The climate of the Tropical Deserts is one of extremes; superlatives characterize the weather elements. Temperatures are the highest on earth, and precipitation is the lowest (see Figure 8.7). The air is so warm and dry that rain is often evaporated before it reaches the ground. The desert can be a fiery furnace by day and uncomfortably cool at night. Each day the sun blazes down from usually cloudless skies, and the heat shimmers from the rocks and dunes. Seasons have little meaning—the months pass in a relatively monotonous succession. Where highlands are absent, winds sweep unhampered across the open spaces; with even a slight breeze the air is filled with sand and dust. The burning winds of the desert are so unpleasant and such a definite characteristic of the region that special names have been applied to them. The *Simoon* and *khamsin* are examples of Saharan winds and the musical sands are found in the Kalihari.

Temperatures. Average temperatures are always high. Averages in high-sun periods range from 85° to 95° and over; noon readings of 105° to 110° are common. The highest shade temperature ever recorded was 136.4°, at Azizia in the Sahara.[3] The heat continues during the low-sun period, and average temperatures range from 60° upward and aften reach 80°. Yearly annual ranges are from 20° to 30°. The clear skies at night promote rapid radiation of heat and diurnal ranges fluctuate from 25° to 50° with extremes as high as 60° and 70°. Surface heating is rapid; temperatures soar as soon as the sun rises and cool just as rapidly when the sun sets. Frost may even form during the cool season of the year in inland positions or

[3] Death Valley in California, with a reading of 134°, has the record in the United States.

FIG. 8.7. Typical climatic graphs of Tropical Desert stations.

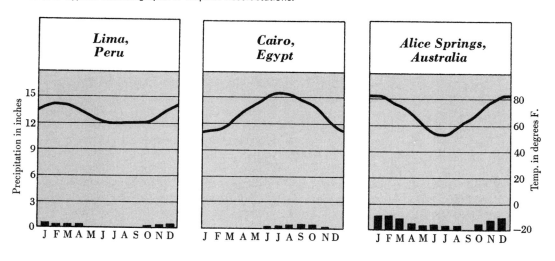

in high elevations. Shelter and warm clothing are needed for comfort from dusk until the sun warms the landscape in the morning. The character of the surface also affects heating and cooling. Sandy areas have more rapid temperature changes than rocky areas, since the rocks absorb more heat during the day and consequently cool more slowly at night. Sands become abnormally hot during the day. Emile Gautier, in his book, *Sahara, the Great Desert,* mentions that in the battle of Metarfa, fought in the sand dunes of the Sahara, the lightly-clad native foot soldiers, unable to hold a prone position because of the intense heat of the sand, remained standing in spite of orders and were all killed.

Except in certain coastal areas, the air is very dry and relative humidities are extremely low, with daytime averaging approximately 25 percent during the hotter season. Despite the high temperatures, however, man is less enervated by the dry heat of the desert when water is available than by the extreme humid heat of the Humid Tropics.

Precipitation. Desert precipitation can be characterized as low and erratic. Yearly totals are under 10 inches, and, except on the margins, it is difficult to indicate any season of maximum. The great expanse of the Sahara has an average under 5 inches, but averages have little meaning, since some years or series of years may be entirely dry; on the other hand, several inches may fall in a single day. Rain often occurs as short, violent, convectional showers, which may create more damage than benefit. Light showers are not too effective, because the moisture is soon returned to the air by the excessive rate of evaporation.

Along the desert coasts of the Atacama and Namib, where upwelling and cool currents chill incoming air, dense mists are common. Fogbanks form offshore over the cold water and are blown to the land. The persistent mist in some cases provides enough moisture for a green cover on the coast hills.

SURFACE FEATURES

To the observer from humid lands the desert is a world apart; landscapes and surface features are entirely different. The gradational forces work under conditions of slow and predominantly mechanical weathering, meager vegetation, low average but high intensity rainfall, and intermittent but rapid runoff. The characteristics of the landforms produced are thus unique to the drylands. Several distinctive features and types can be recognized.

Perhaps the most apparent is the fresh appearance of the landforms. Slow weathering and erosion result in widespread rock features in bold outline. Scanty vegetation makes visible the minute detail; rock debris mantles the base of every cliff and sharp rise. Deserts give the impression of youth.

An outstanding characteristic of the topography is the prevalence of closed basins, large and small. Many deserts have great areas with interior drainage with the erratic runoff directed toward the center of basins. Only a few streams succeed in transversing the entire width of the major deserts; these include the Nile of the Sahara, the Tigris-Euphrates of southwest Asia, the Indus of the Thar, the Orange of southwest Africa, and the Colorado of the Sonoran. In addition, a number of smaller streams rising in the Andes Mountains flow across the Peruvian portion of the Atacama Desert to the Pacific Ocean. In all cases these exotic rivers rise in well-watered highlands from which they derive sufficient volume to carry across the drylands to the sea. Most other streams are intermittent, flowing only during and for a short time after a rainstorm. When volume is sufficient, water may be carried to the center of the basin, forming a *playa* or temporary lake. These stream beds, called *wadis* in the Old World Deserts and *arroyos* in the Americas, are characteristic features even in the driest deserts. Their banks are commonly perpendicular and their flat beds are rocky and sediment-covered. For a short time during a rainstorm they may be brimming full with raging torrents having tremendous erosive power; obstructed by their own deposition, they constantly shift their courses.

Although rains are few and far between, water action is the main force shaping the desert surface. During a storm, water hurriedly moves over the land, down the slopes, some in definite channels and some in sheet floods, carrying sediments into the basin. Lack of volume and velocity causes much water to reach only the margins before dwindling and sinking beneath the surface. Thus the greatest deposition is at the basin margins where the streams form alluvial fans that coalesce and build inward toward the center. In time the basin becomes filled, forming a flat plain. Often the wind alters the surface by removing fine materials, leaving a *pebble desert* or *erg*.

As the bordering highlands are worn back, rock plains called *pediments* develop between the base and the alluvial fill. In some cases entire mountains apparently have been worn away, forming larger rock plains characterized by a thin and discontinuous mantle of rock fragments. This landform, termed *hamada*, has its largest expanse in the Sahara.

FIG. 8.8. Physiographic diagram of Australia.

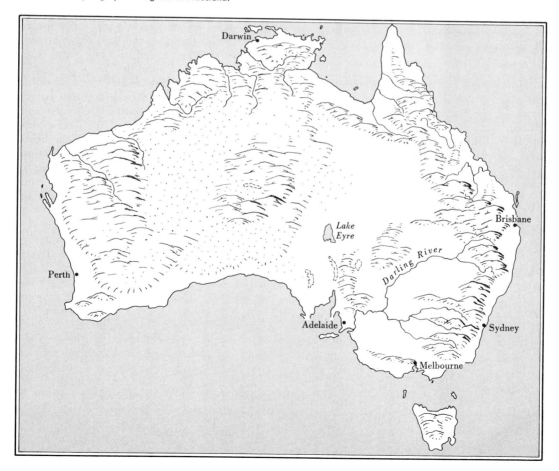

FIG. 8.9. Erg desert. The sand dune scene is in Saudi Arabia. Notice the wind-made ripple marks on the dune and the wadi bed behind the dune. (Standard Oil Company of New Jersey.)

Most deserts have some sand-covered portions. Large areas of sandy, or *erg,* deserts occur in the Saharan, Arabian, Thar, Australian, and American drylands. The largest erg is in Libya and includes an expanse of sand approximately 200,000 square miles in size. Here the work of the wind is most apparent, as frequently the surface is a billowing sea of dunes (Figures 8.8 and 8.9).

The large deserts display all the features mentioned, with many variations. It should be clearly understood that for drylands as a whole, features other than sand are usually dominant.

WATER SUPPLY

Water means life in the desert. Knowledge of where and how a continuous supply can be found is of utmost concern to the desert dwellers. These drylands can be inhabited only so far as this great limiting factor will permit.

Exotic river. The most significant water source is the exotic river that rises in rainy areas beyond the margins and has sufficient supply to carry across the desert. Utilization of many such streams for irrigation agriculture began in the early history of man. The Nile is the classic example, with its irrigated valley supporting the bulk of the Egyptians since the beginning of settlement. Like most of the exotic rivers, the Nile has a season of flood during which its waters, if uncontrolled, inundate riverine areas and deposit loads of silt to revitalize the soils of the valley floor. For the many centuries prior to 1900, the farmers along the Nile practiced a basin type of irrigation—fields surrounded by dikes acted as a reservoir, taking in water and silt during the seasonal overflow. Water was thus held until the flood flow subsided and the fields were soaked; then the dikes were opened, fields drained, and the crops planted. The width of the agricultural belt paralleling the river depended upon the size of the yearly flood. Since 1900 the Aswan Dam and other engineering works have been constructed on the Nile; flood control has been effected and a more uniform yearly flow of the river maintained. Water is now distributed to the fields at all seasons through a system of irrigation canals and ditches. The area under irrigation has been increased and the cropping season lengthened, but the periodic deposition of fertilizing silt has been lost.

Irrigation works also date back to the beginning of settlement in Iraq, where water is taken from the Tigris and Euphrates Rivers as well as from their combined channel, the Shatt al Arab, which flows from their confluence above Basra 120 miles to the Persian Gulf. The Indus River, which crosses the Thar Desert, has had a similar history and today supports very large irrigation works. Desert irrigation with water supply taken from the Peruvian streams had reached a high degree of development before the Spanish era. Today, irrigation agriculture is practiced in many of the fifty-odd river valleys that cross the narrow coastal lowland. Irrigation in the Imperial Valley of California is based upon water diverted from the Colorado River, and in Arizona upon water from the Gila and Salt Rivers.

Wadi bottoms. In the Sahara and southwest Asian deserts, the seepage groundwater below the bed of a *wadi* is sometimes reached through wells. A more or less continuous string of settlements may develop where the subsurface water is abundant and easily tapped. For example, the great Wadi Dawasir with its extensive branches in southwest Saudi Arabia has several such stretches.

Springs. A natural spring may occur even in the desert land. These concentrated natural outflows from underground may result from a variety of conditions involving the position of the groundwater table, the rock structure, and the configuration of the land. In many cases they are caused by the local groundwater table being exposed on the side of a valley.

Artesian water. An artesian water supply depends upon the existence of a certain type of geologic structure (see Figure 8.10). In this structure an inclined pervious layer, or *aquifer stratum,* such as sandstone, is tapped between impervious layers. The aquifer exposed in an area of sufficient precipitation fills with water. Having no outlet, the water in the lower portion of the incline is under pressure. A fissure in the overlying strata will result in a flowing spring, or water will rise in a well bore or even flow from it as long as head or pressure is maintained. The greatest development of artesian water occurs in the eastern part of Saudi Arabia where a number of wells have been drilled by the Arabian American Oil Company.

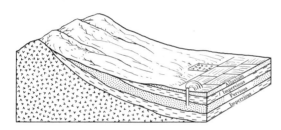

FIG. 8.10. Artesian structure.

FIG. 8.11. Diagrammatic cross section of a kanat. This ancient, laboriously hand-constructed water supply system brings life to thousands of sun-parched acres in the drylands of North Africa and Asia. It is found to a lesser extent in South America. In construction, the water supply is located by first digging the horizontal tunnel on a gentle grade from where water is desired to the source. Vertical shafts are spaced along the line to bring the excavated material to the surface. The greatest distribution is in Iran, where the system has tended to promulgate a semifeudalistic organization of agriculture because construction and maintenance costs are beyond the means of the peasant class. (George B. Cressey.)

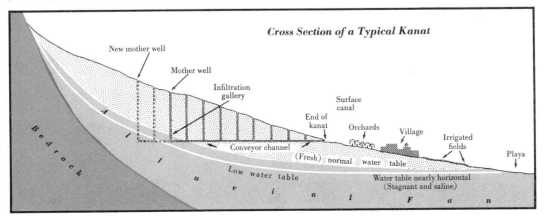

162 THE TROPICAL REALM

Kanats. In southwest Asia, particularly in Iran, the *kanat* is an important ancient system used to tap the water table and to collect subsurface seepage in mountain valleys. Collection tunnels direct the water into a gently sloping central underground tunnel or "horizontal well," which conducts it to better soils fringing alluvial fans or to the plains (see Figure 8.11). Water may be transported as far as 25 miles. The kanats are not lined and thus are subject to obstruction by caving; openings are provided at intervals along the course so that workmen can enter the channels to clear them. This ancient system has the unique advantage of radically reducing evaporation. Kanats are also used in North Africa; in the oases of Marrakech, for example, they are of basic importance, even in city water supply.

Other sources. The advance of technology is making possible the use of the sea as a freshwater supply through desalinization. A great deal of effort has been devoted to the development of low-cost processes. By the mid 1960s approximately two hundred plants were in operation; a plant supplying Kuwait, for example, has a capacity of 15 million gallons a day. With nuclear plants in the offing, it would appear that this source of water supply may be increased in significance along littoral areas.

NATURAL VEGETATION

Few places in the desert are entirely devoid of plants, and even in the most barren areas there may be a host of dormant seeds that will spring into life at the first sign of

FIG. 8.12. Desert vegetation in the American Southwest. The cacti store water in their thick stems and are able to withstand long dry periods. (United States Forest Service.)

moisture. Carpets of vegetation are rare, and plants are scattered as if they had been carelessly sown. Growth is low; however, a tall saguaro cactus occasionally towers over the lower shrubs (Figure 8.12). At certain seasons many plants produce delicate blooms in various shades of red and yellow, making the desert appear as fertile and colorful as any garden.

In general, plant life endures the desert through one of three characteristics. They tolerate or evade, or they go dormant during the drought season. The most apparent desert plants are *xerophytes,* species adapted to drought and high rates of evaporation. Xerophytes show a remarkable adjustment to harsh environment; their methods of defense against arid conditions are extremely interesting. Root systems are well developed. Some plants have taproots that go deep in search of water and others have amazing lateral systems many times the size of the plant. Water is stored inside stems and roots; often a thick bark, narrow leaves with waxy or hairy surfaces, or the complete absence of leaves retard evaporation. Some plants are equipped with sharp spines or have unpalatable flavors that discourage grazing animals. The drought-escaping annuals have seeds that can wait years for a rain—when enough moisture is present, they sprout, flower, and go to seed in an incredibly short time. Perennials will remain brown and appear lifeless until rain stimulates new green growth. The perennial grasses are most common in the transitional portion—where the deserts merge with the grasslands and, although annual totals are low, there is a regular season of rain.

Many types of grasses are found in most deserts, usually present in solitary tufts. Areas that appear barren may show, on closer inspection, large varieties of pygmy plants. Thornbushes, especially the acacia, are common in the Old World Deserts. The highly alkali-impregnated soils of the Atacama probably make it one of the barest of all the Tropical Deserts. The Sonoran is noted for its great variety of succulents—fleshy and thorny plants such as giant saguaro, cholla, other cacti, and yuccas. Their shapes give a weird appearance to this desert's landscape.

NATIVE ANIMAL LIFE

Desert animal life is sparse and restricted to species capable of making adjustments to the arid environment—namely, the ability to subsist on the scanty vegetation and to survive with a minimum of water. Dew, in some cases, is the only water source. Desert animals tend to acquire camouflaging colors ranging from pale yellow to red. The majority are small; the desert fox and a species of antelope are the largest in the Sahara. Burrowing, running, and jumping animals are common. There are few birds, and even insects are scarce except around oases or animal habitats.

The camel is the animal most often associated with deserts. Domesticated for a beast of burden, his use is chiefly confined to the African and Asian deserts. The Romans introduced the camel from Asia to North Africa about 200 A.D., making nomadism possible for the first time, thus stimulating an activity that has continued. Camels were introduced into Australia in the 1860s and were used to carry mail and supplies to ranchers and miners. The United States Army experimented with this animal in the Southwest during the mid-nineteenth century, but the War Between the States disrupted the continuance of the project.

The camel is especially well-adapted to an *arid* environment; his ability to withstand heat and to go for several days without water has made him invaluable for caravan traffic. The feet and knees are padded against the friction of rock and sand, and the nostrils can be closed during sand storms. The nose and lips are tough enough to browse among the coarse desert forage.

SOILS

Desert soils have evolved under conditions of low rainfall, high evaporation, and sparse vegetation. They have been subjected to little leaching and are rich in soluble mineral nutrients which in places have become concentrated in excessive amounts in the surface layer. With little or no plant-life cover, these soils are low in organic matter; moreover, they have retained largely the color of their parent material and range from gray to red. With good management, some can be made fruitful.

Under irrigation, desert soils will produce a wide range of crops, since the application of water can be adjusted to the length of the growing and maturing periods. The principal needs are organic material to improve soil structure and the elements nitrogen and phosphorus, which may be added through fertilization or the growth of legumes and the plowing under of green manure crops to increase organic nutrients.

Occupance in the Tropical Desert

Immense stretches of the Tropical Deserts are entirely without human life, yet within these seas of waste there are scattered islands, many too small to appear on a world map, where a permanent water supply has favored settlement. In these oases, densities often rival those of the Monsoon Lands. For example, the 32 million people of the United Arab Republic (Egypt) are mainly crowded into the narrow belt, usually not over 14 miles wide, which extends the length of the country along the Nile River; within this fertile ribbon, rural populations average nearly 2,000 persons per square mile. Within the remainder of the Great Sahara, the total population probably does not exceed 8 or 10 million people.

Throughout the drylands the presence of man is determined chiefly by the availability of a permanent water supply, and population density depends upon the amount of the supply. The bulk of the people are oasis agriculturalists, small groups are nomads depending upon their small herds, and lesser numbers are engaged in mining or commercial activity in the "Ports of the Desert." A principal variation is the retirement and tourist developments in the United States southwest. It bears repeating that life of all kinds is found on the desert in direct proportion to water supply.

Despite the seemingly hostile character of the desert environment, man has been an occupant of parts for many centuries. The antiquity of civilization in the Middle East and Egypt are well documented in the annals of history. Great contributions to science, art, architecture, engineering, and philosophy stem from these ancient centers. Three great monotheistic religions also originated here. In modern history, the contributions have been modest; however, significant changes are under way in some parts.

NOMADIC HUNTERS

The Bushmen, the nomadic hunters of the Kalahari, rate with the Australian aborigines (discussed under Wet-Dry Tropics) as the most primitive people of the deserts and the world. It is believed that this disappearing race once occupied most of South Africa but was pressed into the most inhospitable parts of the desert by stronger, more advanced groups. The Bushmen are dwarf people, seldom exceeding five feet in height. Now numbering but a few thousand, they are scattered in clans of a few families each and are totally dependent upon the desert environment for their livelihood. Mere existence requires considerable ability, much hardship, and privation.

Game is their principal food. Clever hunters, the Bushmen have a keen knowledge of the habits of the wild animals that they kill with bow and poison-tipped reed arrows or throwing sticks. Their only other important item of equipment is the digging stick, made of a sharpened spike of hardwood inserted in the hole of a round, flat stone. This implement is used by the women to secure succulent roots of desert plants. Lizards, snakes, frogs, worms, and ants and their eggs also are sources of food, consumed especially when game is short. The Bushmen make no provisions for the future, thus there are times of feasting and times of near-starvation.

Within fixed tribal limits they change location frequently to follow the game supply; therefore, the Bushmen cannot accumulate many possessions or build substantial shelters. Dwellings are of the rudest construction and are abandoned at a moment's notice when lack of game or water compels a move. Caves, shallow holes, or a few branches stuck in the ground and covered with skin or grass are usual types. Household utensils are limited to ostrich egg shells and a few crude earthen pots. Food is eaten raw or partially cooked over an open fire started with fire sticks. Clothing is of the simplest form, composed of animal skins. Men wear breech clouts, triangular pieces which pass between the legs and are secured around the waist with a string. Women wear a piece of skin hung from the waist and reaching the knees. At night a cape, consisting of several skins sewn together, is worn to give protection from the cooler temperatures.

The Bushmen have changed little as the result of contact with other cultures and have sought to preserve their own way of life. Up to the present, the opportunities offered by their environment have been too limited to tempt preemption by other people. At the same time, their habitat and way of life rigidly restrict their numbers.

GRAZING

Grazing is an extensive form of occupance with few evidences or permanent works of man. Its greatest distribution occurs in the grass and shrublands between the dry margins of crop agriculture and barren deserts. Generally these areas provide low animal-carrying capacity, and grazing activities are further conditioned by problems of seasonality of precipitation and available forage as well as limited potable water supplies. For the most part the land base has not attracted other forms of occupance. Moreover, the grazing use is highly susceptible to decline. The fragile balance that exists in this environment can be easily disrupted by overgrazing. When the most palatable species are depleted commonly there is an increase in undesirable species, and/or soil loss and severe erosion.

In the total relationship of man and nature, grazing is a step up the scale over those economy types dependent upon hunting and gathering. The grazers have a symbiotic, rather than a parasitic, relationship: The herders are dependent upon their livestock, and the livestock are dependent upon the natural vegetation. Even in the most rudimentary forms, the herders must engage in managerial functions of moving the herds to new grazing grounds, assuring water supplies, protecting the herds, aiding with the birth of the young, and so on.

Most grazing systems depend, at least to a modest degree, on exchange; however, the scale of difference is wide. In the rudimentary nomadic forms of the Old World drylands, the dependent people commonly exchange surplus animals or animal products with oases farmers and with merchants in commercial centers for food such as grains, tea, and dates, for cloth, and for various items of equipment. Some restrict direct use of their animals to renewable products such as milk, blood, hair, and wool, killing for meat only on very special occasions. In the most highly commercial forms market linkages are, in many situations, quite complex, with stocker and feeder producers, feed lots or feed farms, transportation firms, packing or processing plants, and brokers and retailers included in the total scope; thus the grazing activity may be but one stage in preparing the animal product for the consumer.

The major concentration of the rudimentary forms, as implied, is on the margins of the Old World deserts, stretching as a great belt across North Africa, through Southwest Asia, and on into Central Asia. The traditional use of this vast area has been a system of migratory grazing closely attuned to the meager vegetation, limited water, and low and erratic precipitation. The system is, further, a response of small numbers of people to limited outside contacts, limited or no competition from other land uses, limited resource-converting techniques for identification and use of other elements of the environment, and—for the most part (until the recent rise of nationalism)—uncontested usufruct practices.

Although there are variances in many of the details, the general practices are similar from group to group. Their lives revolve around the needs of their animals which are the chief manifestation of wealth. Mobility is the overriding characteristic—quest for animal forage is the stimulus. Low carrying capacity demands that the functional population group—commonly a patriarchal family of 20 to 50—and its herd be small. The necessity of movement requires portability of shelters, equipment, and possessions. Seldom will a given campsite have accessibility to forage that will supply animal requirements for more than a few days, or at most a few weeks. Similar to the gathering groups of the humid tropics, however, these pastoral nomads do not wander aimlessly; they follow time-tested routes and are in the same general area at the same time each year. In general, they follow the flush of vegetation and thus are well aware of the pattern and whims of precipitation. The yearly migration cycle of a particular group may cover several hundred miles.

Groups vary in the kinds of animals they keep. Some specialize in single-stock herds; others may have mixed herds. Almost universally animals are nondescript, although some camel raisers have pride in animals they have developed through selective breeding. Whereas preference may play an important role in kinds of animals kept, the quality and kind of forage available also has an influence. Cattle, for example require the better forage supply and regularly and readily available potable water. Goats, on the other hand, are browsers and can subsist on shrub as well as grass and can take a portion of their water needs from morning dew. Camels and goats can exist in conditions of poor graze, and camels can go several days without water.

In recent years the grazing groups have come under closer governmental influences. These have either been directed at a greater degree of control of area, labor, and production, as in the Soviet Union, or at settling the nomads and changing their means of livelihood.

Purely commercial forms of grazing exist primarily in the technically and economically advanced or advancing nations. The major distribution of the commercial phase is in the middle latitude drylands; however, it exists to limited extent on the margins of the tropical deserts of North America and Australia. Although the environmental base has general similarities to that of the rudimentary forms, it commonly has somewhat better natural forage production and is more accessible. Also the commercial system has a number of significant differences in organization and operational procedures—including more advanced techniques of resource use and adjustments to problems of space

FIG. 8.13. Bedouin encampment at a water source. Here a well has been drilled by the Arabian American Oil Company. A number of such wells have been drilled along the oil pipeline across the desert. Here a stable supply has attracted a concentration of Bedouins and their animals. (Arabian American Oil Company.)

and distance. The commercial grazers are under such additional influences as more formal land ownership, the detailed nature and seasonality of the forage land controlled, transportation linkages with the market, and the market demands and the competitive position for the market.

In general, commercial grazing takes place on a fixed land base, owned, formally leased, or rented by the operators. To compensate for low carrying capacity, the functional land unit (ranch, station, or hacienda) is large; size measurement may be in sections or square miles rather than acres or, even more importantly, in terms of the animal-unit carrying capacity—since the grazing base varies in quality from area to area. The grazing lands of a particular unit are commonly divided with fences to enhance both forage and animal management. Other improvements may include developed water sites for animals, irrigated hayfields for supplemental feed supplies, range forage improvements through control of competitive species or planting of more desirable species, and developed roads or trails to enhance circulation. Ranch headquarters, commonly localized by water supplies, good quality forage, or ease of access, generally contain the operator's home, bunkhouse for workers, barns and sheds for storage, and mustering corrals.

As the result of the rational balancing of the factors of market and forage base, along with operator preference, there are many variations in the actual speciality of commercial ranchers. Some concentrate upon calves, some upon feeders, and others upon

FIG. 8.14. Dhows on the Nile. Dhows have been used for thousands of years as a common means of transport for both commodities and passengers. Today they provide a sharp contrast with river steamers, which ply the river to the Aswan Dam during high water. This picture was taken from the left bank of the Nile not far from the tombs of the Egyptian kings who chose this desolate desert area as a safe spot for burial. Across the river are the ruins of Karnak, remarkably preserved in the hot, dry climate of the Tropical Desert.

mature animals as their salable product. Still others produce wool and lambs, breeding animals, or simply wool.

OASIS AGRICULTURE

Old World Deserts. Since the dawn of human history, oasis agriculture has been the basic element in the life and economy of the Old World Deserts. Today it continues as a major support of the majority of the people who inhabit the desert regions of Asia and Africa.

Egypt, quite appropriately, has been called the Valley of the Nile. Its life—human, plant, and animal—depends almost entirely upon the water of that river and the alluviums deposited on its floodplain and delta over the centuries (Figure 8.14). The total cultivated land of some 6 million acres is about 2.5 percent of the country's total area. Because of multiple cropping, an average of one and one-half crops per year is produced, extending the effective cultivated area to an equivalent of 9 million acres. With about 75 percent of the Egyptians directly dependent upon the soil, farms are necessarily small, averaging about 2.5 acres (Figure 8.15). A further problem is the fact that more than one-third of agricultural land is in the hands of absentee owners, whose tenants must pay rent.

Agricultural practices are a peculiar combination of ancient and modern methods. The sickle, flail, and wooden plow, ancient pole and bucket irrigation, and the waterwheel drawn by animals exist side by side with great irrigation engineering works and the disc plow, tractor, and combine. Ignorance of agricultural science on the part of the peasants stands in striking contrast to a number of modern government projects. Taking into consideration all aspects of agriculture, Egyptian production still depends largely upon the *fellah*, or the small farmer with the hoe. Abundant and cheap labor and the predominance of small farm units restrict the use of farm machinery. Although yields per acre are high, the production per individual is very low.

Cotton is the dominant enterprise in agriculture as well as in the total economy of Egypt. Not only is it the most important cash crop, it is also the major item among Egyptian exports, normally occupying about 20 percent of total cropland. Corn and wheat are the ranking bread grains, the former being the staple diet of the farmer and the latter produced for the city dweller. Rice, too, is a major crop; acreage is being expanded in response to high yields and ready markets. Barley and grain sorghums are also important; together all the grains occupy 50 percent of the total cropland. Legumes, horse-beans, lentils, and chick-peas are supplementary food crops. Vegetables are grown abundantly in all seasons, and the production of sugar cane is well-established.

Livestock are limited in number. Pasture space is lacking, and competition for food crops is too great for feed production. The water buffalo cow is most important; this hardy draft animal is immune to most pests and diseases and is a good milk producer.

The oasis of the exotic Nile Valley completely overshadows others of the Sahara. Nonetheless, small dots of irrigation agriculture exist wherever there is a water supply. The major distribution is found along the wadis descending from the Atlas Mountains and radiating from the Ahaggar and Tibesti

FIG. 8.15. Agricultural village in the Nile Valley. Fertile alluvial soils, continuous growing season, and an unfailing water supply make the Nile Valley one of the most productive agricultural areas for its size in the world. This richness, however, is not reflected in the lives of the people. The population pressure on the land is so great that farms are small, and output per person remains very meager. Most of the Egyptian farmers live in small villages similar to this one. (Oliver H. Heintzelman.)

FIG. 8.16. Wooden water wheels. Among the great sights of northern Syria are the 2,000-year-old wooden water wheels in and near the city of Hama, 125 miles north of Damascus on the banks of the Orontes River. Built by the Roman conquerers of Syria, these enormous contraptions scoop water out of the Orontes and deposit it on the land for irrigation purposes. The incessant creaking of the wheels, continuing day and night, have given Hama the name of "The Melodious City." (Arab Information Center.)

Highlands. In all, intensive agriculture prevails with large amounts of energy and time expended to produce the greatest possible yields. The abundant sunshine, controlled applications of water, rich soils, and general freedom from weeds and insect pests result in abundant harvests. A wide variety of fruits, grains, and vegetables is common. Throughout most of the Old World Deserts, date palms are a characteristic feature of the oasis landscape. Probably the oldest known cultivated tree, it is well suited to be "King of the Oasis." "With their feet in water and their heads in fires of the heavens," the tropical desert oasis is the natural habitat. The fruit of this high-yielding tree is both a subsistence crop and a money crop. It provides food for the oasis dweller and his animals as well as an item for exchange. Moreover, the trunk and fronds provide material for construction, furniture, and fiber. Even the stones are crushed and used as livestock feed.

Frequently a two- or three-story agriculture is practiced to achieve the utmost from the land: An annual grain or vegetable crop forms the ground level, an intermediate-height fruit the second, and the skyward-reaching date palm the third level. Some of the more accessible and larger oases produce commercial crops of cotton and sugar. The farmers normally live in villages with flat-roofed houses constructed of either sun-baked mud or stone. Thick walls and roof provide protection against the intense heat of the day and the cool of the night.

The oases of southwest Asia have similar characteristics. Those of Saudi Arabia are, for the most part, small. Of the wadis, Dawasir, with its extensive branches in the southwest part of the plateau, is outstanding; long stretches along this system have adequate subsurface water supplies for agriculture—some of the finest dates in the country are produced there. In the central part of the plateau there are a number of water pits, huge natural wells which range in diameter from 150 to 1500 feet, and from 420 feet upward in depth. Al Kharj and Aflaj are two of the most significant. Crops of dates, alfalfa, and sorghum are presently grown on several thousand acres, and the possibilities for increasing the arable area are excellent. The eastern part of the country offers the greatest potentials for new irrigation (Figure 8.16). The likelihood for artesian wells exists in an area paralleling the Arabian Sea and extending westward for 100 miles. The Arabian American Oil Company has drilled a number of wells in this area, and several large springs exist. The Hofuf Oasis, the largest in Saudi Arabia, has some 25,000 acres under irrigation on which two million date palms are growing (Figure 8.17).

The most notable oasis of southwest Asia is found along the Shatt al Arab and the Tigris and Euphrates Rivers of Iraq. This ancient land produces about 75 percent of the dates of commerce with 30 million date palms, one-third of all in existence, situated along the banks of these streams. The most productive area is along the Shatt al Arab where the palm groves follow the courses of hundreds of canals leading out into the desert, sometimes for only a few hundred yards, but in other cases for distances of five miles. These canals receive water with the normal tidal rise of the Shatt al Arab. Iraqi farmers also cultivate wheat, barley, rice, corn, sorghum, and sesame. In recent years cotton has become established as an important crop. Agriculture has always been the basis of the general economy. There is little doubt that up to the thirteenth century A.D. a much larger area was under irrigation than at present; the ruins of an obsolete irrigation system can easily be traced. Several modern developments have recently added to the cultivatable area, and there

FIG. 8.17. Tower of Babel at Babylon, Iraq. Near this pool of water and date palm grove once stood the famous Tower of Babel, a towering temple of seven brick pyramided terraces, connected by a continuous ramp which lead to the topmost room, a sanctuary for the chief god of Babylon, Marduk. The Tower of Babel dominated the ancient city of Babylon rising from a 300-foot square ground plan. It was completed by Nebuchadnezzar in the 6th century B.C. The Bible refers to the Tower of Babel (Genesis 11:4-9). The site of ancient Babylon is a two-hour drive from Baghdad, Iraq. (Arab Information Center.)

FIG. 8.18. Arab Pioneer, one of many on the Dujaila Land Resettlement Project of the Iraq Development Board, located 110 miles from Baghdad in an area still indicated as uninhabited desert on most of the world's maps. A 39-mile canal here now has watered 160,000 acres into fertility. The once-arid land has been apportioned among 1,200 former share-croppers' families on a homesteading agreement similar to those by which the American West originally was settled. (Arab Information Center.)

remains a potential larger than the present total.

Oasis agriculture in the Thar Desert is wholly in the valley of the exotic Indus River. The irrigated land has been greatly expanded during the past 40 years, and today the Indus Valley of West Pakistan has one of the largest single expanses of irrigated land in the world, on the order of 23 million acres. The bulk is dependent upon large-scale government-constructed dams and diversion canals. Six dams are either in operation or under construction, and at least three more are proposed. The Sukkar Barrage in the Sind is the most famous. Completed in 1932 by the British, it is nearly one mile in length and supplies water for about $7\frac{1}{2}$ million acres through 36,000 miles of main canals. This irrigated area has become a notable producer of such crops as cotton, rice, and wheat. Unfortunately nearly 20 percent of the irrigated area has been severely damaged by salinity and waterlogging. This primarily results from excessive applications of water by farmers and by canal leakage. Since 1959 a counter attack has been under way to return this damaged land to production by sinking numerous wells from which water is pumped and returned to the canals, coupled with improved application techniques.

American deserts. The Incas had established irrigation agriculture in the Peruvian desert before the coming of the Spaniards, but their carefully constructed and maintained water distribution systems fell into disrepair following Spanish conquests. Agriculture continued solely for local subsistence until near the end of the nineteenth century, when white and mestizo landlords began production of commercial crops, particularly sugar. Since the mid '20s, cotton has supplanted sugar as the leading crop. Rice has also become significant, and numerous other crops such as corn, wheat, fruits, and vegetables are produced for local consumption. Today, three million acres are devoted to irrigation agriculture in the valleys of the coastal streams. Government-sponsored development programs include several schemes to divert water from streams east of the Andes, through tunnels, to increase the irrigated area along the west coast.

The waters of the Colorado and Gila Rivers which flow through the Sonoran Desert were long ago used to some extent by the Indians. When white men arrived, this usage was evidenced only by the ruins of canals and settlements. Modern irrigation began about the turn of this century (Figure 8.19). The most famous center is the Imperial Valley, actually the north slope of the Colorado Delta, which through the ages pushed across the rift depression of the Gulf of California, completely cutting off the northern end. Water is carried from the Imperial Dam on the Colorado to nearly a million acres by the All American Canal and the Coachella Branch. The Salton Sea, occupying the lowest part of the depression, 246 feet below sea level, serves as the drainage reservoir for the irrigation system. Continuous heat and sunshine make possible year-round production of a wide range of crops including melons, lettuce, cotton, citrus, alfalfa, and many varieties of vegetables. The latter are grown for sale beyond the area; the advantages of high yields, quality produce, and off-season markets permit profitable shipment to all parts of the nation. The Coachella Valley, north of the Salton Sea, is the major source of dates in the United States; fruit from this district dominates the American quality market. The developments centering around Phoenix and Tucson, Arizona, taking water from the Salt-Gila River system, have made this area a noteworthy producer of cotton.

In recent years, major irrigation projects have been developed in stream valleys of the northwest coast of mainland Mexico (several of which are within the short wet-season phase of the Wet-Dry Tropics). In most of these, cotton is the major commercial crop during the summer; also, an important winter vegetable enterprise has been evolving to exploit markets in the United States. Lower cost labor as well as more reliable temperatures than in Southern California are the principal advantages.

MINING

Minerals contribute to the significance of the deserts and are the only reason for man's presence in certain sections. Their profitable extraction has caused some settlements to develop in barren areas. Water has often to be piped many miles or actually hauled in tank cars, and all other necessities have to be imported to supply workers in these settlements.

Oil. The outstanding mineral resource from a tropical desert today is oil. One area, the Middle East, possesses about two-thirds of the world's proven reserves. Development has been largely by foreign interests, particularly by American and British companies. The major producing areas lie close to the Persian Gulf in Saudi Arabia, Qatar, Kuwait, Iraq, and Iran. Bahrein, a small Arab sheikdom consisting of a group of islands twenty miles off the Saudi Arabian shore, is also significant. Since 1955, oil development has occurred in the Algerian and Libyan Sahara, from which pipelines have been laid to the Mediterranean coast. The output of all these producing areas is steadily increasing; the major portion moves out to serve the industrial countries. Western European countries, essentially without domestic supplies, are particularly dependent upon this region for this highly significant energy source (see

FIG. 8.19. On the right a citrus grove in the Salt River Project—on the left the raw desert. Illustrated here are both the limitations and possibilities related to water in the deserts. (United States Bureau of Reclamation.)

FIG. 8.20. Arabian oil development: an oil well in a sea of sand. (Arabian American Oil Company.)

Figures 8.20 and 8.21). It is doubtful, however, that the Middle East in the forseeable future will reach a world production share commensurate with its reserve share. Lacks in local markets, technology, and capital, and dependence upon foreign markets together with political problems (including world oil politics and strategies) form deterrents.

Nitrate. The importance of the Chilean section of the Atacama once revolved around its abundant deposits of nitrates. In the basins between the low coastal range and the Andes, the evaporation of former lakes left rich deposits of mineral salts. These beds, containing the nitrate-bearing material called *caliche,* vary from a few inches to several feet in depth and extend in an irregular belt 450 miles north-south. Many natural deterrents had to be overcome for exploitation. All workers, materials, equipment, and food had to be imported, and water had to be piped 100 miles from the Andes. Until World War I, mining was crude, employing much hand labor. Production costs were high, but so were profits. Chile held a virtual monopoly on the world supply of this valuable mineral fertilizer and industrial chemical, and for many years the export tax provided about half the total revenue. Following the war, the high-cost Chilean product met serious competition from synthesized nitrates. The industry has recently modernized to cut costs; the nitrate companies have merged, and new treating processes in a few large plants have been established. The major share of the world supply of nitrogen compounds, however, comes from other than mining operations: as a by-product of coking and especially by fixation of atmospheric nitrogen. Interestingly nitrogen is the principal resource won directly from the atmosphere.

Guano. Guano is a unique fertilizer product of South America. On the dry, barren islands and headlands along the coast of Peru, this excrement of numerous bird colonies accumulates in large quantities. The cool Peruvian or Humboldt Current, which passes immediately offshore, results in upwelling and a nutrient-rich water environment favoring abundant microscopic organisms that support a tremendous fish life that, in turn, supplies food for countless birds. No less than a score of species frequent the islands, but cormorants, pelicans, and boobies deposit most of the guano which accumulates rapidly in this arid climate. The guano was formerly exported in large quantities, but today the industry is a government monopoly, and the total production is used to fertilize the oases fields of the Peruvian coastal lowland. The chain of relationship is interesting; the cool current not only accounts for the arid climate but supports the pyramid of life that accounts for the guano, itself a factor in maintaining the high productivity of the oases.

FISHING

During the 1950s the Peruvians began to capitalize upon the potential fish resources in their offshore waters. As part of an economic development program extensive fishmeal processing facilities were constructed, and a fishing-processing activity based upon foreign markets was initiated. By 1965 Peru had advanced from insignificance to the leading fishing nation of the world in terms of tonnage landed—mostly in anchovies.

MANUFACTURING

Manufacturing is limited, for the most part, in the Tropical Deserts by the lack of an adequate combination of raw materials, energy resources, markets, capital, and technology. Most existing industries are concerned with the processing of local agricultural products. There are exceptions: Oil refining is significant in the Middle East; Egypt is advancing the development of sugar refining, cotton ginning, and cement manufacturing; the American Southwest has a limited amount of agricultural processing and general manufacturing in the major urban centers; and San Diego, with its aircraft plants, makes a significant contribution from this area.

Traditional handicrafts still exist in the regions outside of the United States; however, they have been constrained considerably by the increasing availability of machine manufactures through the channels of commerce.

URBAN CENTERS

Cities are functional developments established at locations where they can perform useful services. Product assembly, warehousing, distribution, manufacturing, business activities are usual kinds of urbanizing influences.

The number, size and rate of growth of

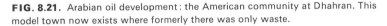

FIG. 8.21. Arabian oil development: the American community at Dhahran. This model town now exists where formerly there was only waste.

urban places depend directly upon the needs of and opportunities offered by their surroundings. Reciprocal access with ease and reasonable cost are fundamental also. The deserts with scanty population, poor transportation, and limited needs to be served naturally support few cities. But just as the margins of the oceans have their ports, so do the deserts. Ancient cities such as Damascus, Syria, Baghdad, Iraq, and Tombouctou, in Mali, developed as the result of favorable situations with respect to caravan routes and have long functioned as assembly and distribution points for items moving from and to their surrounding areas.

Many of the largest of the desert cities actually are ports. Included in this class would be Karachi, in Pakistan; Alexandria, Said, and Suez, in Egypt; Dakar, in Senegal; Lima, in Peru (Figure 8.22), served by the port of Callao, and San Diego, in California. Basra, date port of Iraq, is about 100 miles up the Shatt al Arab but is reached by oceangoing vessels drawing up to 30 feet. Cairo, largest city of the realm, is the chief commercial and administrative center of Egypt. It serves all the Nile Valley as well as much of the eastern Sahara from its favorable river position at the head of the delta, where caravan routes funnel in from the east and west. Abadan, in Iran, owes its growth to oil refining and shipment.

A few cities have developed as the result of religious significance. Mecca, in Saudi Arabia, is an outstanding example. This city, birthplace of the Prophet Mohammed and

FIG. 8.22. City of the desert: Lima, Peru. The plaza shown here is typical of many Latin American cities. (Pan American-Grace Airways.)

FIG. 8.23. Syrian center. The City of Homs, 100 miles north of Damascus, is seen in this aerial photograph. Near the center is the comparatively modern Grand Mosque dominating the community. (Arab Information Center.)

FIG. 8.24. Impact of oil on Arabia. Within two decades, the people of Saudi Arabia have become as familiar with the sight of high-speed streamlined trains as they once were with the slow plodding camel caravans dating from Biblical times. By camel the trip from Damman, a Persian Gulf port, to the inland capital city of Riyadh takes about 90 hours. In a self-propelled Budd car capable of speeds up to 90 miles an hour, the normal time of the 358-mile trip is 9 hours. (Arabian American Oil Company.)

seat of the Islamic religion, is the cherished destination of every person of Islamic faith at least once during a lifetime. During the pilgrimage season its permanent population of about 185,000 is at least doubled.

Phoenix, Arizona, one of the rapidly growing cities of the western portion of the United States, has a population of nearly one million. It serves as a commercial center for a significant agricultural area, is a popular health and winter resort, and is the state capital.

CHANGE IN THE MIDDLE EAST

The Middle East is the desert area of greatest change today. The tapping of its rich oil reserves is bringing capital for modern development. The impact upon the local peoples has been noteworthy. Stable income has attracted many former oasis dwellers and nomads to work in the oil affairs. A number of new centers of settlement have become established. New markets for food have extended the outlets for oasis produce. Royalties received by the governments from the oil are being used to establish schools, to install hospitals and modern sanitation, and to develop irrigation projects, as well as to improve transportation and the general economy. In general, the effects upon the inhabitants have been favorable.

With these developments the Old World lands are assuming new importance in international affairs. Long significant as the early center of civilization and the sites where the great monotheistic religions were born, the region now has added importance. It is a strategic bridgehead between three large land masses; it contains the Suez Canal and is a vital stepping stone in the transworld air-line systems; it is a major producer of petroleum. Although many racial groups are represented, there is an underlying cultural unity with the Islamic faith binding the peoples together. The Arab nations are beginning to act collectively for the benefit of the whole and are through this cooperation becoming a potent force in world affairs. In part this has been spurred by the speed and efficiency with which Israel, created in 1948, has constructed a relatively prosperous economy and a strong state with importance far out of proportion to its tiny size (8,000 square miles and a 1969 population of 2,790,000).

nine

TROPICAL HIGHLANDS

The Tropical Highlands, despite a ruggedness of terrain, are the favorite sites for settlement in the broad zone of the tropics. In contrast to the hot and humid lowlands, the higher altitude climates are invigorating—lower and more stimulating temperature and more healthful living conditions are the major attractions. Generalizations for a regional treatment of the Tropical Highlands are difficult because homogeneity as a regional characteristic is lacking.

Vertical zonation is the key to highland geography. The Tropical Highlands display a vertical layering of climatic conditions, natural vegetation, and land utilization possibilities. Temperatures range from tropical to arctic, climates, from perpetual summer to eternal spring to year-round sunny winter. It is possible in a few hours' climb to experience as many temperatures and landscapes as would otherwise require thousands of miles of latitudinal travel. Natural vegetation proceeds from that of selva and jungle to flora reminiscent of the cold tundra. Cultivated crops ascend the highlands in a ladder-like pattern; the various tropical products are at the bottom of the ladder and the potato occupies the top rung (Figure 9.1).

Location

Highlands are widely scattered in the tropical zone; unlike the climate regions, they follow no definite pattern. Only the more massive areas are shown in Figure 9.2; but small areas are scattered throughout the tropical regions. Mountains and plateaus occupy a major share of Middle America, and the lofty Andes form the backbone of western South America. Highlands are also present in the West Indies and in eastern Brazil.

Large sections of east-central Africa are filled with high tablelands. Mountains also form the backbone of Southeast Asia and are scattered throughout Indonesia, the Philippines, and the islands of the central Pacific Ocean.

FIG. 9.1. Kenya Country. (Ray Bassett.)

FIG. 9.2. The Tropical Highlands.

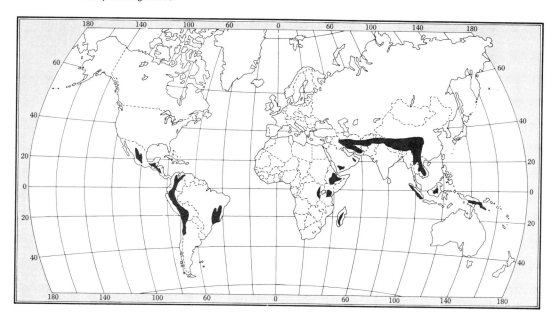

Physical Environment

CLIMATE

There is no Tropical Highland climate as compared to that of the Rainy Tropics, Wet-Dry Tropics, or the Tropical Deserts. Similar patterns of amount and distribution of rain and temperature are distinctive features of these other tropical regions. The many differences in exposure and elevation in the Highlands make it impossible to state averages in these climatic elements (Figure 9.3). One characteristic, however, that all Tropical Highlands have in common is the low annual ranges of temperature. Regardless of elevation of a given station, each month has an average temperature that is a near facsimile of its predecessor. Only daily varieties of weather break the yearly monotony.

The lower pressure of the higher altitudes has a physiological effect on man. Breathing becomes more difficult and lung trouble is often prevalent. Travelers are subject to mountain sickness, the *soroche,* which causes faintness, nosebleeds, insomnia, and loss of appetite. The rarefied air is rich in violet rays, and one either soon acquires a coat of tan or must constantly guard against serious sunburn. Cooking habits must be revised because the boiling point of water is lowered and food takes longer to cook.

Temperatures. Tropical Highland temperatures are controlled chiefly by altitude. Temperatures decrease about 3.3 degrees with each 1,000 feet of elevation. Annual temperature ranges for stations, however, remain about the same between high and low elevations; only actual temperatures differ. For example, Quito, Ecuador, at 9,350 feet, has only a .7° range between the average temperatures of the warmest and coolest months, whereas Guayaquil, on the western coast of Ecuador, with an elevation of only 39.4 feet, has a similar range of .5°.

FIG. 9.3. Climatic graphs of Tropical Highland stations.

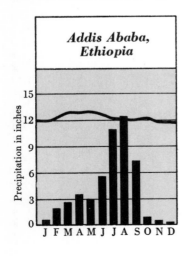
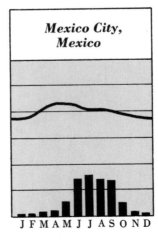
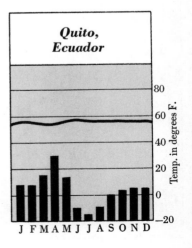

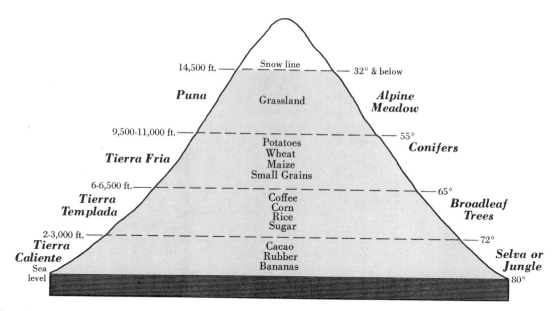

FIG. 9.4. Altitudinal zonation.

Sunshine is intense in the dry, thin air, and exposed surfaces heat rapidly. Noon temperatures are high except in the shaded areas. The Tropical Highlands are sometimes called the "Land of Hot Sun and Cool Shade." The same factors that allow rapid daytime heating operate when the sun sets, and night and early morning are uncomfortably cool. The rising sun is a welcome sight after the chill of the night.

Precipitation. Highlands in general receive more precipitation than the associated lowlands. Seasonal distribution varies with latitude. Areas subject to the constant influence of the equatorial low receive rainfall every month, whereas in Monsoon and Wet-Dry locations there are wide fluctuations. Addis Ababa, the capital of Ethiopia, located at 9.2°N, with an elevation of 8,000 feet, receives about 70 percent of its precipitation in the high-sun months. Quito, on the equator, has precipitation every month. Thundershowers typify the rain; snow falls only in the high elevations.

Slopes exposed to incoming moisture-laden air receive copious amounts of rain in contrast to the dry shadows on leeward sides. Windward sides of mountains with sufficient elevation receive precipitation even in desert areas such as the highlands of Arabia; this makes agriculture possible on slopes and provides a source of water for irrigation below.

Zonation. The Tropical Highlands are divided into four broad altitudinal belts based on temperatures, vegetation, and agriculture (see Figure 9.4). The first belt is the *tierra caliente,* or hot lands, rising from sea level to altitudes of 2,000 to 3,000 feet; the environment is that of the associated lowlands. The second zone, the *tierra templada,* or temperate lands, extends to 6,000 or 6,500 feet and is often called the "Coffee Zone." Although coffee thrives at this elevation, sugar, cotton, and other crops are also grown. The *tierra fria,* or cool lands, follows and reaches to 10,000 or 11,000 feet. This is a belt where wheat, barley, and root crops can be produced—crops suggestive of agriculture in the middle latitudes. The *puna* or *paramos,* the cold lands, is the zone reaching as high as 15,000 feet. The puna is the limit of settlement; crops are unimportant, but some grazing exists on the lower margins. Permanent snow occurs at about 15,000 feet.

The four zones do not have a stable elevation, becoming increasingly lower with distances from the equator. Furthermore, the division lines are seldom parallel; wide variations exist due to factors influencing temperatures, such as exposure to winds and orientation to the sun.

SURFACE FEATURES

Highlands everywhere are the most spectacular of the earth's surface features because of their bulk, height, and steepness of slope. They differ greatly, depending upon their earth materials, manner of formation, and age (see Middle Latitude Highlands chaps.). Two primary divisions are recognized: mountains and plateaus.

Mountains are conspicuous elevations having small summits and the preponderance of their areas in steep slopes. Isolated mountains are common in all portions of the earth, but most are parts of larger groups that display considerable differences in size, shape, elevation, relief, origin, and age. There are small mountain groups like those of Oman and Yemen on the Arabian Peninsula, and in contrast there are great mountain systems covering large areas like the Andes and the Rockies. In all cases local relief is over 2,000 feet; on the basis of local relief, mountains with elevation differences of 2,000 to 3,000 feet are classed as low, from 3,000 to 4,500 as rough, 4,500 to 6,000 as rugged, and above 6,000 sierran.

On the basis of their size and arrangement, it is common to divide the larger mountainous complexes into ranges, systems, chains, and cordilleras. A mountain *range* consists of a long, narrow ridge or of closely-spaced ridges that are similar in age and origin. The Cascade Mountains of Oregon and Washington and the Eastern (Oriental), Central, and Western (Occidental) ranges of the Northern Andes are examples. A mountain *system* includes a group of ranges somewhat related as to origin, structure, alignment, form, and age. Examples are the Rocky Mountains and the Northern Andes. A mountain *chain* is a long, narrow mountainous belt including ranges and systems that may be more or less independent in origin and age and without similarity of structure or form. The mountains of Central America and of southern Europe are often called chains. *Cordillera* is an inclusive term used to refer to a group of ranges, systems, and chains that are fairly compact and cover a vast area. It is applied to the mountains of western North America and to the Andes of South America (see Figure 9.5).

Plateaus, not so easily distinguished as mountains, are tabular uplands having local relief of more than 500 feet. They usually stand well above bordering areas on at least one side. According to position on the continents and relationship to other landforms, they are of three major classes: *piedmont* plateaus, which lie between mountains and neighboring seas or plains, such as the plateau of Patagonia in Argentina; *continental* plateaus, which rise rather abruptly on all margins from bordering lowlands or seas, like that occupying a large part of Africa; and *intermontane* plateaus, which are more or less surrounded by mountains, as in Mexico.

Latin America. The most continuous and largest area of highlands in the tropics fills much of Middle America and extends southward in the Andes of South America. The highlands of Middle America are diverse and difficult to characterize. They are neither simple in geology nor in surface configuration, ranging from broad plateaus to towering active volcanoes. The highlands of the Mexican portion have essentially two elements, a central plateau, which is by no means uniform, and dissected mountainous borders. The mountains, higher on the west, converge in the south in an east-west aligned volcanic zone, including the great peaks of Orizaba and Popocatepetl. Highlands continue through the Central American countries, broken occasionally by intermontane basins. The Andes are the highest continuous mountains in the world, running the length of the continent of South America. This great barrier, rising virtually out of the Pacific Ocean, is 100 to 400 miles wide and has a crest that is seldom under 12,000 feet high. Many peaks reach above 18,000 feet; and Mt. Aconcagua in Argentina (22,835 feet) is the highest peak in the western hemisphere. This massive system is made up of several ranges that are complexly interrelated and, in places, formed in echelon. From Venezuela southward into Argentina, between the principal ranges there are many high intermontane basins. The most notable are the major basins of Bolivia, often known collectively as the *Altiplano*.

The ancient Brazilian Highlands, in contrast to the youthful Andes, have been subjected to weathering and erosion through countless ages. Most of the area stands at elevations between 2,000 and 4,000 feet and is generally hilly or plateau-like in character, although some peaks do reach 8,000 feet and more, particularly in the southeast. The steep

FIG. 9.5. The Buenaventura Highway. Rugged terrain and heavy vegetation make road building extremely difficult and costly. This highway is winding down the slopes of the Western Cordillera into the Cauca Valley of Colombia. Notice the farmstead in the central portion of the picture. Where soils exist, slopes as steep as these are sometimes farmed. (Standard Oil Company of New Jersey.)

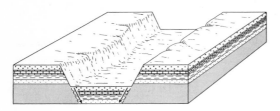

FIG. 9.6. Diagrammatic sketch of a rift valley.

and severely dissected eastern escarpment appears like a range of mountains from the sea.

Africa. The continental plateau of Africa reaches its greatest elevation in the east-central portion of the continent. Starting from the knot of the Abyssinian Mountains, the highlands run southward through the lake region. They have a relatively uniform surface and are of volcanic origin, built up by layer upon layer of lava poured out upon the plateau. The region is traversed lengthwise by a series of rift valleys, part of the great system that extends for 4,000 miles from Lebanon at the east end of the Mediterranean Sea southward through the Red Sea and into central Africa. Linear lakes occupy the two main branches in Africa with Lakes Albert, Edward, and Tanganyika along the western line, and Lakes Rudolf and Nyassa along the eastern line; Lake Victoria lies in the tableland between the two rift valleys.

This great rift system was produced by down faulting; as the result of tremendous compressional and tensional stresses, parallel faults occurred, and portions of the earth literally slipped downward. The resulting trench-like structure is termed a *graben* or rift valley (see Figure 9.6). Such features are present in other parts of the world, but nowhere are they so spectacular.

Southeast Asia and adjoining islands. The principal characteristics of Southeast Asia and the adjoining islands have been discussed in the chapters dealing with the Rainy Tropics and the Monsoon Tropics. It will be recalled that mountains are the most widespread landforms of Southeast Asia, with many ranges spreading southward from eastern Tibet in a fan-like pattern through Burma, Thailand, and Indochina; one range continues farther southward to form the backbone of the Malay Peninsula. Elevations tend to be highest in the north where deep gorges and steep slopes often dominate; however, the Annam Mountains of Indochina rise to over 10,000 feet as do the mountains along the India-Burma border.

The geology of the islands adjoining Southeast Asia is extremely complicated. The western islands of Indonesia, including Sumatra, Java, and Borneo, stand on the comparatively shallow continental shelf of Asia, whereas the easternmost islands, including New Guinea, are on the shallow continental shelf of Australia. Celebes and the Moluccas are essentially mountaintops that rise from the deeper seas between. The Philippines are similar to the central islands of Indonesia—mountains in the process of rising from the sea. Actually, two mountainous areas form the main outlines of the archipelagoes, or island groups; one, the Sunda Arc, reaches from Sumatra eastward through the Lesser Sunda group (Bali eastward through Timor); the other, the Sulu Arc, reaches southward through the Philippines into the Moluccas. Both contain many extinct and active volcanoes; Indonesia is one of the most active volcanic areas on earth today.

NATURAL VEGETATION

Vertical zonation characterizes Tropical Highland natural vegetation; broad zones ascend the slopes, roughly paralleling types are found in a horizontal range from the equator to the poles. The forest sequence includes tropical rain forest and jungle in the low levels, followed by zones of broadleaf evergreens, deciduous broadleafs, and ending with a belt of conifers. The forest zones are often disrupted by patches of grass and bare areas; occasionally there is a mixture of types. Spotty distribution is a result of local conditions such as exposure to sunshine, dry-shadow locations, or bare rock surfaces. Alpine meadows begin at about 13,000 feet and stretch to the snow line. On the wet windward side of mountain slopes the forest zone will attain higher elevations resulting in a smaller amount of alpine meadow. The reverse is true on the dry leeward side, where the alpine meadows reach their greatest extent. Rainfall also influences types of vegetation on lower slopes. Where definite wet and dry seasons are experienced, savanna and scrub forest may displace the selva as the dominant lower cover.

NATIVE ANIMAL LIFE

Native animal life follows a pattern similar to other elements of the physical environment in that zonation features distribution. Lack of food and reduced temperatures, however, limit many species to the middle and lower elevations. Fauna in the tierra templada are quite similar to types living in the adjacent lower levels.

Two wild members of the camel family, the vicuna and guanaco, are distinctive in the high Andes. Both are smaller in stature than a camel and do not possess a hump. Vicuna-hunting during the Inca regime was restricted to the rulers and today law protects the animals from extinction. The alpaca, raised for fine fleece, and the surefooted pack animal, the llama, are the domesticated animals of the Andes.

Occupance in the Tropical Highlands

Study of the distribution of the world's population reveals that man generally prefers the lowlands as a dwelling place; at least 90 percent live at elevations below 2,000 feet. Difficulties of cultivation, transportation, construction, and isolation usually make highland environments unattractive to large-scale settlement; nevertheless, few of the larger areas are without people. In the tropics an irregularity in the "general rule" exists— the Tropical Highlands are often inviting. The most pleasant and healthful climate is found in the tierra templada zone, which has attracted concentrations of population. The more invigorating highland climates have stimulated the development of the highest orders of civilization within the tropics. This is certainly true in Latin America, where the Inca and Aztec civilizations flourished and where the Spanish and Portuguese colonizers soon recognized the more favorable qualities produced by higher elevations.

The center of Brazilian life and economy is on the plateau in the east and southeast portions of the country, where the preponderance of that nation's 90 million people lives. Similarly the majority of the peoples of Middle American and Andean countries have chosen the favorable basins, valleys, and plateaus of the highlands rather than the hot and humid lowlands. The greatest concentrations of cities and developments are located at moderate elevations. In the Andes there is frequently a zonation of peoples as well as climate, vegetation, and crop possibilities. The occupants of the tierra caliente zone are largely native migratory agriculturists and Negro descendants of former slaves. A few areas have been developed for plantations. The accessible basins and neighboring gentle slopes of the tierra templada are occupied by native and mixed breeds as well as people of European background who are engaged in commercial agriculture. The tierra fria and the puna are sparsely populated by native Indians pushed into these higher elevations by European settlement; their livelihood is based upon the production of hardy crops and the pasturing of animals. The population is by no means evenly spread but is patchy in response to the distribution of land suitable for agriculture and to transportation and accessibility. In general, there is a considerable gulf between the social and economic well-being of the native and mixed-race people and those of European backgrounds. In recent years this gap has been the basis for considerable discontent. The peasants are beginning to seek more equitable opportunities and greater voice in social and economic order. The Rockefeller Foundation, through technical assistance and funds, is attempting to aid in this movement.

In contrast to the long-settled Latin American Highlands, concerted effort to develop the African regions other than by native peoples has been recent. Delay was due to such factors as the recent penetration of the continent by Europeans, a lack of the lure of stored native riches, the problem of accessibility, and the fact that colonizing powers had richer lands to develop elsewhere. These highlands, usually 3,000 to 6,000 feet above sea level, have a moderate,

FIG. 9.7. Rice terraces in Indonesia. Although population pressure on the land is great, the people of the Asian region live primarily on the plains and in the valleys. In some places population density is so great that intensive cultivation has been forced up the hillsides. This scene is on densely settled Java. The surface here has been completely reshaped by man to permit water control and rice production. (United Nations.)

pleasant climate, and many sections are favored with wet and dry seasons and youthful volcanic soils. Although in the minority because of the commercial infrastructure they have established, the Europeans have been the significant instrument of change. The British had recognized the development possibilities and had been giving increased attention to commercial agriculture during the late stage of the colonial period. Today, Kenya in particular plays a significant role as a supplier of agricultural raw materials to industrial Great Britain.

The highland corners of the southern fringes of the Arabian peninsula comprise Yemen and Oman, which have long been fruitful "islands" in a sterile desert region. These portions are sufficiently elevated to intercept regular precipitation from the monsoons. Yemen on the west receives summer rain in large enough quantities to support 5 million people and their agriculture. In marked contrast to the arid wastes of the lower deserts, flowing streams for centuries have been diverted to irrigate carefully terraced slopes that produce cereals, fruits, and the famous mocha coffee. Oman, somewhat further north on the eastern corner, receives less rainfall and that during the winter season, thus it supports fewer people. Oman, in contrast to Yeman, receives its precipitation during the northern hemisphere winter.

Little developmental attention has been given to the highlands of the Monsoon Tropics of the Asian mainland. There, the mountainous areas are sparsely populated by primitive tribes who make little imprint upon the land except as their fire-cleared fields cause alterations in the nature of the vegetation. In some areas hill stations have been established for the use and comfort of Europeans during the hot season.

The highlands on the islands bordering Southeast Asia, especially Java in Indonesia (Figure 9.7) and Luzon in the Philippines, support notable populations. The all-important activity is the production of rice for food. For example, the rice terraces of the semiprimitive Igorot people in the Central Mountains of northern Luzon are famous. These giant terraces, covering 250 square miles, cling to the steep sides of the valleys with retaining walls 20 to 30 feet in height. It is believed that more than 2,000 years were required to build them. They are an admirable engineering feat, even by modern engineering standards.

Settlement in the Tropical Highlands is based primarily upon agriculture and mining. In the more progressive countries, processing of raw materials is establishing manufacturing as an important part of their economies.

AGRICULTURE IN LATIN AMERICA

Three distinctive forms of agriculture are recognized in the Latin American Highlands. The first and least complicated is that practiced by the Indians of the tierra fria in the Andes, where climate rigidly restricts cropping possibilities. Here people live in crude mud or stone huts and cultivate small fields in the valley bottoms or on the slopes, which are sometimes terraced. Unresponsive to government attempts to introduce efficient methods, the people practice agriculture in a traditional and simple manner. Crops are adjusted to altitude; in the highest zone of agriculture, the potato is the principal crop, grown to approximately 14,000 feet in Peru; grains are produced between 10,000 and 13,000 feet. Most groups also keep animals, with the native llama (Figure 9.8) and alpaca significant in Bolivia and Peru; mules, sheep, and goats are more important farther north. A few Indian people living above the crop line are dependent entirely upon a pastoral economy; sheep are their principal animals.

The second and slightly more advanced form of agriculture is practiced in the tierra templada by the Indians and by mixed breeds—the mestizos, white and

FIG. 9.8. Peruvian Indian farmer with pack llamas. This sure-footed relative of the camel is the principal beast of burden in the Andes. (Martha Carbonne.)

Indian; mulatto, white and Negro; zambo, Negro and Indian; and a mixture of all three. These groups comprise the major segment of the Latin American populations. Corn is their principal food, supplemented with beans, often grown in the same field. Other crops include chile pepper, peas, and small grains. Some produce a limited amount of commercial products. Holdings are small, usually occupying poor sites, equipment is simple, and living standards are very low. The old civilizations exist beside the new, little altered by 400 years of contact.

In direct contrast is the third form—commercial agriculture under the administration of people with European background. Holdings are large and equipment and methods are relatively modern. Coffee is the most prominent and distinctive crop. Brazil is the great world producer, with centers chiefly in the states of São Paulo and Minas Gerais where the hot, moist growing season and the cooler, dry harvest season, together with large expanses of gently sloping land with porous dark-red loam soils, provide an ideal environment. Although there is some production on small farms, the characteristic unit is the *fazenda,* or coffee plantation, resembling the English manor of the fourteenth century—each tends to be a more or less completely integrated economic unit. Some operators produce other crops or graze cattle in valleys where there is danger of coffee-damaging frost.

Coffee is also a major commercial crop in the highlands of the Northern Andean countries, especially in Colombia, as well as in Middle America and the West Indies (Figure 9.9). Often the greater part is grown on small farms. Several other export crops are produced, including cotton, sugar, and tobacco.

AGRICULTURE IN AFRICA

Parallel forms of agriculture are found in the African Highlands. In the high, dry grasslands natives live by pastoral activities; lower, in areas where climate, slope, and soil are suitable, permanent agriculture has become established. The nature of native farming varies from one sector to another; some groups, such as those found at 3,500 to 5,000 feet on the Uganda Plateau, have developed a relatively advanced culture, living in villages of neat, round, grass huts surrounded by fields of yams, peanuts, manioc, grains, vegetables, and fruits. Others plant crops and then follow herds about the grasslands, returning to their fields only for the harvest. During this century the impact of white settlement has in part encouraged some commercial production.

White settlement has been almost wholly for the purpose of commercial agriculture. The completion of a railroad from the coast across Kenya to Uganda at the turn of the century initiated development, making possible the export of crops. The largest white colony is found in the tierra templada zone of Kenya; only about 50,000 of an 11 million population, however, are white.

The colonial administration here allowed European people to gain control of an area the size of the state of Maryland. Since the beginning, coffee has been the chief crop; it is grown on plantations using native labor. Other crops include sisal, tea, cotton, sugar cane, tobacco, pyrethrum (a kind of chrysanthemum used for the production of insect powder), and wheat. In the years immediately following World War II, large-scale plantings of peanuts were attempted, but with little success.

MINING

Precious minerals were the first attraction of the Latin American Tropics; exploitation has continued, and in the modern era the minerals of the highlands contribute significantly to the economies of a number of countries. Mining is the entire basis for settlement in some localities. The most noteworthy minerals of the Andes are copper and tin. Copper mining was carried on by the Indians before the coming of the white people, but this century has witnessed its major development. Several countries mine copper, but Chile, with one-fifth of the world's reserves, is outstanding. The working of the largest deposit at Chuquicamata, 10,000 feet above sea level, supports a community of 25,000 people. Tin deposits are located in Bolivia, with the principal mining districts in the mountains at elevations of 14,000 to 18,000 feet. This country has also been famous for silver production in the past.

Mexico has been a leading mineral-producing land since the arrival of the Spaniards. Today important quantities of silver, lead, gold, zinc, and copper are mined. Brazil possesses one of the largest high-quality iron reserves in the world in the central Belo Horizonte district in Minas Gerais, but exploitation is restricted by lack of quality coal

FIG. 9.9. Colombian coffee grown under shade. The trees are kept pruned to facilitate harvest. (J. Granville Jensen.)

FIG. 9.10. Steel mill at Volta Redonda illustrates the spread of modern industrialization to tropical countries. This mill is the largest in South America. (Brazil Government Trade Bureau.)

Careful mineral surveys have not yet covered large areas of the Tropical Highlands, and there are possibilities of further discoveries. Although minerals are essential for modern development, in the long run mining constitutes a weak foundation for continuous support of communities, since minerals are exhaustible resources.

MANUFACTURING

The Tropical Highland regions of Latin America have developed industry as an important segment of their economic activities. Industrial plants here, using local raw materials, are manufacturing goods chiefly for home consumption in an attempt not only to bolster their own economies but also to lessen their dependence upon other areas. The leading single industrial center of Latin America is the district around São Paulo, Brazil, where thousands of factories ranging from small workshops to large modern plants produce textiles and clothing, machinery, cement, and a variety of other products.

and the distance of the reserves from the sea and major foreign markets; however, a steel center has been established at Volta Redonda, despite the difficulties (Figure 9.10). Manganese is also present in the same area and is being mined largely for the United States market.

The extensive stretches of volcanic rock which bury the possible mineral-bearing crystalline rock over much of the African Highlands have hindered exploration, except where eroding streams have exposed the underlying formations. Where the volcanic cover is lacking, as elsewhere in the humid tropics, the solid rocks are obscured by a deep mantle of decomposed rocks and soils that also hinders mineral discovery. Despite these conditions some spectacular mineral developments have occurred in recent decades, and doubtless there will be others. The mineral development of Tanzania has been especially noteworthy; diamonds are the leading product, gold is significant, and tin, lead, and mica are also being mined.

One of the large steel mills in Latin America is located at Volta Redonda, in the Paraiba Valley inland from Rio de Janeiro. Iron ore is brought from the Belo Horizonte district and coal from the southern provinces of Santa Catarina and Rio Grande do Sul. Location of the plant of Volta Redonda places it near the center of the growing urban markets at the median cost point between coal and iron ore. Medellin, in the central Andean highlands, dominates industry in Colombia. Highland Mexico contributes materially to Latin America's industry; and development is progressing under nationally sponsored programs. The diverse products of Mexico's factories include iron and steel, textiles, pottery, food products, and a host of others. Major industrial concentrations are centered near Mexico City.

The Brazilian Plateau and the highlands of British East Africa have especially good possibilities for continued development; both have large areas adaptable to modern agriculture, plus room for more people. Agricultural raw materials as well as minerals and forests should encourage expansion in industry. The state of São Paulo, Brazil, especially exemplifies the trend toward industrial awakening. A promising area also exists in the central Andean range of Colombia, where Medellin and Manizales, with low cost hydroelectric power, mineral resources, improved transportation facilities, and a progressive citizenry, are establishing notable industries today.

URBAN CENTERS

The highlands of the Latin American Tropics support most of the largest cities of the realm (Figures 9.11 and 9.12). São Paulo, at an elevation of nearly 2,700 feet, is now the largest city in Brazil and one of the fastest growing urban centers in the world; it is rapidly surpassing a population of 6 million. Mexico City, of similar size, is situated in a basin (Valle de Mexico), 7,500 feet above sea level and is the center of government, commerce, industry, and culture. Caracas, in a productive valley at 3,000 feet, is the outstanding city of Venezuela, with a population of 1.9 million. La Paz, Bolivia, with over 600,000 people at an elevation of 12,000 feet in the Andes, is the highest major city in the world. Quito, Ecuador, with a population of 405,000, is 9,350 feet above sea level; Bogotá, cultural center and capital of Colombia, with a population of over 2.2 million, is situated at an elevation of 8,700 feet. At least a dozen additional cities exceed 100,000, and there are many in the 25,000 and 100,000 class.

The African Highlands have as yet few large cities. Addis Ababa, capital and commercial center of Ethiopia, has a population of 650,000. The population of Nairobi, Kenya, is about one-half million; and that of few other centers approaches 100,000.

FIG. 9.11. São Paulo. A view of Avenida Anhangabau, the busiest traffic artery of downtown São Paulo. (J. Granville Jensen.)

FIG. 9.12. The skyline of Nairobi, Kenya. This modern city, the capital of Kenya and commercial center of East Africa, symbolizes the continuing westernization and economic advance of Black Africa. (William C. Rense.)

PART THREE

THE MIDDLE LATITUDE REALM

Our analysis of the earth as the human habitat now focuses upon the regions of the middle latitudes. In contrast to the regions of the tropical realm, those of the middle latitudes have temperature seasons—and it becomes appropriate to recognize spring, summer, autumn, and winter. Temperature characteristics are coequal with precipitation characteristics as criteria for setting the regions apart.

The total range of climatic differences in the middle latitude realm comprises many climates, ranging from nearly tropical to nearly polar and from extremely wet to extremely dry. For our purposes, however, there are eight major types: Dry Summer Subtropics, Humid Subtropics, Long Summer Humid Continentals, Short Summer Humid Continentals, Dry Continentals, Marine West Coasts, and Subarctics. As in the tropical realm, the highlands will be separated out for independent appraisal. Within the total scope of the middle latitude realm there are also vast differences in natural vegetation and soils, but, as in the tropical realm, their relationships with climate are such that the regional framework provides a key to their understanding as well as to the understanding of other characteristics of and variations in the physical-biotic system.

In the realm as a whole, there are more extensive and varied evidences of man as an active agent of modification, organization, and earth use than in the tropical realm. In the middle latitudes, man has achieved his greatest advancements in resource-converting techniques, in adjusting space to suit his needs, and in economic and political organization and development. Change has gone on at an accelerated pace during the twentieth century. Yet there are vast differences in the the degree and character of occupance: for example, the Asian Humid Subtropics teem with farmers, and farming looms large in national economies; but Japan provides an outstanding contrast—she has become a commercial industrial-urban society. In the United States manufacturing, commercial, and service activities are more important than agriculture in the employment of labor; yet the United States is the leading world producer of agricultural commodities. In Western Europe the high man-land ratio rivals that of the Orient, but in Western Europe secondary and tertiary activities are major employers of labor. Canada, Australia, and most of the southern hemisphere countries of the realm, in contrast, have modest populations and considerable open space.

In the chapters that follow, these and other characteristics of the man-earth system will be examined.

ten

DRY SUMMER SUBTROPICS

The Dry Summer Subtropics are almost universally known as the mediterranean regions, a name derived from the lands bordering the Mediterranean Sea, where the unique climatic pattern of mild, rainy winters and hot, dry summers is most extensively developed.[1] Similarity is a distinctive feature of mediterranean lands. Mountain slopes or plateau escarpments form picturesque backgrounds for every region, and all face the sea. Low shrubs and brush mantle hillsides, interspersed with scrubby gray-green or bluish broadleaf evergreens. Cultures are complex and varied, but farming is the basic activity in all regions. Although there are differences in scale, types of agriculture are similar, and crops are almost identical; a definite mediterranean characteristic is the olive-citrus-vine-wheat quartet. Sunshine, sea, and scenery are the bases for a flourishing resort activity in all regions of the Dry Summer Subtropics.

The imprint of new cultural patterns is still fresh in the mediterranean lands of the Americas, South Africa, and Australia, whereas the actual borderlands of the Mediterranean Sea have a rich, colorful past and have made countless contributions to western civilization. Many of the classics in art, literature, and science stemmed from the early Greeks. The Romans made many contributions to the basic principles of justice, law, and order. Roman roads were epic in land transport as were the developments in water transport and irrigation technology, and the Mediterranean Sea was a nursery school for navigators. Three of the great monotheistic religions were nurtured here. The impact of the Mediterranean Basin lands on the western world is clearly stated by Ellen Churchill Semple: "All the world is heir of the Mediterranean. All the world is her debtor. Much that is finest in modern civilization traces back to seeds of culture in the circles of the Mediterranean lands and transported thence to other countries."[2]

[1] Dry Summer Subtropics and the term "mediterranean" are used interchangeably. When Mediterranean is capitalized, it refers to the sea or basin.

[2] Ellen Churchill Semple, *The Geography of the Mediterranean Regions* (New York: Holt, Rinehart & Winston, Inc., 1931), p. 3.

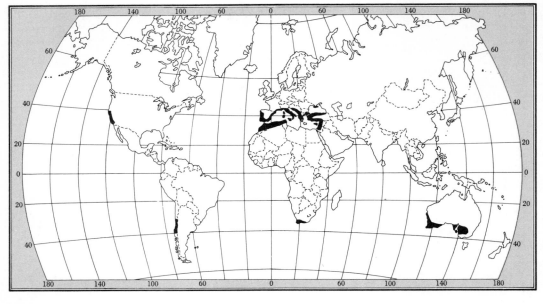

FIG. 10.1. The Dry Summer Subtropics.

Location

The Dry Summer Subtropics border the tropical zone on the western margins of the continents, usually between latitudes 30° and 40°, with extensions inland varying with continental outline and surface. All continents are represented (Figure 10.1).

The greatest areal development is in the lands fringing the Mediterranean Sea. All the European countries touching the Mediterranean Sea have areas included within this type; so do Portugal on the Atlantic west coast, Turkey, Syria, Lebanon, and Israel on the east shore of the basin, and the Barbary Coast states of Africa on the south shore.

FIG. 10.2. The Mediterranean Basin.

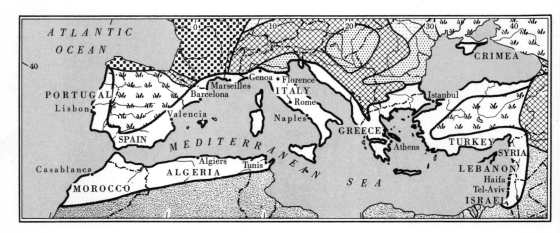

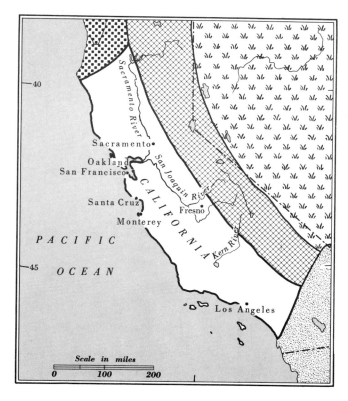

FIG. 10.3. California.

FIG. 10.4. Central Chile.

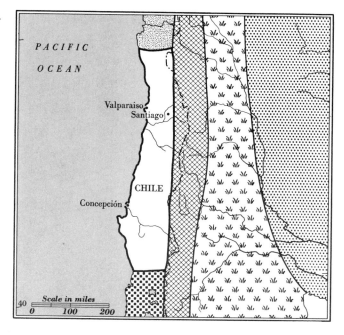

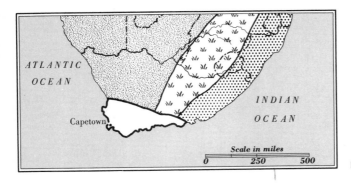

FIG. 10.5. The Capetown district.

The region extends through the Bosporus Straits, along the Black Sea coast of Turkey, and includes southern Crimea of the USSR (Figure 10.2).

The North American region is restricted wholly to California west of the Sierra Nevada Mountains, but excludes the desert to the southeast and the cool, moist north coast (Figure 10.3). The middle one-third of Chile, from about Coquimbo (30° S) to the Rio Bio Bio, comprises the South American region (Figure 10.4). On the southwest tip of Africa there is a small region known as the Capetown District (Figure 10.5). Two areas occur in Australia, one on the southwest fringe including the city of Perth, the other bordering the Great Australian Bight on the east and extending through the southern margin of the Murray Basin (Figure 10.6).

FIG. 10.6. Australian Dry Summer Subtropics.

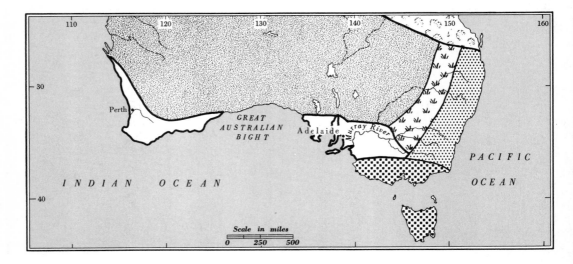

Physical Environment

CLIMATE

The Dry Summer Subtropics are transitional areas between arid and humid climates; the equator margins of the regions are fringed by deserts, and the humid lands are normally poleward. Winter is the season of rain; aridity rules during the summer. Seasonal changes are the result of alternating controls. During the dry-summer season the mediterranean regions are dominated by the dry, settling air of the subtropical high and the trade winds. In winter the high-pressure belt and the trades shift equatorward, and the regions come under the influence of the rain-bringing cyclonic westerlies.

A high percentage of sunshine is a mediterranean trait. Summer skies are clear and blue, and even in the rainy winter there is still a high percentage of sunny weather. Sea breezes are common during the summer, moderating high daytime temperatures. Summer sea breezes in the Mediterranean, however, are rarely cool due to the warmth of the sea. Furthermore, the sirocco, a hot, dust-laden, parching wind, occasionally sweeps in from the Sahara. A desert wind, known as the Santa Ana, occurs in southern California. There are several local winter winds in the Mediterranean Basin, such as the mistral and bora, which cause considerable discomfort to the inhabitants. The cold, dry mistral ranges from Barcelona to Genoa but is especially pronounced in the lower Rhone Valley, where farmers plant screens of cypress trees or build windbreaks to protect crops from the cold blasts. The bora, a similar winter wind, occurs along the Adriatic coast of Yugoslavia.

Temperatures. Winters are mild with average temperatures between 40° and 50°.[3] Frostfree periods range from nine to twelve months. Frost is most likely to occur when nights are cool, clear, and calm, when the earth heat is rapidly radiated into the atmosphere. When the cool layer of air near the surface reaches a temperature of 32° or lower, the water vapor usually condenses into white frost, or tiny ice crystals. Frost is especially hazardous to citrus fruits and out-of-season vegetables. Citrus-growers utilize the slopes, which are safer from frost since dense cold air flows down the hillsides and settles in the valley bottoms (see Figure 10.7). The most common weapons for combating frost are small heaters scattered throughout the orchard. Another method, but more expensive, is the use of wind machines that attempt to mix the cold surface air with warmer air aloft.

Summer average monthly temperatures range from 70° to over 80° and compare with desert recordings. Like the deserts, too, relative humidity is low and heat is dry. The clear skies are ideal for rapid heating, and noon temperatures frequently soar above 100°. Temperatures drop sharply in the evening, and summer diurnal ranges are high.

FIG. 10.7. Air drainage. Cool air, like water, tends to flow down slopes and settle in low places. Thus there is less danger of frosts on slopes than in valley bottoms.

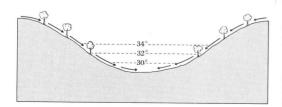

[3] In the winter season on the French Riviera an overcoat is found to be comfortable evening garb, and furnaces in villas are started after the sun sets.

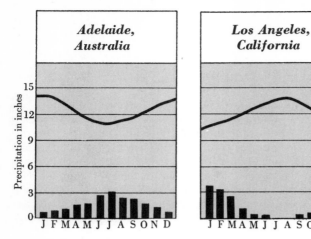
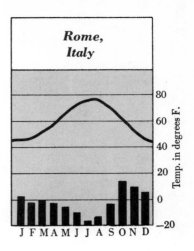

FIG. 10.8. Climatic graphs of Dry Summer Subtropic stations.

Precipitation. Dry Summer Subtropic precipitation is moderate, with yearly totals averaging from 10 to 30 inches. The distribution is uneven, with the greater part of the precipitation falling during the cool season (Figure 10.8). Rain is usually a summer rarity. Light showers followed by sunshine are typical, but occasionally there are periods of cloudiness with rain occurring as a persistent drizzle. Snow is rare on the lowlands, but during winter it usually falls in the mountains, where a deep snow cover is especially significant since it provides irrigation water for the summer dry season. Thunderstorms, except in the higher elevations, are uncommon. Fogs occur on those coasts that have paralleling cold ocean currents; parts of the California coast are some of the foggiest areas in North America.

Mediterranean precipitation is unreliable; annual amounts and time of arrival both fluctuate. Low amount, poor distribution, and unreliability of precipitation place distinct limitations on possible land uses and make irrigation necessary for the growth of many crops. It is fortunate that rain coincides with the cool season; a summer maximum would lose much of its effectiveness due to the high evaporation rate.

Location and local differences. Location in respect to latitude, water bodies, deserts, and mountains causes many local differences in mediterranean temperatures and precipitation. Coast locations have cooler summers as well as milder winters than interior stations. Several California stations illustrate these conditions—the average January and July temperatures of Santa Cruz on the coast are 50° and 63° as compared to 45° and 79° at Merced in the interior. Desert margins are drier and have shorter seasons of rain in contrast to poleward locations. Chico receives 24 inches, whereas Fresno, over 185 miles to the south, has an average of 10 inches. Exposed slopes receive amounts considerably higher than the neighboring lowlands. Mountains in many areas also form effective barriers as protection against invasion by cold continental air.

SURFACE FEATURES

Plains occupy a relatively small portion of the Dry Summer Subtropic regions. Rugged topography frequently dominates the landscapes, restricting the amounts of land suited to cultivation (Figure 10.9).

The greatest complexity of landforms is found in the Mediterranean Basin, most of which is surrounded by mountains (see Figure 10.10). The three southward-jutting peninsulas, Iberian, Italian, and Balkan,

FIG. 10.9. Judaean Hills, Jordan. Often mentioned in the Bible, the Judaean Hills are solemn and impressive even in their aridity of midsummer. The winding road recalls the Bible story of the Good Samaritan who cared for the wounded stranger on the road to Jericho. (Arab Information Center.)

FIG. 10.10. Physiographic diagram of the Mediterranean Basin.

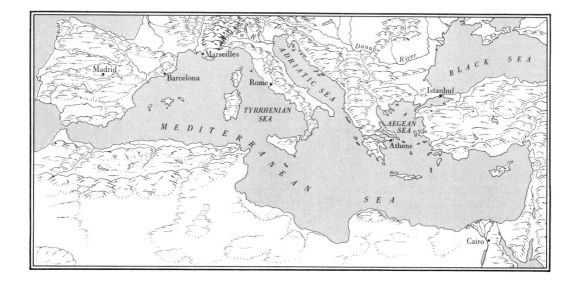

FIG. 10.11. Monaco as seen from France. This view illustrates the dearth of level land along much of the Mediterranean coast. Settlement, as a result, has been forced up the slopes. Monaco is a 370-acre independent principality on the French Riviera. The site was chosen by the Phoenicians as an outpost and concentration point because of its excellent harbor and easily defended position. Today this small state, especially its main city, Monte Carlo, is renowned as a gambling center with international attraction. It rates as a major lure of the Riviera. (Oliver H. Heintzelman.)

along with the Turkish Peninsula, consist of complex groups of highlands bordered by small and disconnected plains. Most of the plains are depositional, having been built by mountain-fed streams dropping their sediment loads in quiet, protected bays. The largest include the lower Rhone Valley, Arno Valley, and Guadalquivir Valley. Within the region there has been a widespread development of karst topography, particularly in southern France, southeast Italy, the Dalmatian Coast of Yugoslavia, and in Israel.

The coastline of the Mediterranean Sea is of interest for its influences on man's early activities. The shores of the Aegean Sea, formed by subsidence of mountains and valleys at right angles to the sea, are rugged with numerous islands, inlets, and protected harbors. Difficulties of land travel in these regions led to the early development of water transport and navigation. Elsewhere mountains parallel the sea, often rising steeply from the shore, forming few protected harbors and leaving little room for man. However, coasts such as the Rivieras of France and Italy, have beautiful settings and have become resort areas of world acclaim.

Man has severely altered the surface features of the Mediterranean Basin. Through the centuries, population crowding on the plains has forced wider use of the hilly and mountainous areas than is common in most other climatic regions (Figure 10.11). In places the slopes have been terraced to make agriculture possible. Removal of the natural vegetation, by fire, cutting forests, and overgrazing, has exposed the slopes to rainfall, causing many serious erosion problems. The results have been denudation of soils from slope land and an increase in flood and silt problems on many river plains. This has reduced the habitability of some lowlands by producing swampy and malarial conditions.

In the regions of North and South America the landforms have a north-south orientation, and in both there are found three parallel units—high bordering mountains, central valleys, and low coastal uplands. The Central Valley of California, lying between the Sierra Nevada and the coast range, is the largest and most important lowland area of the entire realm (Figure 10.12). Its alluvial surface was built principally by streams of the Sierra Nevada. Drainage is provided by

FIG. 10.12. Physiographic diagram of California.

the Sacramento River system in the north, the San Joaquin in the central portion, and the Kern and King Rivers, which flow into the enclosed Tulare Basin, in the south. The Central Valley of Chile, between the towering Andes and the low coast range, also forms a long corridor but differs from its counterpart in California by being divided into separate basins by spurs of the Andes. The drainage pattern differs in that the major streams are transverse, descending from the Andes and directly crossing the basins and coast range to the ocean.

The Capetown District of the Republic of South Africa is separated from the continental plateau by the rise of the land in a series of steps that form the Little and Great Karoos. The coastal zone is not a plain but an area of broken topography characterized by rugged hills and small fertile valleys.

Only in the Australian regions are mountains absent from most scenes. In the southwest, hills are even lacking, but an escarpment separates the coastal plain from the western plateau. The lowland character of the Murray Basin is broken in the west by the Flinders Range, which extends northward from Adelaide and the Gulf of St. Vincent, and by the Mount Lofty group of hills immediately east of Adelaide.

NATURAL VEGETATION

Vegetation types are similar throughout all the mediterranean regions, although species differ from continent to continent. Most of the species have xerophytic characteristics such as thick bark, small, stiff, shiny leaves, thorns, waxy surfaces and other devices to protect against excessive transpiration during periods of high temperatures. The most common type consists of low evergreen shrubs and brush thickets known as *chaparral* in California and *maquis* in the Mediterranean Basin. This type is more widespread than in the earlier period of human occupance, being more tolerant than forests of low-moisture conditions and fire. Where found today, forests are composed mainly of broadleaf evergreens; conifers grow on the higher and wetter mountain slopes. The broadleaf trees are widely spaced, with large trunks and gnarled branches, typified by myrtle, holly, holm oak, and cork oak. Tall trees are rare, except for the stately stands of eucalypti in southwestern Australia.

The natural vegetation of the Mediterranean Basin has been altered by overgrazing, fire, and by centuries of timber cutting without provision for reforestation. Many slopes are almost bare. Alterations have been so great that it is difficult to deduce the nature of the original cover.

Grasses and forbs are common in California. Annuals were introduced by the early Spanish and in many places dominate today. The seasonality of precipitation is clearly marked where grass is found; in winter the landscape is green, but in summer the parched blades are dry and brown.

SOILS

There are decided differences in the soils from locality to locality in the mediterranean regions, resulting from diversity of landforms and parent material as well as from variations in rainfall. On slopes, which are especially common in the Mediterranean Basin, the soils are frequently shallow and poorly developed. In the more rainy sections, leaching is advanced, and the abundant moisture and warmth cause the residual soils to be a reddish color. In the leeward and lower rainfall areas, leaching is less pronounced and fertility tends to be high. Many of the best agricultural lands are found in localities with alluvial soils that are particularly productive under irrigation.

Occupance in the Dry Summer Subtropics

The Dry Summer Subtropic regions comprise about 1 percent of the world's landmass and are the homeland of 4 percent of the world's population. The complex nature of the topography results in an uneven distribution of population; concentrations are generally centered on the plains, but even some mountain sections are crowded. The old lands of the Mediterranean Basin, occupied for thousands of years, have densities comparable to the settled lowlands of India and China. The indigenous peoples of the other Dry Summer Subtropic regions made relatively little imprint upon the land, and their numbers were small and cultures simple when white settlers arrived. Fire alteration of natural vegetation, however, was widespread. Effective settlement in these new mediterranean lands dates back little more than a century but, despite their later development, these areas are rapidly filling. The state of California has made remarkable strides in population growth and today ranks first among the states of our nation. Approximately 90 percent of the population of Chile is located in the Dry Summer Subtropic portion, which forms the heart of the country. Chile, a state of peculiar spatial form, over 2,600 miles from north to south, with an average width of only about 100 miles, has a northern one-third restricted by desert and a southern one-third restricted by cool, moist climate and rugged terrain.

Almost all mediterranean lands lack the raw materials and energy resources for major industrial development. Resources for the extractive industries—especially mining, and forestry—are also limited. Climate, soil, and water are the attractions and agriculture is the core element in all the economies. In no other regions of the world is farming so diverse and yet so specialized. Agriculture was, is, and will probably continue to be the dominating activity of man in most of the mediterranean regions.

California presents several variations from these generalizations. The American asssessments of the developmental attractions of this region are in the framework of a highly sophisticated economy, high standard of living, advanced transportation system, mobile population, etc.

AGRICULTURE IN THE OLD MEDITERRANEAN LANDS

From the earliest settlements, man in the Mediterranean Basin has depended upon agriculture for his livelihood; he has shown remarkable ingenuity in selecting crops which completely utilize the "niggardly given gifts of nature"—the light and poorly distributed rainfall for general agriculture and the high proportion of land in steep and uncultivatable slopes. As Miss Newbigin pointed out a half century ago, Mediterranean man has, "whenever possible, replaced the natural vegetation by a series of crops which make use of every drop of water, every square foot of soil, and yield him a complete dietary."[4] The suitability and success of the crops that he has been growing since ancient times is indicated by the fact that these same crops form the base of agriculture wherever man has settled in regions of Dry Summer Subtropic climate.

Conditions for farming are not altogether unfavorable. The long growing season, long

[4] Marion I. Newbigin, *The Mediterranean Lands* (New York: Alfred A. Knopf, Inc., 1924), p. 63.

FIG. 10.13. Grape harvest in France. This activity is often a family enterprise and grapes are one of the major crops in the Languedoc area of the Midi. (French Embassy Press and Information Division.)

hours of sunshine, high temperatures of summers, and mild, moist winters, together with the naturally fertile soils of many of the farming areas, are desirable for certain crops. Indeed, when irrigation is possible, mediterranean areas with smooth topography have agricultural possibilities rivalled by no other climatic region. The advantages of irrigation for extending the range of crops and increasing the supporting capacity were recognized by the ancients who built reservoirs and aqueducts, some of which are still in use. Unfortunately only a small portion of the region can be irrigated; much is too steep, and in many cases sufficient water is not available.

The most dense agricultural populations in the world, outside of the rice lands of Asia, are found in the Mediterranean Basin. Italy, for example, has an average density of about 460 persons per square mile; however, population is far from being evenly distributed. The greatest numbers are found in the most habitable lowlands, where more than 600 persons per square mile are common. The lowest densities are in the mountains and drier sections, but even there 100 to 150 persons per square mile is not unusual. Population pressure has caused use of the poor lands wherever possible. Everywhere settlement is near the limit of capacity of rural activities. About 22 percent of the Italians are farmers, and an even higher portion of the population of other countries of the basin is engaged in tilling the land: 30 percent in Spain, 40 percent in Portugal, and 50 percent in Greece and Albania. Farms, generally privately owned but sometimes rented or sharecropped, are necessarily small; most are under 20 acres. And at least one-third are no more than 5 acres. Fragmentation of holdings is common; tillers live mainly in small villages. Farming must be intensive to support the large families that are typical throughout the Basin. In some areas there are large estates, especially in Spain and Italy, which utilize local peasant labor. In Italy there is a movement toward land reform; and some of these estates have been expropriated and the land redistributed to the formerly landless laborers. In Yugoslavia and Israel communal systems of farming exist. Yugoslavia has not gone as far as other communist countries, however, in collectivization. The kibbutz, or cooperative farm of Israel, is unique in that it exists in a free society. The land is held and worked by all the families in common; meals are served in a central hall. In both Yugoslavia and Israel there are also private farms.

Through the ages, three general classes of crops have been developed to secure maximum use of water and land. These include (1) crops utilizing winter rainfall, such as wheat, the most significant cereal, barley, and beans—these are planted in the fall or winter to take advantage of cool-season moisture and mature before the encroachment of the long, hot, dry season; (2) drought-tolerant crops whose large root systems and other special adaptations permit growth even during the dry summer season—these include grapes and tree crops such as olives, nuts, and various fruits; and (3) crops grown under irrigation, including a great variety of vegetables and many fruits.

Crops are also adjusted to slope. Annuals occupy the lowlands, and tree crops and vineyards are usually planted on alluvial fans and on hill slopes, which are sometimes terraced (Figure 10.14). These locations provide good air drainage that minimizes the danger of frost for fruits; furthermore, erosion dangers are lessened as tree crops require a minimum of cultivation and have large root systems which aid in holding the soil in place. Citrus fruits are restricted to the warmer portions; vineyards are often given the choice south-facing slopes that produce superior quality grapes. Olive and fig orchards occupy the less favorable sites, both being hardy, drought-tolerant trees. Chestnuts are sometimes the breadstuff in the poorer mountain agricultural areas. In a few localities, particularly on the terraces of the French and Italian Rivieras, flowers are a colorful crop.

A two- or three-story agriculture, com-

FIG. 10.14. Terraces in Calabria (on the "toe" of Italy). Here the "niggardly given gifts of nature" are utilized to the utmost. Through shaping the surface by terracing, the slopes and meager soil resources have been made fit for agriculture. Notice the conical hill in the background that is circled from top to bottom. Drought-resistant olives and fall-planted wheat occupy most of the land; where irrigation is possible, multiple crops of vegetables are grown. Citrus fruits and grapes add diversity on some of the better lands. (Oliver H. Heintzelman.)

mon in a few localities, secures the maximum use of land and water. Annual crops are planted between rows of grapes, and the vines frequently climb fruit or olive trees. Subsistence needs are the strong motivation for the farmers; however, a portion of the citrus, nuts, olives, grapes, vegetables, and tobacco flow out into world commerce. All farm familities have demand for commodities that can not be produced on their farms.

The pressure upon the cropland for food restricts the integration of livestock with crops on many farms. Generally there are few dairy or beef cattle, but large numbers of sheep and goats, used for meat, milk, and fiber, are found in the nonarable areas throughout the hills and mountains. These animals, able to live on the scant vegetation, have contributed directly to the serious erosion problem by overgrazing until slopes become bare to the winter rains. Swine also are common in the areas of oak forest, where they feed on the forest mast.

FIG. 10.15. Shasta Dam. This structure is a key element in the Central Valley Project. The Sacramento River floodwaters are held for release during the dry-summer season for irrigation benefits downstream and in the San Joaquin Valley, to which water is transferred by canal. By means of the dam, electricity is also generated, recreation facilities are provided on the artificial lake, and navigation is improved in the lower reaches of the river. (United States Bureau of Reclamation.)

AGRICULTURE IN THE NEW MEDITERRANEAN LANDS

In contrast to the strong subsistent accent of the Mediterranean Basin, the other Dry Summer Subtropic regions are engaged in commercial agriculture. This difference is explained in terms of their development during the Industrial Age, their lack of encumbering populations and ancient culture and associations with nations of advanced commercial economics. The crops are usually similar, but the farms are larger, more modern, mechanized, more specialized, and the products are grown principally for markets outside the region.

California. California offers the prime example of commercial agriculture in the new mediterranean lands. This state has advantages over other regions in possessing the largest continuous lowland suited for agriculture, in having a sufficient water supply for big-scale irrigation (Figure 10.15), and in having access to extensive domestic markets and the latest developments of resource-converting and space-adjusting technology. Specialty agriculture, consisting primarily of horticultural crops, is characteristic.

The general patterns of mediterranean crops were set by the Spanish missions in California toward the close of the eighteenth century. During the short period of Spanish reign, the bulk of the land was used for cattle, with fruits and vegetables commonly grown in small plots near the missions. In the modern era, with the development of nearly 8 million acres of irrigated cropland, extensive grazing and grain production have decreased, and specialty crops with higher unit value have become the foundation of California agriculture, thereby allowing California, despite unfavorable distance, to compete for eastern and midwestern markets. The agriculture of this state has been aptly described by the phrase "variety in abundance." Most crops produced in the intermediate zone can be found in the Central Valley. In recent years no less than fifty commodities have each returned 10 million dollars or more annually to farmers. Altogether the agricultural enterprises return about 4.4 billion dollars to California farmers, placing the state first in the nation in total farm income. Nearly one-third of the state's commercial farms have sales in excesss of 40 thousand dollars annually.

Crops are usually localized in well-defined areas where the details of climate, soil, slope, and water supply are most suitable. For example, citrus groves are found on the alluvial fans of the Los Angeles Lowland (although urban expansion is claiming almost all of the citrus land) and along the slopes fringing the southeast portion of the Central Valley, where air drainage is good and frost danger is low. Acres of peaches are planted in the San Joaquin Valley; grapes are grown in many localities—Fresno is the raisin center; the Napa Valley is a wine grape center; olives are produced largely in the drier sections, especially in the Corning area; a variety of vegetables is found on the delta lands of the Sacramento and San Joaquin Rivers; rice is prominent in the landscape of the central Sacramento Valley; cotton has become a big crop in the southern end of the valley. Alfalfa, producing five or six cuttings a year, has encouraged the fattening of beef cattle, and large urban markets have favored dairying. Truly agriculture is big business in California. Truly, agriculture is a big business in California.

Southern hemisphere regions. Middle Chile, the Capetown District of South Africa, and the regions of Australia are significant agricultural areas in their respective continents. Development, as in the case of California, is historically recent, but again mediterranean crops form the basis of the economy.

In the Central Valley of Chile physical conditions for agriculture are excellent, with fertile alluvial soils, favorable relief, long growing season, and ample water for irrigation. Summers are slightly cooler than the average for California or the Mediterranean Basin regions. Development has lagged behind other similar lands for several reasons: (1) Large estates, known as *haciendas,* still contain well over half the cultivated land of central Chile. The estate owners, preferring the life of country gentlemen, have concentrated on stock rearing rather than on intensive plow agriculture. Consequently, much potential cropland is only extensively used. (2) Modern methods have only recently been introduced to Chile, and in many cases primitive systems still prevail. (3) The isolation of the region, even from South American markets, has restricted sales of commercial production. Wheat is the principal crop. Corn, not a true mediterranean crop, is grown extensively by the very low-income tenant farmers and is their most important food grain. A variety of fruits and vegetables is found, but acreages are limited because of small markets; only grapes are produced in sufficient quantities for export, and these leave the country as wine.

South Africa was one of the first areas of the Dark Continent to be colonized by Europeans. Colonization from the beginning was for the purpose of permanent settlement. There are many parallels in early history between this area and the United States. The colonization of both began in the seventeenth century; early difficulties with the natives of Africa may be likened to those with the Indians in America; the revolt of the Dutch settlers against the British may be compared to the American revolution; and the need for labor supplies in both led to use of slaves, which became the basis for similar race problems. There is continued evidence that the majority of the white South Africans favor white supremacy in political affairs, not wishing to yield political power to the black South Africans. Also the commercial infrastructure and white domination of economic enterprises could not exist in the present form without the low-paid unskilled and semiskilled black labor. Another aspect of the population problem of South Africa is the division in the white population between the British South Africans and the Afrikaners or Boers who are of Dutch descent, speak a derivation of Dutch, and outnumber the people of British stock.

FIG. 10.16. Rural landscape in the Paarl Valley of South Africa. Orchards and vineyards dominate land utilization; the scene is reminiscent of parts of California. The white strip in the far background is a newly constructed highway. (South Africa, Government Information Office.)

FIG. 10.17. California fishing fleet. The artificial port of Los Angeles boasts the greatest fishing industry in the world. Nearly 2,000 fishermen are employed on the fleet of large vessels that catch fish for packing and for the fresh-fish trade. (City of Los Angeles.)

The Capetown District bears more similarity to European and American agriculture and settlement than any other part of Africa (Figure 10.16). Wheat is the principal crop and is grown on large farms operated by Europeans ranging in size from 200 or 300 acres to several sections (a section equals one square mile, 640 acres). Owing to limited rainfall, about half of the wheat land lies fallow each year to store moisture for the succeeding crop year. This district is also one of the chief fruit-growing areas of Africa. The Dutch began producing grapes at an early date in the area around Capetown, and during the first half of the nineteenth century they shipped large quantities of wines to England. Plant diseases and insect pests have retarded this phase of the industry, and although grapes for wine are still important, the Capetown District as an exporter is overshadowed by other areas. Production for fresh markets is also worthy of mention—grapes and other fruits ripen here in March and April when northern hemisphere countries are still without fresh supplies. There is some export of fine table grapes to Great Britain in the "off season," along with peaches, plums, and citrus fruits.

Orange production is a young, but growing, activity. Improved transportation to western Europe, irrigation development, and the advantage of "off season" markets have been the main factors in its establishment.

Wheat and sheep are the important enterprises of the Australian regions. Both provide commodities well suited for long-distance export. Farming, in the main, is still extensive with units ranging from one to two sections or more in size. In the Murray Basin where large-scale irrigation works have been developed, there is more intensive agriculture, characterized by smaller land holdings, tree fruits, grapes, and dairying. Markets and labor supply are the main limitations for full development of agriculture at the present time. The total population of Australia is only slightly more than twelve million, thus the home consuming capacity is relatively small.

FISHING

Fishing is of local significance in almost all the mediterranean economies and has developed as an important commercial activity in many. A string of fishing villages with small fleets occupy inlets and bays along the Mediterranean Sea coasts, where fishermen have combed the waters for several thousands of years. Sardines constitute the chief commercial catch, but Greece and Mediterranean North Africa have long established sponge fisheries. In recent years South Africa has begun to develop a commercial phase: for example, frozen lobster tails from that country are now appearing on the American market. Large catches of tuna have made California an American leader (Figure 10.17). Much of this catch is made beyond mediterranean waters by California fishing fleets ranging the open Pacific; "clippers," boats of special design, sail as far south as the equator in search of tuna. In recent years this has lead to international incidents because of controversy over the width of the territorial sea.

The growing interest in recent decades in the resources of the sea has resulted in concerns for political, social, and economic significance. The 108 sovereign states that face the sea along some 200,000 miles of coastline (ranging from the United States with 11,650 miles to Iraq with 10 and Monaco with 3) have particularly strong interests. Ideally every part of the global sea (all the oceans are, in fact, joined together by wide passages) should be governed by internationally accepted rules. This would contribute toward stability in international relations as well as cooperation in the use and management of the ocean resources and shipping lanes.

The breadth of the territorial sea, however, continues as a controversy among the world's nations. Claims vary from 3 miles by the United States and most maritime nations of Europe and the commonwealth, through 12 miles by the Soviet Union and many others, to 200 miles by some Latin American countries. Major problems of offshore sovereignty are summarized in the question: "What state holds jurisdiction over what part of the sea and to what degree?" Much work is yet to be done before the sovereignty of the sea has full agreement and laws of the sea can be devised, respected, and enforced. Not the least of the background tasks is hydrographic mapping.

FOREST INDUSTRIES

The sparseness of the forest cover has limited lumbering industries. There is some milling, but the timber comes from the wetter mountain slopes that border the region. One exception is noted in southwest Australia, where valuable stands of jarrah and karri, eucalyptus species, are exploited. The cork oak furnishes the most valuable forest product in the Mediterranean Basin, with concentrations in the western portion. Commercial cork comes from the outer bark of the tree, which is stripped about every ten years for a total of eight to fifteen peelings during its lifetime (Figure 10.18).

MINING

Although not widespread, mining is a noteworthy activity in several areas of the Dry Summer Subtropics. Quantities and varieties conducive to great industrial developments are lacking in the Mediterranean Basin, but a number of minerals are significant. Bauxite is mined in the southern Rhone Valley of France and near the coast at several points in Yugoslavia. Italy is a leader in mercury production. Rich iron deposits are worked in Algeria, and the ore is shipped to western Europe; phosphates in quantity are also mined and exported. Turkey is one of the leading producers of chromium and also mines coal (Figure 10.19), molybdenum, lead, and zinc. Spain has considerable mineral and wealth but, except for copper in the south, most deposits are on the plateau and in the mountains outside the mediterranean fringe. The quarrying of building-stone is of

FIG. 10.18. Cork oak grove in Spain. Notice that the trunks and large limbs have been stripped of bark, which is used for commercial cork. These trees have a life span of 150 to 300 years. (Oliver H. Heintzelman.)

FIG. 10.19. Open-pit mining scene at Zonguldak on the Black Sea coast of Turkey. This is the nation's leading coal-producing area. Surface mining costs less than any other form of mining and provides the advantage of more complete recovery of the mineral. (Turkish Information Office.)

some importance throughout the Mediterranean Basin. Shortages in timber have caused stone to be used extensively. In addition the statuary marbles of Carrara, Italy, are world famous.

California is outstanding for petroleum, ranking as a leading producing state, with major fields in the Los Angeles area and the southern part of the Central Valley. Understandably, as a principal area of population and economic growth, California is a major producer of many nonmetallic building and industrial materials: sand and gravel, pumice, cement, gypsum, etc.

The major mineral wealth of Chile, the nitrates of the desert and the copper of the Chuquicamata area in the Tropical Highlands, have already been discussed. It is chiefly for these two minerals that the country is noted, although modest amounts of coal and iron are also mined. The chief iron mines near Coquimbo are controlled by the Bethlehem Steel Corporation; the ore, handled entirely by machinery from mine to ship, is exported to Sparrows Point, Maryland, for smelting. Chile, a leading coal mining country of South America, mines about 5 million tons annually. This fact indicates that the entire continent is short in domestic supply of the greatest source of industrial energy. The coal mined in Chile, primarily near Concepción, is of low grade, but its domestic use is forced by government policy through the implement of high tariffs on foreign imports. Some high grade coal is imported for blending with domestic resource for use in metallurgical industries.

The two most notable metallic-mineral areas of Australia are closely associated with the eastern Dry Summer Subtropic region. Broken Hill, in west central New South Wales, has one of the great lead-zinc ore bodies of the world. Production has been continuous since 1883, and the end is not yet in sight. One of the continent's largest iron deposits is west of the head of Spencer Gulf, with the bulk of the production coming from Iron Monarch. The ore is hauled by rail to the port of Whyalla, 34 miles away, for shipment by boat to steel mills at Newcastle and Port Kembla on the east coast.

INDUSTRY

Factory towns characterized by belching smokestacks, clusters of industrial plants, and raw material stockpiles typical of the manufacturing belts of western Europe and eastern United States are not common in the Dry Summer Subtropics. Heavy industry is curtailed by a scarcity of coal and iron; shortages in fibers limit textile production. Chile, however, has a steel industry near Concepción, utilizing domestic resources and imported high-grade coal. Much of the regions' industry is associated with the products of the orchards, vineyards, and fields. The Mediterranean Basin countries and California are especially important in the processing of agricultural commodities. Varieties of fruit are dried or canned, and fresh or frozen fruit juices are canned and bottled. Olives are pressed for oil, bottled, or tinned. Tons of grapes are crushed by vintners who have gained world-wide reputations for the quantity and quality of their wines. Many varieties of vegetables are canned or frozen for market. In the Grasse area of southern France, acres of flowers are processed for essential oils used in the perfume industry that is centered there (Figure 10.20).

The sea also contributes to the food-processing industry. Sardine canneries are important in the Mediterranean Basin; Monterey, California, is a fish-canning center. California tuna canneries concentrated in Los Angeles and San Diego[5] prepare much of the United States' pack.

California stands out as the major industrial area of the mediterranean lands. Petro-

---

[5] San Diego falls into the Tropical Desert classification in this book.

FIG. 10.20. Grasse, France—the perfume center. A long tradition in this industry has made Grasse synonymous with perfume. Raw materials such as musk, moss, ambergris, oils, and flowers are drawn from all parts of the world. (French Embassy Press and Information Division.)

FIG. 10.21. Menton—resort center on the French Riviera. Menton is the easternmost town on the French Riviera. Although once an attraction for European nobility and sophisticates, it now caters to moderate-income clientele, leaving the luxury trade to its more glamorous neighbors, Nice and Cannes. (Oliver H. Heintzelman.)

leum for energy and petrochemicals, local and nation-wide markets with substantial buying power available by excellent transportation connectivity, and the sunny climate have stimulated the largest and most diversified industrial development on the Pacific Coast. The motion picture industry originally concentrated in Hollywood to take advantage of the sunny climate and the variety of California scenery. Favorable flying weather and outdoor working conditions were important factors in the establishment of four leading aircraft companies. The Hollywood label places the stamp of approval on sports clothes, and Southern California competes with New York and Paris as a style-setting center. Tons of petroleum are refined and thousands of automobiles are assembled. Iron and steel, rubber, pottery, ships, electronics, ballistic missiles, and furniture are also included in the industrial diversification. Fontana, near Riverside, is a notable iron and steel center, receiving coal by rail from deposits near Sunnyside, Utah, and iron from Eagle Mountain in the San Bernardinos. Several steel mills operate in the San Francisco Bay area. Industries are concentrated chiefly in the populous Los Angeles Basin and the San Francisco Bay area.

TOURISM

People of the mediterranean regions "sell" their sunshine, winter climate, scenery, and beaches. The rugged shoreline, narrow beaches, and sun-facing slopes of the Rivieras of France and Italy, the Dalmatian Coast of Yugoslavia, and the Crimea of the USSR are the playgrounds of Europe. Scattered along the Mediterranean coast are clusters of small settlements, as well as larger centers such as Nice and Cannes, all directing their energies to the tourist and resort trade. Provision of lodging, food, personal services, tours, curios, and distinctive consumer

FIG. 10.22. The French city of Carcassonne possesses a well-preserved example of fortifications characteristic of the Middle Ages. It guarded the gap from the Mediterranean to the Acquitaine Basin of Western France. Vines dominate the foreground. (French Embassy Press and Information Division.)

URBAN CENTERS

Since the days of ancient Greece, a large percentage of the people of the Mediterranean Basin have lived in concentrated settlements. The early development was a response in part to commercial activity and in part to the need of massing for protection. Most of the early centers became no more than hamlets, villages, or small towns. The centers that grew during the ancient period and flourished through the Middle Ages contributed much to the plan and function of later urban centers throughout the world; here cities first developed the specific functions that are now peculiar to them—service as centers of commerce, manufacturing, culture, education, and political activity.

Istanbul, Athens, Naples, Rome, Florence, Genoa, Marseilles, Barcelona, Valencia, Casablanca, Algiers, Tunis, and Beirut are but a few of the ancient cities which today are busy metropolises. In Italy alone there are a dozen cities with a population of over 100,000. In contrast to the long-established centers that dominate the urban activity of most of the nations that fringe the sea, most of the city needs of Israel are served by two new centers, Tel Aviv and Haifa. Both have grown during this century to become major ports and commercial centers. There is little dispersed settlement through the Basin, and even the farming populations are concentrated in villages scattered through the countryside. Many cling to the slopes, using land not suited to agriculture; in the earlier periods these sites offered some protection.

In the new lands each region tends to have one or two dominating centers. Santiago is the great city of Chile and Valparaiso is its port. Capetown is both port and urban center of the Cape Province of Africa. Adelaide is the main city serving the Murray Basin, and Perth serves southwest Australia. In California there are six major cities, Los Angeles, San Francisco and Oakland on San Francisco Bay, San Jose in the Santa Clara

goods for this trade, although annually cyclic, is a major source of income for these areas. Part of the economy of the little principality of Monaco revolves around its famous gambling casino. California and "America's Playground" are synonymous to many people in the Western Hemisphere. This state has been especially active in promoting its recreation opportunities; it attracts thousands of visitors as well as many retired persons who establish permanent residences there. As yet tourism is a minor activity in the southern hemisphere regions.

Valley, and Sacramento, and Fresno in the Central Valley, that are among the fastest growing urban centers of the entire world.

RESOURCES FOR GROWTH

Agriculture will remain the bulwark of the economies of the Dry Summer Subtropic countries. The possibilities for development, however, are distinctly limited. Mediterranean Basin lands are presently farmed to near capacity—little can be added to their arable acreage. Any material change will have to result from multifarious approaches such as erosion control, reclamation of marshlands, intensification of farming, reforestation, increased fishing activity, extension of industries utilizing hydroelectricity, local raw materials, and ancient skills, and greater

FIG. 10.23. San Francisco occupies a 46.6 square mile fingertip between the Pacific Ocean and one of the world's greatest natural harbors. The hub of a nine-county complex, the city is linked to northern California by the Golden Gate Bridge (upper left) and to the east shore by the Bay Bridge (lower right). Marin County's Richardson Bay is visible beyond Alcatraz and Angel Islands (upper right). (San Francisco Convention and Visitors Bureau.)

FIG. 10.24. Ruins of Paestum. Paestum is an example of the glorious past of the Mediterranean Basin. This colony, located on the Gulf of Salerno in southern Italy, was founded by the Greeks in the 6th century B.C. Such ruins are significant tourist attractions. To the local inhabitants they lend reality to stories of the time when Greek, Roman, Byzantine, Saracen, and Norman invaders came, saw, conquered, and left their marks in stone. (Oliver H. Heintzelman.)

capitalization of the "glorious past" as well as climate and scenery to attract tourists in increased numbers. Probably the greatest potentials lie in the improvement of farming techniques, and although programs are under way to raise the standard of living of the peasants, progress is extremely slow where rural populations are dense, traditions are long established, and education and capital low.

California, the "Giant of the West," is the most rapidly developing mediterranean region. Present trends indicate no diminishing of pace. Problems undoubtedly will arise, especially in shortages of water supply. Technology, unified effort, and available capital are major assets for combating this limitation, as evidenced by transport of water from the Colorado River to the Los Angeles Basin. Additional future supplies may be obtained from the sea or from outside the region, particularly by reservoir storage in the Sierra Nevada Mountains. Other problems are presented by the pace of population growth and concern the development of orderly settlements, transportation facilities, and other services. Deterioration of urban environments, urban scatteration on prime croplands, increasing demands for recreation space, air pollution, and traffic jams are but a few of the problems that have attended the rapid growth of California in the commercial, industrial, urban age. All beg solution.

Isolation has been the major deterrent of the southern hemisphere regions. The worldwide trend toward greater national self-sufficiency has been the impetus for recent attempts to reduce dependence upon outside areas. Because of present underdevelopment, Chile probably has the greatest relative potential for future growth.

eleven

HUMID SUBTROPICS

The Humid Subtropics, along with the Dry Summer Subtropics, are the final outposts of the tropics. They represent transitional zones between the continuously warm climates and those where winter cold becomes a definite characteristic. In Africa and Australia, however, they fail to serve this role because of the limited poleward extent of these continents.

The mild, rainy climate of the Humid Subtropics favors the cultivation of a wide variety of crops, thus agriculture dominates land utilization where surfaces are reasonably smooth. Sharp contrasts occur, however, from region to region in farming types and crop accents, but more especially in scale. Commercial farming, with an emphasis on cotton was established early in the Humid Subtropics of North America. Commercial farming also predominates in the southern hemisphere regions. Food crops for hungry millions take precedence in the Orient, where farmers have tilled the soil for thousands of years. Where land is arable, man is everywhere; his roots are deep in the earth, and landscapes are completely altered by his works.

Although the Humid Subtropics are outstanding agriculture regions the exploitation of forests, minerals, and sea provides considerable diversification. Three of the regions have had significant industrial awakenings in the past seventy-five years. In the twentieth century, Japan rose from a feudal state to become the industrial giant of the Orient. The area of the United States Humid Subtropics is now developing its storehouse of energy resources and raw materials and has entered an era of industrialization. The coastal zone of New South Wales is the principal industrial belt of Australia.

Location

The Humid Subtropic regions lie on east sides of continents, generally between latitudes 25° and 38°. The interior borders are usually marked by increasing aridity and are

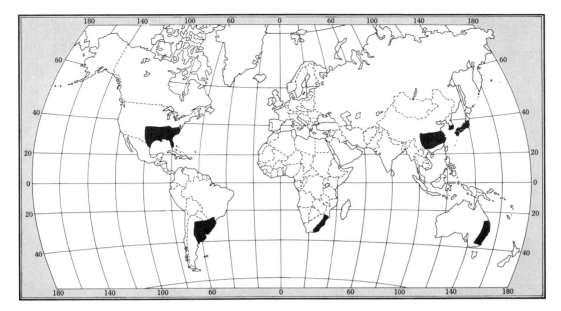

FIG. 11.1. The Humid Subtropics.

FIG. 11.2. The Humid Subtropics of the Orient.

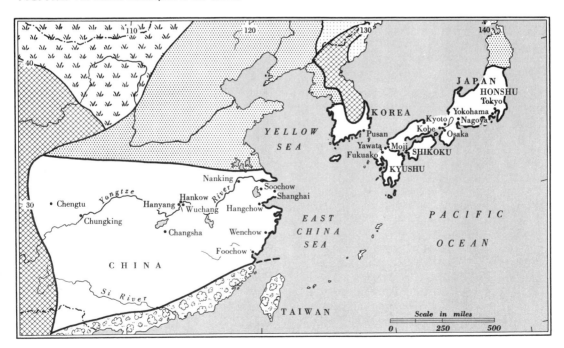

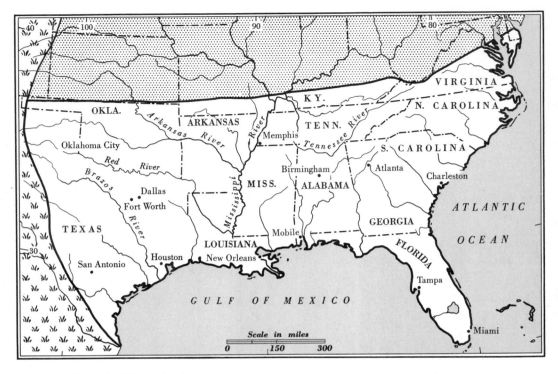

FIG. 11.3. The United States South.

set at the 20-inch rainfall lines. All continents are represented in these regions (see Figure 11.1).

The largest region is found in Asia, where the Humid Subtropics occupy virtually all of China between the Tropic of Cancer on the south, the Central Mountains north of the Yangtze Valley, and westward to about the 100th meridian. Also included are southern Korea, the Japanese Islands of Kyushu and Shikoku, and the portion of Honshu south of latitude 38° (Figure 11.2).

In North America the region includes the southeastern quarter of the United States, southward from the line denoting the annual frost-free season of 200 days and eastward from the 20-inch annual rainfall line that bisects the panhandles of Oklahoma and Texas and bows slightly eastward to terminate near the mouth of the Rio Grande. Hereafter this area (Figure 11.3) will be called the United States South. The Humid Subtropics of South America center upon the middle and lower parts of the Paraná Basin and the Rio de la Plata. The region contains the pampas of northeastern Argentina, all of Uruguay, most of the southern projection of Brazil, and the southern portion of Paraguay (Figure 11.4).

The African region is restricted by the Drakensbergs to a relatively narrow belt comprising the Natal Coast along the eastern border of the Republic of South Africa (Figure 11.5). In Australia the Eastern

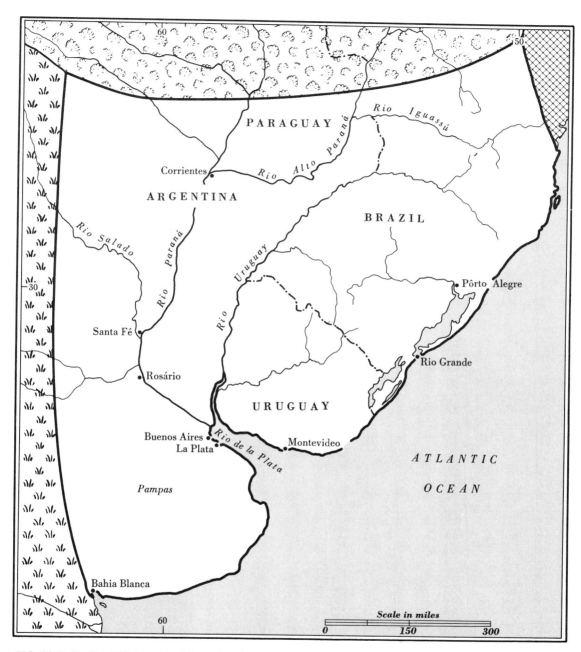

FIG. 11.4. The Humid Subtropics of South America.

223

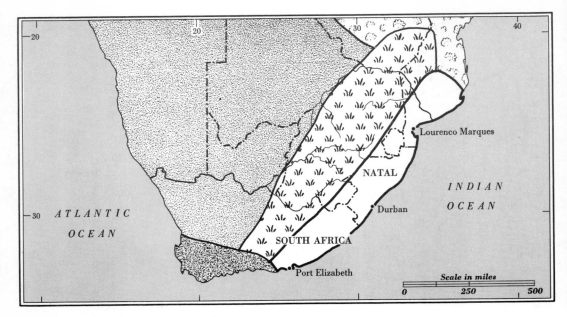

FIG. 11.5. The Humid Subtropics of Africa.

FIG. 11.6. The Humid Subtropics of Australia.

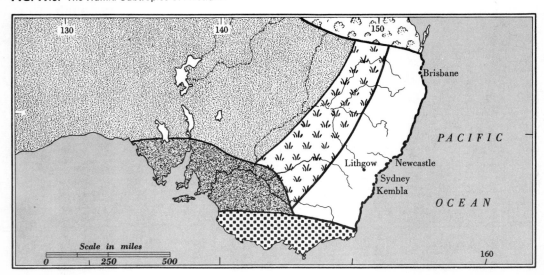

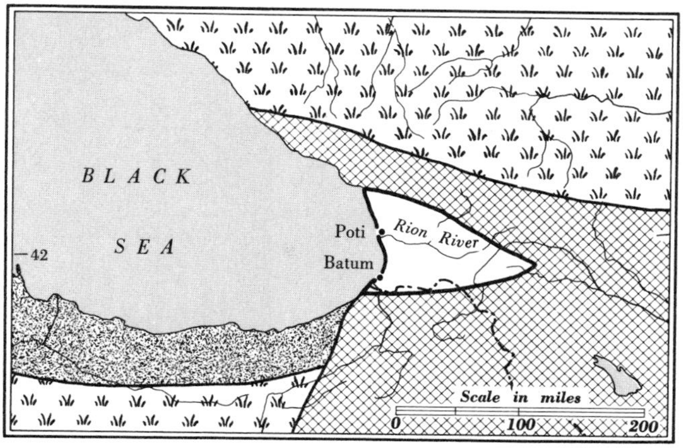

FIG. 11.7. The Humid Subtropics of the Soviet Union.

Highlands restrict the Humid Subtropics to a coastal location from about Brisbane, Queensland, southward along the margin of New South Wales to Cape Howe (Figure 11.6).

A small area not conforming in continental position is found at the eastern end of the Black Sea in the USSR. With its base on the sea, it is pinched into a wedge shape by the converging mountains (Figure 11.7).

Physical Environment

CLIMATE

The Humid Subtropics have hot, moist summers and generally mild winters. All regions, except the small area in the USSR, occupy east coast positions in similar latitudes, but the character of their continental backgrounds accounts for climatic differences. The regions in the northern hemisphere are contiguous to sources of cold continental air in winter, whereas those in the southern hemisphere fringe tropical landmasses, and cold air from the Antarctic is modified by movement over oceans. The great differences in the heating and cooling of the northern hemisphere continents and their surrounding seas create monsoon tendencies in the Asian and North American regions, especially pronounced in the former.

Temperatures. Summer temperatures are high with averages of 75° to 80°; daytime readings are often above 90° and occasionally exceed 100°. Humidities are also high and, with the oppressive heat, make summer weather particularly uncomfortable. Nights remain warm and sultry, offering little relief from the heat. Summers parallel the climate of the Rainly Tropics. Winters are mild with average temperatures between 40° and 55° (Figure 11.8). In Asia and North America, cold continental air is a close neighbor, manifested by wintry blasts. No barriers block the southward movement of cold air into the southern portion of the United States, and it streams into the region via the Mississippi Valley. These cold waves, sometimes called "Northers," occasionally bring below-freezing weather even to Florida. China has a series of mountain ranges that create blocks to polar air from the cold interior of Asia; how-

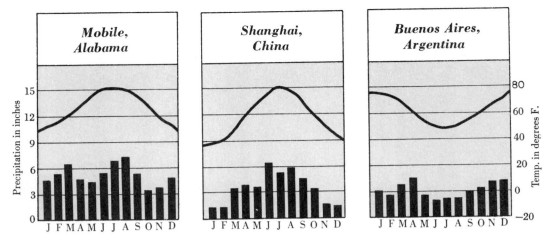

FIG. 11.8. Typical climatic graphs of Humid Subtropic stations.

FIG. 11.9. Hurricane damage on the Gulf Coast. This is a portion of Biloxi, Mississippi after Hurricane Camille struck. Camille killed more than 255 people, destroyed nearly $1.5 billion in public and private property. (National Oceanic and Atmospheric Administration.)

ever, the Lower Yangtze Plain is exposed and experiences lower winter temperatures than are common for the Humid Subtropics as a whole. Lacking cold continental backgrounds, all three of the southern hemisphere regions have higher average winter temperatures. Growing seasons are long, ranging from 200 to 260 frost-free days in the United States and Asiatic regions to a continuous growing season on the Natal Coast; the equatorward margins of all regions occasionally have entire frost-free years.

Precipitation. Rainfall is abundant and occurs throughout the year, with most regions having a summer maximum. Averages range from 20 inches, on the subhumid borders, to about 60 inches. Summer precipitation is chiefly a result of convectional storms; the American Humid Subtropics average between 40 and 60 thunderstorms a year. Weak cyclonic storms account for a part of the summer precipitation. Heavy rains in late summer and early fall are a result of tropical cyclones—known as hurricanes in the Atlantic and typhoons in the Pacific. Despite a prevalence of summer rain, there is a high percentage of sunshine. Winter precipitation is mainly cyclonic. Instead of the sunshine-shower weather of summer, winter rain more often falls as a persistent drizzle from dull gray skies. Less rain falls in winter, but there are generally more completely overcast days. Snow is a rarity in the southern hemisphere regions but is not uncommon in the American South and the Orient, especially near the northern boundaries.

Local phenomena. Several local phenomena are important in the climatic picture of the Humid Subtropics. Coastal areas of the northern hemisphere regions are subject to violent tropical cyclones in the late summer and early fall; hurricanes blow out of the West Indies through the American coastal states (Figure 11.9); typhoons create destruction on the south China coast and sweep into southern Japan. Cold winds occur occasionally in the southern hemisphere regions; the *pampero,* a violent wind of short duration, strikes the pampas of Argentina in summer, and the "Southerly Bursters" cool the Australian coast. Hot northwest winds from the interior of Africa descend upon the Natal Coast during spring and summer. Tornadoes, small violent storms, occur in the spring and summer, chiefly in the western part of the North American region. They hang from large cumulo-nimbus clouds like funnels and spin with tremendous speeds. The portion that touches the surface may be from 300 to 1500 feet in diameter. When a tornado strikes, property destruction is almost complete.

SURFACE FEATURES

The Orient. There is little similarity in surface features from one Humid Subtropic region to another. Mountains fill the bulk of the Oriental regions, although the plain of the Yangtze River is one of the great lowlands of the world. This powerful stream, with its numerous tributaries, has laid an alluvial surface which reaches 600 miles inland and varies from 20 to 200 miles in width. The delta is building seaward at the rate of one mile in 70 years. Water features are an outstanding characteristic of this plain; canals, rivers, and lakes are prominent in the landscape.

South China is a land of broken topography—the major portion is in slope. Streams draining into the Yangtze, the ocean, and the Si River have carved it into a maze of hills and valleys. Level land does not exceed 10 percent of the area.

The most important section in western China is Red Basin of Szechwan Province, the gathering area for the principal headwater tributaries of the Yangtze River. This basin, named for its brick-red sandstones, is rimmed by mountains but is by no means of uniform surface; sharp, dissected anticlines

FIG. 11.10. (top) Terraces in south China. The press of a huge population has made necessary the careful terracing of many of the slopes to make more land available for food production. (Theodore Herman.)

FIG. 11.11. (bottom) Typical landforms of Japan. Modern industrialization and urbanization in this country often occur at the expense of agricultural land, since they are attracted by smooth surfaces and access to water transportation. A major copper smelter is shown in this scene. (Mining World.)

and gentle synclines trending northeast to southwest are numerous. The only major level area is the Chengtu Plain, formed by the alluvial fan of the Min River.

South Korea is thoroughly hilly. Mountains paralleling the east coast drop sheer to the Sea of Japan but descend gently westward and send out low spurs, which separate a number of significant basins with lowlands in the west and south.

Japan is a mountainous island country; everywhere highlands form the background. The largest lowland is the Kwanto or Tokyo Plain of eastern Honshu, with an area of about 5,000 square miles. Most of the plains are peripheral, formed by rivers depositing their sediments in quiet bays. Altogether these plains comprise no more than 15 percent of the country.

United States South. The United States South lies dominantly within the physiographic province known as the Atlantic-Gulf Coastal Plain. The outer margin is remarkably featureless; broad, flat-bottomed valleys with bordering bluffs rising 20 to 50 feet in height provide the only significant relief features. Along the coast, which merges imperceptibly with the sea, there are vast stretches of swamps. Inland, elevation increases, drainage is better, and the surface, longer exposed to the forces of degradation (which wear away the land), displays a greater variety of relief owing to the difference in resistances of the sedimentary materials. The harder members form low, hilly ridges parallel to the coast in Alabama, Mississippi, and Texas, and the weaker strata have been lowered to form broad vales. The ridges, known as *cuestas,* have a long slope seaward and steeper slope on their interior sides.

The Mississippi River bisects the region, and its broad floodplain, 50 to 75 miles wide, and large delta is one of the most productive areas of the nation (Figure 11.14). On the west the region makes contact with the Great Plains and on the north merges with the Interior Lowlands and the Appalachian Highlands. The low Ouachita and Ozark Highlands break the smooth transition to the interior. The transition to the Appalachians, however, is less perceptible; the piedmont is, for the most part, a broadly rolling surface.

FIG. 11.12. The Florida Everglades. This is the "Sea of Grass" on the rim of Florida's vast and sprawling Everglades. This view shows cabbage or sable palm islands, and an expanse of sawgrass, which is not grass but a sedge. In other sections, the Everglades is a forest of cypress providing a home for Florida's Seminole Indians and refuge for huge flights of colorful native birds and rare wild flowers. (Florida State News Bureau.)

FIG. 11.13. A view of the Piedmont, taken on the Blue Ridge Trail between the Virginia and North Carolina state line. This marginal land, formerly cultivated, is now being used for pasture. Many acres of such land have been planted to pine in recent years. Rapid growth of trees and a ready market for pulpwood are major stimulants. (Standard Oil Company of New Jersey.)

FIG. 11.14. A view across the Mississippi River near Natchez. The flat floodplain of this great river varies from 25 to 75 miles in width. Notice the convectional storm in the background. (Standard Oil Company of New Jersey.)

South America. The South American region has relatively uniform relief. The only significant hilly area is in southern Brazil and the northern one-third of Uruguay. This extension of the Brazilian highlands rarely exceeds 1,000 feet elevation but is nonetheless more rugged than the undulating plain of southern Uruguay. The plain of the Paraná River comprises southern Paraguay; and a lowland known as the Argentine Mesopotamia occupies the area lying between the Paraná and Uruguay Rivers. This merges with the flat and featureless pampas of northeastern Argentina, the most important segment of the region (Figure 11.15).

Africa and Australia. The regions of Africa and Australia are similar. Both are compressed between highlands and the sea, and in each the slopes rise steeply from narrow and discontinuous coastal lowlands. Special mention should be made of the broad and fertile Hunter Valley in the Australian region (see Figure 11.17). The Hunter River separates the New England Plateau from the Blue Mountain Plateau, forming an important passage inland from Newcastle to the interior.

USSR. The region of the USSR is the triangular-shaped lowland of the Rion River and the lower slopes of the bordering Caucasus Mountains on the north and the Arme-

FIG. 11.15. A physiographic diagram of southern South America.

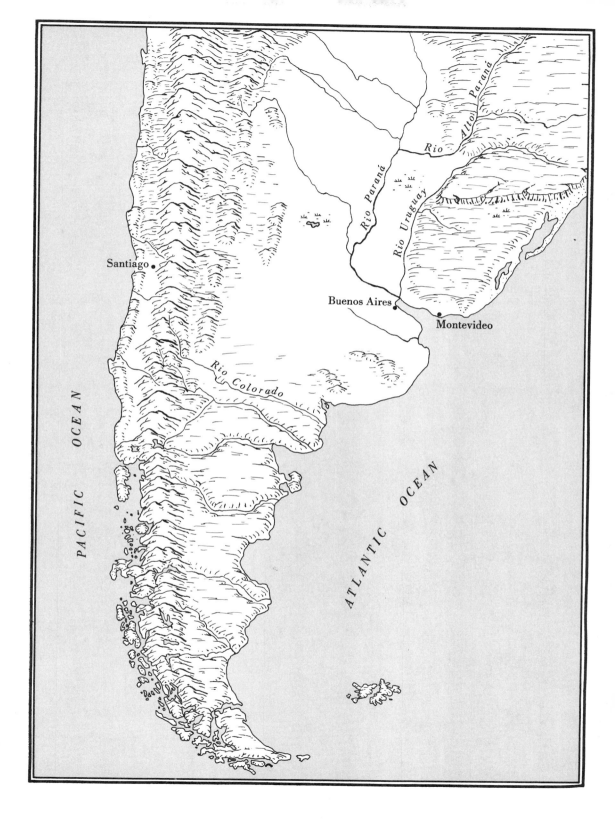

FIG. 11.16. A landscape in Natal. Note the rolling nature of the countryside. The crop on the right is the drought-resistant crop, kaffir corn. (South Africa, Government Information Office.)

FIG. 11.17. The Hunter Valley of Australia. Aberdeen Angus cattle fatten on improved pastures flanking the Hunter River, which winds through the foothills of the New England Plateau. (Australian News and Information Bureau.)

nian Highlands on the south. The area along the coast consists mainly of marshes and lagoons, although a large-scale reclamation project has been under way to make this portion suitable for agriculture.

NATURAL VEGETATION

Forests that constitute much of the natural cover are favored in the Humid Subtropics by an abundance of moisture and a nearly year-round period of growth. In the long-occupied land of the Orient, forests have been cut over many times, and virgin forests remain only in the inaccessible areas. Broadleaf evergreens prevail on the equator margins, with occasional understories of bamboo thickets; northward there is a replacement by mixed forests of conifer and deciduous broadleaf species.

In the United States South conifers occupy the sandy, less-fertile coastal plains as well as margins of the Piedmont, with longleaf, shortleaf, loblolly, and slash pine predominating. The pine forests grow in open stands with a low undercover of coarse grass and shrubs. Dense forests of hardwoods, such as red gum, and tupelo, thrive in the poorly drained river bottoms; bald cypress is also found in poorly drained areas. More open hardwood stands occupy the better-drained bottom land sites. Hickory, chestnut, oak, and poplar form a cover in the rougher Appalachian Uplands. Toward the drier western margins the forests are gradually replaced by woodland and then by grasses, with trees restricted to the river banks.

The Paraná Basin of South America has a mixed forest dominated by the Brazilian pine (a close relative of the Chilean Araucaria pine). There is also found a variety of low evergreen trees and shrubs, including the maté tree (South American or Paraguay tea). Coarse marsh grasses abound in the poorly drained areas. Starting in southern Uruguay and stretching to the Rio Colorado

FIG. 11.18. A live-oak forest. The forests on the southern border of the United States South are evergreen. The Spanish moss hanging from the branches is gathered for use in packing and cheap upholstery. (United States Forest Service.)

FIG. 11.19. Virginia bald cypress growing in grassy lake near McNab Arkansas. (A. W. Nelson, United States Forest Service.)

FIG. 11.20. A stand of longleaf pine. Notice the open nature of the stand and the flatness of the terrain. (United States Forest Service.)

spectacular seasonal color pattern: black in the spring from the burning of old growth, then green, turning to brown, and culminating in silver when the grass spikes flower.

Subtropical forests are the characteristic cover on many of the wet coastal slopes in Humid Subtropical Africa. Some of the better forests of Australia are found along the east slopes of the coastal ranges. The forests north of Sydney contain red cedar and several varieties of Araucaria pine; south of Sydney, pure stands of eucalyptus are dominant.

NATIVE ANIMAL LIFE

Animal life is often a picturesque part of the physical environment, especially where man's numbers are few, but in the Humid Subtropics man is nearly everywhere. Consequently the native animals of these regions are either the small wily species, or are found chiefly in areas generally avoided by the farmer—the mountainous lands, the forests, and the swamps. Native animals, except for birds, are scarce in the closely settled lands of the Orient. The United States South, with its many square miles of forest and swamp, still has a variety of native animals. Game birds, as well as others, deer, fox, squirrels, opossums, and raccoons are found in the forest. Mink, otter, and especially muskrat inhabit the great marshes along the Gulf Coast. Muskrat trapping in the Louisiana swamps is a major winter activity for thousands of trappers. The greatest concentration of wild life is found in the Everglades of Florida. This 5,000-square-mile region of water birds including herons, cranes, ibises, egrets, and others; snakes, frogs, and alligators are also abundant. Crocodiles, manatees or sea cows, and sea turtles are found in the same areas. Living on the drier islands in the area are typical forest animals. The Everglades is one of the most colorful animal refuges in the United States.

is a broad expanse of grassland, the Argentine pampas. This level to undulating plain appears like a sea of grass broken only by solitary umbrella-shaped ombu trees. The grass cover of the pampas is unusual in a climate type normally associated with forests. Many scientists have offered explanations; some believe that repeated fires by Indians favored the grasses and prevented the establishment of a tree cover. The pampas, before the modern period of agriculture, had a

A few small animals live in the forests on the northern rim of the Humid Subtropics in South America; birds are plentiful on the pampas. The South African region contains many animals that are typical of the drier grasslands to the west, such as species of antelope and the scavenging hyenas. Snakes are common, and crocodiles and hippopotamuses are found in streams near the coast. Marsupials are associated with the Australian region; the major representatives are the kangaroo, wallaby, and flying opossum. Here, too, are found many unusual birds such as the lyre bird and emu.

SOILS

Red and yellow soils, named for the characteristic color of their subsoils, are typical of the Humid Subtropics, but exceptions are common. The red soils are dominant on sloping land, and the yellow soils prevail on the more level, flat, or imperfectly-drained sites. The prominent factors in their development are the warm and humid climate and the associated forest vegetation. The climate promotes intense chemical breakdown, strong oxidation, and leaching reminiscent of the humid tropical soils. The forest returns relatively little organic material. The soils tend to be acid and low in organic matter and plant nutrients. Though low in inherent fertility, they are easily tilled, respond well to fertilization, and are potentially fruitful. Good management practices, however, are essential to continued cultivation.

Many of the better agricultural soils, particularly in the Orient and the Mississippi Bottoms, are of recent alluvial origin. In the areas of lower rainfall, such as the western edge of the United States region and the South American plains, the soils have developed under a grass cover, have been subjected to only slight leaching, and are rich in organic and mineral matter; consequently, they are highly productive. Excellent soils have also developed on limestone in several places in the United States South; those of the Black Belt of Alabama and the Black Waxies of east Texas are especially important. These soils, known as *rendzina,* are examples of intrazonal soils (great soil group classification: clacimorphic suborder). They are mature soils that have characteristics that reflect the influence of parent material over the normal effects of climate and vegetation. Not all the exceptions are on the favorable side. Along seaward fringes, youthful soils on marine sands are frequent; these are inherently infertile but are often easily cultivated and responsive to fertilization and irrigation. Poorly drained organic soils are found in areas of coastal swamps. These are also intrazonal soils (hydromorphic suborder). The soils of the Argentine pampas are unusually fertile, having developed under grass vegetation and moderate precipitation. The smooth topography of the region enhances its utility for agriculture.

Occupance in the Humid Subtropics

The Humid Subtropics are the most populous of all the regions. The long-settled rice lands of eastern Asia abound with people. Japan, with an area of 143,000 square miles, supports 101 million, the greatest concentration being in the narrow belt extending from the Kwanto Plain westward along the borders of the Inland Sea to north Kyushu. The Chinese are crowded into the agricultural lands of the Yangtze Plain and the narrow shoestring valleys among the hills; the Red Basin stands out in the west as an island of black on a population map. Population density in the China region usually amounts to

FIG. 11.21. (top) Even the rivers of China are crowded. This scene in Soochow Creek in Shanghai shows the permanent homes of many Chinese. Boats are means of livelihood as well as homes. (Theodore Herman.)

FIG. 11.22. (bottom) Preparing the paddies for rice in a south China valley. The men on the treadmill are lifting irrigation water from the canal to the fields in the traditional manner. (Theodore Herman.)

1,000 persons per square mile of agricultural land, and 2,000 or more is common.

The United States South has the greatest rural densities of any major region of the United States, containing nearly half of the total agricultural population. Similarly, the Humid Subtropical regions are some of the most highly populated sections of South America, Africa, and Australia.

AGRICULTURE

The Humid Subtropics are rivalled only by the Monsoon Tropics in the numbers of farmers they support. The abundant moisture supply, warmth, and long growing season combine to produce excellent conditions for plant growth. A great variety of crops is possible, including subtropical fruits, tea, fibers, vegetables, and grains. The combinations and emphases vary from one region to another depending upon man's culture history, needs, numbers, techniques, institutional restraints, and markets. In the Orient farming is mainly of subsistence order, but in the other regions commercial agriculture has prevailed from the beginning of white settlement.

The Orient. Agriculture in the Orient is farming in its most intensive form. In few other portions of the world does man live so close to the earth and depend so dominantly upon the products of cultivated soil. The small holdings, two or three acres in size and commonly fragmented into half a dozen plots, must be made to yield the complete support of the farm family. Thus the people toil from dawn until dark giving infinite care to the crops, which are grown predominantly for food.

Rice, the most desired crop because of high yields and food value, dominates the landscape during the warm rainy season, growing on every field which has suitable impervious subsoil and water supply for irrigation. The paddies, seldom larger than a

FIG. 11.23. Transplanting rice in the Orient. The rice is germinated in seedbeds to give the new crop an early start while winter crops mature and fields are prepared. (Theodore Herman.)

FIG. 11.24. Rice harvest in the Orient. With a small hand sickle every stem of the rice is carefully cut and gathered into bundles for drying and threshing. (G. Martin.)

fraction of an acre, are level and surrounded by dikes, for rice must grow most of its life standing in water (Figures 11.22, 11.23, and 11.24). Other summer crops include corn, sweet potatoes, sobyeans, sugar cane, sesame, and vegetables; cotton is grown in portions of the Yangtze Plain where the autumn is relatively dry. Following the harvest in late summer or early fall, the fields are planted to winter crops of wheat, barley, beans, or other hardy vegetables. This double cropping is practiced wherever physical conditions permit. When vegetables are grown, their shorter maturing period allows the production of several crops during the yearly cycle. Little land is unused; even the dikes surrounding the paddies are usually planted. Such intensity of production is possible only with heavy fertilization; all waste vegetative matter, animal and human excrement, ashes, and sediments from the canals are used to keep the growing capacity of the land high (Figure 11.25). The multiple cropping index total country basis for Japan is about 1.6 and for China about 1.4; in the humid subtropic portion these figures would be significantly higher.

Little land can be spared for pasture or feed crops, and except for swine and chickens, which can live on household refuse, animals are scarce; even draft animals are restricted to large farms. The large dependent population simply means that the conversion loss from vegetable calories to meat calories can not be offered; the people have to feed at the primary production level of the food chain. For the same reason, commercial crops

FIG. 11.25. Collecting human excrement from cesspools to be used for fertilizer. (Theodore Herman.)

FIG. 11.26. A Japanese farm village. House roofs are of thatch and tile. The thatch roofs are steep to shed the heavy rain. Drying racks for rice are in the right foreground. Deciduous trees are in the background. (G. Martin.)

are limited—silk and tea are the only noteworthy commercial products of oriental agriculture. Sericulture (production of raw silk), important for many centuries, is an admirable adaptation; the mulberry trees frequently occupy slope lands, and the feeding of the leaf to the silkworms requires great amounts of careful, patient labor. Competition from synthetic fibers has seriously reduced the world markets for silk since World War II. Tea, similarly, is a slope-land crop utilizing areas that would otherwise be in natural vegetation.

There have been major changes in the organization of Chinese agriculture under the communist regime. Following 1949, collectivization was carried out rapidly, then in 1958, as part of the "Great Leap Forward," a program was devised to consolidate the 750,000 collectives into some 24,000 communes. The communes were to be super-sized farm units that would completely obliterate the time-honored family-oriented farm structure. There were to be dormitories for workers, nurseries, schools, health facilities, homes for the aged, as well as processing plants for farm products and other light manufactures based on local resources. Land and practically all other items of personal property were to become commune property. The peasant was to receive income in the form of wages based on a labor point system, which was to pay the individual according to his skill and contribution. The system, it was believed, would permit complete governmental control of production and disposal of agricultural commodities, would accelerate the rapid modernization of all aspects of agriculture, and would benefit the government's total development design.

A series of three low farm production years, however, coincided with the period of the commune initiation (1958–60). As a result it was necessary for the government to backtrack beginning in 1961. Although weather conditions improved, three years of "Cultural Revolution" in the late 1960s continued to curb the commune movement. The collectives continued as the main organization form through the decade. During the recent period peasants on the collectives have been allowed to maintain private plots and sell commodities raised on a limited free market. It is believed that about 80 percent of the hogs and 95 percent of the poultry were being raised on such plots, which accounted for 5 to 7 percent of the cultivated land in 1966. Incentives, such as additional fertilizers, cotton cloth, and other bonuses have been part of the government push to bring production back to former levels. Whereas it is believed that these measures have helped in this endeavor, there is little indication that the regime will go much further in granting farmers independence. Rather, China appears to have employed the more lenient measures in response to the pressures of emergency.

Farming in modern Japan, also, now differs significantly from the traditional pattern, especially in terms of resource-converting techniques. Use of improved seeds, commercial fertilizers, and generally more advanced cultural practices are now typical in Japan. In recent years mechanization has advanced rapidly; hand tractors with power take-off equipment are in widespread use. These developments are reflections of the general growth of industrialization and commercial orientation. A further indication is the substantial reduction in the role of farming in the national economy—less than 30 percent of the labor force is employed in farming. Moreover, there has been a rise in part time farming, made possible by mechanization which reduces labor needs and necessary because of rising standards of living. The Japanese farmer is increasingly being integrated into the commercially oriented economy of the nation. He is beginning to respond to the growth in market for other than grain and starch foods—vegetables, fruits, meat, milk, eggs—and he is showing increasing concern for achieving the greatest possible money returns.

There has been a general upsurge and diversification of the economy of the United States South. The one crop, cotton, that dominated the economic, social, and political affairs of the region and resulted in much abuse of cropland while other latent resources were underused (from the time that Eli Whitney invented the cotton gin until the end of the Depression years), has lost its regency. The assets of varied land resources, water, minerals, seacoast, and climate are being welded into an increasingly stronger and more varied economy that features forestry, mining, fishing, manufacturing, and tourism as well as farming.

During the last thirty years, southern agriculture has seen major changes. Cotton acreage has been reduced (Figure 11.27a), new enterprises and mechanization have been added, tenancy has been reduced, and the fruits of American agricultural research have been, in general, accepted and employed by the farmers. The changes had their beginnings in the 1930s, stimulated by cotton overproduction problems, government programs, the work of the Tennessee Valley Authority, and an improving backlog of research findings. World War II served as a fillip to mechanization and diversification, with the concomitant loss of labor to the military services or defense industries and the rise in demand and price for other agricultural commodities besides cotton. The movement for change generated during that time was sufficiently strong, so that the South in recent years—like the remainder of the nation—has continued to sophisticate its economy. Although remnants of the past persist, such as some areas of rural slums and a racial problem, the advancement is widespread and portentous of continued improvement.

Cotton acreage today is only a quarter of that of the 1930s and there has been a change in the distributional pattern. The decline has been most marked in the southeastern states. The Black Belt of Alabama, for example, once a famous production area, now pro-

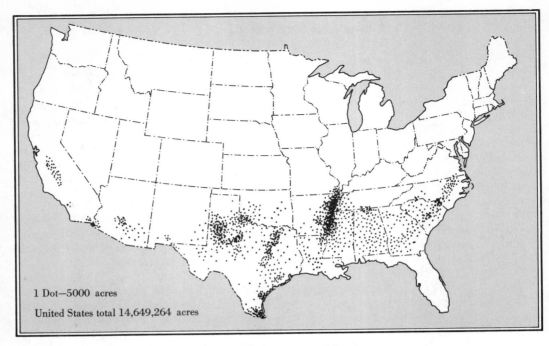

FIG. 11.27a. Cotton harvested, average 1959. (U.S. Department of Commerce, Bureau of the Census.)

FIG. 11.27b. Cotton, acreage 1939. (U.S. Department of Commerce, Bureau of the Census.)

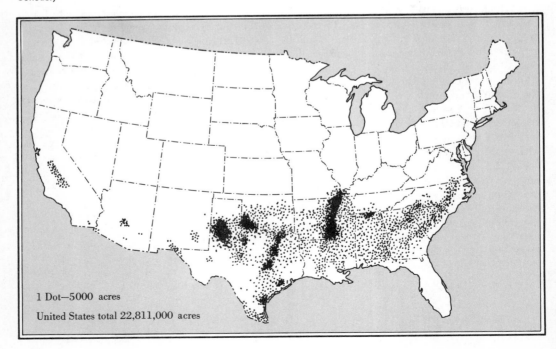

duces little cotton; the heavy soils that are wet into the spring, resulting in late planting and a greater incidence of pests, along with the problems attending the costs of grass removal from the fields of this natural grassland, have been factors favoring a shift to livestock. There has been a general movement westward of production; the Mississippi floodplain and the High Plains of Texas (where irrigation is practiced) are now the outstanding areas (see Figure 11.27b). Despite the reduction in cotton acreage as the result of improved seeds, fertilizers, and cultural methods, there has been relatively little drop in the regional crop. Yields per acre have been as follows: 1866–75, 163 pounds; 1931–35, 191 pounds; 1941–45, 262 pounds; 1961–65, 491 pounds; and during the last five years of the 1960s the yields ranged from a high of 527 pounds to a low of 436. Even in the major areas cotton does not dominate the cropland use as in the past. Other cash crops of significance include corn, peanuts, tobacco, and soybeans. A number of localities favored by climate and soil have established vegetable and truck-crop production as specialties; the development of canning and quick-freezing industries has aided in expanding the marketing possibilities. Fruits are grown in some sections; the citrus of central Florida and the peaches of the Carolinas and Georgia are notable. Sugar cane is a

FIG. 11.28. (top) A citrus grove in central Florida. Thousands of acres are devoted to citrus production in this section of the state, where fertile limestone soils, sunshine, rain, and minimum frost danger aid in the production of fruits of the finest quality. This is the leading orange-producing district in the United States. (Florida State News Bureau.)

FIG. 11.29. (bottom) Mechanical cane cutter. About one-quarter of a million acres of land are devoted to sugar cane in the Mississippi delta. The cane in this picture will be used for planting stock. When planted, the bamboo-like stalks sprout new roots from each point. (Standard Oil Company of New Jersey.)

FIG. 11.30. Rice harvest on the coastal plain of Louisiana. This scene provides a striking contrast to that shown in Figure 11.24. In the United States rice-producing districts, large farms, along with labor and other input costs, make such machinery both possible and necessary. (Allis-Chalmers.)

FIG. 11.31. Mechanical cotton picker. Agriculture in the United States South has undergone rapid mechanization in recent years. (Mississippi Agricultural and Industrial Board.)

traditional crop on the Mississippi delta lands and is now also grown in the Everglades; rice is noteworthy on the low coastal plain of western Louisiana and northeast Texas as well as on the heavy soils adjoining the Mississippi floodplain of Arkansas. The major tobacco-producing districts of the United States lie along the northern margins of this region, with centers on the Piedmont and inner coastal plain of the Carolinas and Virginia and in the Nashville Basin of Tennessee. Tobacco is also produced in the states of Florida, Georgia, and Louisiana.

The livestock enterprises represent the most important additions to Southern agriculture. A number of factors have combined to promote the development, including (1) cotton acreage reduction, (2) incentive payments given farmers for conservation farming, such as legume production and the planting of permanent pasture on lands subject to erosion, (3) a greater use of hay-forage–cash-crop rotations, (4) improvements in cattle, with stockmen upbreeding their herds, and (5) pasture improvement with greater attention to improved pasture plants and management. Cattle rearing appears to be firmly fixed, and the advantages of year-round grazing, minimum needs for shelter, low production costs, and adaptation to the evolving farming systems point toward future growth. Development of urban markets is encouraging increases in dairying and poultry raising, especially broilers. Both activities are now established, and the possibilities of further expansion are good.

Clearly the United States South has taken on new importance as an agricultural region, and there is little doubt that this portion of

FIG. 11.32. Beef cattle grazing on planted pastureland in Louisiana. This represents a new element in southern agriculture. Many improved pastures will carry one cow per acre through a ten-month grazing season. (United States Department of Agriculture, Soil Conservation Service.)

FIG. 11.33. Pampa north of Buenos Aires. The popular Black Angus cattle are grazing on improved pasture. The trees in the background have been planted and usually indicate a estancia headquarters. (J. Granville Jensen.)

the country will assume an increasingly significant role in the economy of the nation.

South America. The South American Humid Subtropics are the outstanding southern hemisphere agricultural region. Crop production centers in the Argentine pampas, Latin America's leading agricultural region, where fertile soils, smooth topography, climate, relatively small rural population, and proximity to seaboard favor commercial agriculture. The large-scale operations characterized by landholdings averaging several thousand acres in size, mechanization, and extensive crops exemplary of the commercial sector contrast sharply with the paddy-garden agriculture of the Orient. Wheat, corn, sunflowers, and flax for seed are the leading commercial crops of the pampas, and most of the production enters the channels of world trade. Millions of acres of alfalfa are grown to feed local cattle. Crops are adjusted to rainfall. The corn and flax are concentrated in the more humid northern sector; alfalfa is produced in the same district but extends somewhat farther westward and southward; and the wheat zone forms a crescent extending 600 miles from Santa Fe in the north to the city of Bahia Blanca on the coast in the south. In the zone surrounding Buenos Aires truck cropping and dairying are an important response to the urban market.

Crop production is overshadowed by livestock grazing in other portions of the South American region. In Uruguay, for example, 90 percent of the total area is classed as agricultural land and some 90 percent of the agricultural land is in pasture. With Argentina in the lead, attention is being given to improving the quality of livestock, both through breeding and pasture improvement. Today much of the fresh beef in world trade is produced in this region. Cattle dominate in most areas, except in Uruguay where sheep for wool are of equal importance.

There is also a subsistence sector in the Latin America region, producing for family needs. Generally there is not areal competition; the subsistence units tend to be concentrated along the streams in the poorer areas. This results from the land granting pattern that evolved in the early white settlement; the best lands were claimed in that process. In terms of counted farm units, it could be said that a small number of large *estancias* and a large number of small farms characterize farmholdings. In Uruguay, for example, 57 percent of the farmland is held in units of 2,500 acres or more which comprise 4.4 percent of the units. Nearly 30 percent of the farms are less that 25 acres in size and comprise 0.7 percent of the farmland. As to be expected, the small holders are mixed crop farmers.

Africa and Australia. Commercial agriculture in southeast Africa revolves around the production of sugar cane, pineapples, and other tropical fruits, citrus on the coastal lowlands, and cattle in the rougher inland districts. This region, which includes Zululand, has a large African Negro population, and a considerable portion of the cultivated land is given over to raising food crops, mainly corn and kaffir corn.

Southeast Australia, though restricted by mountains and sea to a narrow lowland fringe, is one of the most important districts of the continent. Dairying and beef cattle production are the chief phases of agriculture, and much of the land is devoted to pasture and forage crops. Sugar cane is a major cash crop in the northern district. The Brisbane area is the principal source of commercial tropical fruits. Oranges are grown in the vicinities of Newcastle and Sydney. Much of the land is still in extensive use.

USSR. The small Rion Valley region is more significant to the agricultural economy of the USSR than its size would indicate. With the exception of the small Lenkoran lowland of Azerbaidzhan, this is the only area of the country where tea and citrus fruits, or industrial crops such as tung trees, geranium, camphor, ramie, and the like can be grown. Tea and citrus fruits cover the greatest acreage, but recently grape vineyards have expanded, as have areas devoted to tobacco, almonds, and corn. Sericulture is practiced in eastern margins.

FISHING

Fishing activities are present in the offshore waters of all Humid Subtropic regions, although the importance of the economy as well as the type of catch varies. Fish are especially important in oriental countries, where they form a significant item in the diet. The numerous sheltered coves in the indented China coastline south of Shanghai are clustered with small fishing vessels that supply local markets. Although Chinese fishing has an old history, the present communist government has devoted considerable attention to reorganization and expansion; as a result, China has become one of the top five fishing nations of the world in terms of tonnage landed.

Japan has an outstanding fishing industry and today leads the world's nations in all phases of the activity. Fish products constitute major export items. Sardine, cod, salmon, mackerel, and tuna, as well as crab, figure high in the export trade. Other species, many considered inedible by nonoriental standards, are consumed locally and often eaten raw. Even seaweed is gathered for food, fodder, and fertilizer. Japanese fishermen, although restricted from certain waters, have surpassed the prewar catch. Nearly three million Japanese are dependent upon the activity for all or part of their livelihood. Every Japanese village fronting the sea has some type of fishing activity. Japan could not support her millions without the bounty of the sea. Aquiculture, the cultivation of fish and other aquatic life, is a very old activity in the Orient. Paddies, streams, ponds, canals, and sheltered bays along the coast are utilized to provide food and raw material supplies.

Much of the fishing industry of the United

States South is concentrated in the Gulf of Mexico, with red snapper, pompano, grouper, and sheepshead making up much of the food species caught for market. Large catches of menhaden, seldom used for food, are taken for use in the manufacture of fish oil and meal. Several types of shellfish are harvested with a special emphasis on shrimp and oysters. Numerous fleets are employed in shrimp fishing; the Texas shrimp industry leads all other states. Chesapeake Bay, at the northeast margin of the region, is the site of the major oyster fishery. Tarpon Springs on the west coast of Florida is the headquarters of the United States sponge fleet. This fishery has recently suffered a decline due to a sponge disease and to competition from synthetics.

The three southern hemisphere regions all have small fishing fleets that provide seafoods mainly for local consumption. South Africa, however, is gaining a favorable reputation for the high-quality lobster it is sending into world trade. Lack of markets in the southern hemisphere is the principal limitation upon present expansion in all three regions.

FOREST INDUSTRIES

Forest industries are a relatively minor activity in the Orient. The forest-base has been vastly reduced through the long history of human occupancy, losing area to cropland and being generally degraded in quality through the generations of excessive cutting where accessible to people. Fast-growing bamboo is used extensively as an Oriental substitute for wood and other materials. Charcoal for fuel is an important Japanese forest product; its value exceeds that of lumber. The dwarf-like character of many trees of regrowth forests makes them unsuitable for lumber. The Japanese are now following an orderly plan of forest conservation, realizing that trees can be a crop on land unsuited to agriculture as well as a significant factor in preventing soil erosion and floods.

The United States South has many natural advantages for forestry. The southern pines grow rapidly in the warm, moist climate and can be harvested for pulp in about twenty years, although at this age lumber yields are low. Logging operations can be carried on all year, facilitated by relatively level terrain and lack of dense underbrush. The region also benefits from proximity to the large eastern United States market. Today it is the outstanding pine-producing area in the nation and is taking over leadership in the pulp industry. In recent years with the major growth in the pulp and paper industry in the region, many thousands of acres of land have been established as pine plantations. Here is produced approximately one-half of

FIG. 11.34. Laying an oil pipeline. This is an important means of transporting petroleum products to market from most of the oil fields of the world. Similar pipelines are used to transport natural gas. (Shell Oil Company.)

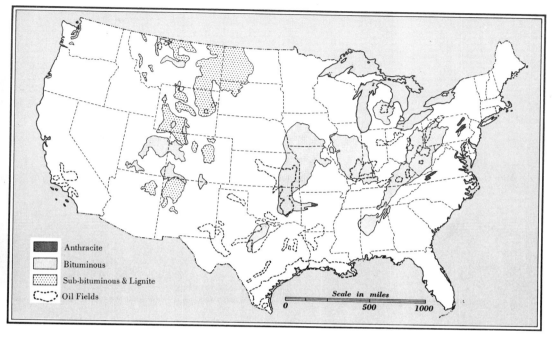

FIG. 11.35. The coal and oil fields of the United States.

the world's naval stores—turpentine and resin distilled from the gum of longleaf and slash pines. Paints, soaps, and medicines are other products using pine gum. The hardwoods are forming the basis for increasingly important veneer, flooring, and furniture industries.

The Paraná pine (Araucaria), an important lumber species of southern Brazil and Paraguay, is exploited for marketing in the large tributary urban centers, such as Buenos Aires and Montevideo. Maté leaves are gathered in Paraguay and southern Brazil for making the popular South American tea. There are small forest operations in the Humid Subtropics of Africa and Australia. The forests of southern Queensland furnish raw lumber and materials for plywood and veneer; the eucalyptus are important for structural timbers.

MINING

The mineral possibilities of China have not yet been fully determined. Large coal deposits are known to exist in the Red Basin, but transportation difficulties have limited mine production to supplying local needs. This same area is the country's most important salt producer. A few small iron deposits are worked along the Yangtze River; tin is mined in Yunnan Province of southwest China; and antimony and tungsten are extracted from the hills of south China.

Japan has a dearth of minerals; those represented include coal, copper, gold, and sulfur. Northwestern Kyushu contains the leading producing coal fields of the southern part of this island nation. Most Japanese coal is of mediocre quality, poorly suited to metallurgical purposes.

FIG. 11.36. Sulfur-loading facilities at Port Sulphur, Louisiana. This new city, twenty miles downstream from New Orleans, was not established until 1933, when local sulfur development began. (American Waterways Operators.)

FIG. 11.37. Phosphate mining near Bartow. Eighty percent of the nation's supply is mined within a few miles of this central Florida town. (Florida State News Bureau.)

The United States South is the outstanding mining region of the Humid Subtropics (see Figure 11.35). Fuels are particularly important; the largest oil and gas province of the nation is situated in the western half, coking coal is found in eastern Oklahoma, and the Appalachian coal fields reach into north central Alabama. In addition, lignite and sub-bituminous deposits are numerous in Mississippi, Louisiana, and Texas, but exploitation is as yet minor. Notable iron reserves are being worked in the Birmingham District of Alabama, and 95 percent of the domestic bauxite comes from near Little Rock, Arkansas. The major native sulfur production of the world is centered on the Gulf Coast of Texas and Louisiana (Figure 11.36). Most of the nation's phosphate rock comes from deposits in Florida and Tennessee (Figure 11.37). Two-thirds of America's kaolin is produced in Georgia.

The South American region is without major important mineral deposits. Brazil's modest coal fields are in the southern states of Santa Catrina and Rio Grande do Sul. African and Australian regions are both coal-producers. The African deposits extending from central Natal into Transvaal are the main source for that continent. The bulk of Australia's high-quality coal is produced in the Sydney-Newcastle area.

The Chiatura manganese deposits of Soviet Georgia are closely associated with the Rion Valley of the USSR. These deposits have been mined for eighty years, and Chiatura is one of the major manganese-producing centers of the world. The mineral is shipped from the ports of Poti and Batumi on the Black Sea to other parts of the USSR and the world.

MANUFACTURING

Japan, since the mid-nineteenth century, has become the industrial leader of the Orient, despite basic weaknesses. Although the "Island Empire" has limited natural resources, it has some coal, and water for

FIG. 11.38. All phases of the petroleum industry are found in the United States South. This large refinery is located on the Mississippi, where it is accessible to cheap water transportation. (Corps of Engineers, U.S. Army.)

power, and it has a large, adaptable labor supply, in addition to a strong will to succeed. By the turn of the century considerable attention had been given to industrialization. In the period between World Wars I and II the country became a competitor for markets throughout the world, and the "Made in Japan" label ran the gamut from expensive chinaware and efficient textile machinery to cheap ornaments and toys. Commerce became the lifeblood of the industrial economy —raw materials had to be imported and, due to the small buying power of the home market, finished goods had to be sold abroad. Thus Japan became a trader nation with ships moored in every major port of the world.

The disastrous effects of World War II crippled Japan's industry, alienated her foreign markets, and lost her territories furnishing raw materials. Furthermore, the growth of economic nationalism restricted access to many of her former markets. But the country made a rapid recovery, and the assets of experience, "know-how," and economic and technical aid from the United States have elevated Japan to an even more prominent position in the industrial world. Today Japan is a commercial, industrial, urban nation, a significant trader in world commerce.

The Japanese industrial belt extends 800 miles from Tokyo to Nagasaki with major industrial nodes located in the Kwanto Plain, Nagoya, the Osaka-Kobe district, and on the coast of northern Kyushu. Industries based in large, modern factories produce iron and steel, textiles, chemicals, tools, optical equipment, ceramics, rubber goods, petroleum products, motor vehicles, electronic equipment, and a host of articles for the variety store trade. She also assumed leadership in shipbuilding. All the manufacturing of Japan is not associated with large factories—small plants are still important. Many commodities for local markets are produced in small workshops usually having fewer than ten workers. Some export goods such as electrical equipment, bicycles, rubber footwear, and ceramics are also made in small plants.

Manufacturing in the Humid Subtropical region of China is concentrated in the Yangtze Valley. Cottage or workshop industries are typical for the manufacture of everyday needs and even some goods for the luxury trade, such as silks and lacquerware. Three centers of modern industry are Shanghai, Wuhan, and Chungking. All have iron and steel mills, but Wuhan, favored by good transport and proximity to energy and raw material resources, is particularly important with production capacity recently expanded. Wuhan is also a machine manufacturing center. Shanghai is a textile center and Chungking is the manufacturing center of the Szechwan Basin.

The traditional agrarian economy of the United States South has been drastically altered and a combination of favorable ingredients has stimulated an industrial revolution. Mines, forests, fields, and the sea provide a diversity of raw materials. Coal, petroleum, natural gas, and water produce power. Labor and capital are available; markets are widespread but easily reached by water and rail transport. Cheap land for factory sites, industrial water supplies, low taxes, low-cost plant heating, and other factors are added inducements.

The textile mills of the Piedmont lead the nation in the production of coarse and medium cotton goods. The Piedmont contains hosiery mills and electronics and furniture factories; tobacco manufacturing is concentrated on its northeast margin. The South is a leader in rayon; pulp and paper industries are significant; the region produces most of the kraft paper and has made a start in newsprint production. Other industries include shipbuilding, petroleum refining, chemicals, aluminum (Figure 11.39), cottonseed and food processing, and fertilizers. The Tennessee River Valley is the locus of considerable new industrial development; included are plants manufacturing chemicals

FIG. 11.39. Aluminum plant. Natural gas for the generation of required electricity and ease of access to the Surinam bauxite deposits are the attractions for the industry here. (American Waterways Operators.)

and textiles and those processing foods and aluminum. The iron and steel plants at Birmingham, Alabama, have the three basic requirements for smelting—iron ore, coking coal, and limestone—in juxtaposition. Industrial diversification is typified by the state of Texas, which has steel mills, light-metal plants, aircraft manufacture, automobile assembly plants, and petrochemical industries, which use natural and refinery gases as raw materials to produce synthetic rubber and plastics. The blast of the factory whistle has become a familiar sound in the South.

Manufacturing in the Humid Subtropics of the southern hemisphere is relatively minor; by world standards these areas lack many of the industrial bases and are principally concerned with producing raw materials for processing in other countries. Industry in the South American area is chiefly associated with processing agricultural commodities such as wheat and meat for export and textiles for local markets. The Natal Coast of Africa manufactures some consumer goods and also has a small iron and steel industry. The major manufacturing belt of Australia, as well as of the southern hemisphere Humid Subtropics, extends from Brisbane to Sydney with iron and steel plants at Newcastle, Port Kembla, and Lythgow. Food, shoes, textiles, and other goods are produced for home consumption.

TOURISM

Tourism is an important activity of most Humid Subtropic regions. The Japanese resort beaches, wooded islands, mountains (especially Fuji), hot springs, and temples and shrines are frequented by thousands of visitors and vacationists. The coast areas in southeast United States draw hundreds of thousands of people each year, particularly from the heavily-populated regions of the midwest and east. Florida has become the recreation center of the South; its revenue from the tourist business is about one billion dollars yearly. In addition to genial winter climate, the region has many interesting cities and historical attractions and features festivals, sport fishing, and athletic events to attract visitors. Leading tourist and recreational centers in the southern hemisphere include Durban in South Africa, the Uruguay beaches near Montevideo, and the Australian coastal area from Brisbane to south of Sydney. Improvements in overseas transportation along with lowering costs are beginning to attract an international clientele. Many popular health and seaside resorts of the USSR are located along the Black Sea littoral.

URBAN CENTERS

Although long primarily lands of rural populations, some of the largest cities of the world have developed in the Humid Subtropic regions, and recent industrialization promotes urban growth.

The Orient. The Yangtze Plain alone has a score of centers with over 100,000 population, at least a half-dozen of which exceed

FIG. 11.40. Shanghai, the gateway to the Yangtze valley. The "pedicab" has replaced the rickshaw in many of the Oriental cities. (Pan American World Airways.)

FIG. 11.41. New Orleans, the great river city of the United States South. This commercial center owes its importance to its situation near the mouth of the Mississippi, where it intercepts trade from the Mississippi Basin to the north and tropical products to the south. (American Waterways Operators.)

500,000. The dominating city of the plain, and of China, is Shanghai, whose admirable situation on the delta of the Yangtze makes it the natural outlet for the richest region of the country; as a result, it has become the leading commercial, industrial, and cultural center, with a metropolitan-area population in excess of 12 million (Figure 11.40). Nanking, 200 miles to the northwest, has over one million population. Wahan—the conurbation of Hankow, Wuchang, and Hanyang—630 miles from the sea, is in an excellent position to serve the upper portion of the plain. The population is about 3 million. Others exceeding a half-million include Hangchow, Soochow, and Changsha. Chungking, about 2.5 million, and Chengtu, nearing 1 million, are the principal cities of the Red Basin. Each of the delta plains along the southeast coast tends to have one large, dominating city; the most important are Wenchow and Foochow.

The main concentration of cities in Japan is found in the thin belt extending from the Kwanto Plain to northern Kyushu. Here are centered the major activities of the nation, including agriculture, industry, and commerce. Within this belt there are four principal nodes of development; from east to west they are: Tokyo-Yokohama, Nagoya, Osaka-Kobe-Kyoto, and northern Kyushu. Tokyo is the largest city with about 12 million people. Yokohama, its ocean port, has nearly 2 million. Osaka, leading industrial center, has a population of 3.2 million. Kyoto, a city of more than 1 million people, contrasts sharply with the modern industrial-commercial Osaka and Kobe. It was the

FIG. 11.42. A panoramic view of Sydney Harbor as seen from an office building in the city's central business district. In the foreground is the main artery of the city which converges on the Sydney Harbor Bridge (the "Coat Hangar"). (D. Lawrence Anderson.)

imperial capital and is today the center of arts and crafts. Nagoya (2 million) is a modern industrial city. There are no great metropolitan centers in the northern Kyushu industrial node comparable to those of the other areas. Fukuoka with over 500,000 is the largest, Yawata is the main steel center, and Moji and Nagasaki the main ports. Seoul is the capital and largest city of South Korea with a population of nearly 4 million.

United States South. Recent industrialization and commercial development have had tremendous influence upon city growth in the United States South. As long as the region was dominantly agricultural with cotton the only major commercial crop, most cities were small and few in number. During the past 30 years urban developments have been more rapid than in any other portion of the United States, and probably more rapid than in any other area in the world for a comparable period of time. Many cities have increased by 50 percent, and several have doubled in size; today 30 cities exceed

100,000. The largest are as follows: Houston, Texas, nearly 2 million, having grown from 292,000 in 1930; Dallas, Texas, about 1.5 million, having increased from 260,000 in 1930; Atlanta, Georgia, 1.4 million; New Orleans, 1.1 million; Miami, Florida, about 1.2 million. Birmingham, Alabama, Fort Worth, Texas, and Oklahoma City exceed .5 million.

Southern hemisphere. There are fewer urban centers in the southern hemisphere regions, but each has one or two large cities. Buenos Aires, on the Rio de la Plata estuary, has 8 million people and is *the* great city of the southern hemisphere. A favorable situation, with a rich agricultural hinterland and land and water transportation, has stimulated it to become the major port and the commercial, industrial, and governmental center of Argentina. Montevideo, with over 1 million people, dominates Uruguay even more completely, handling virtually all the nation's commerce as well as serving as the federal capital. Rosario, Argentina, 200 miles upstream from Buenos Aires on the Paraná River, is the third city of the region with nearly three-quarters of a million population.

Durban, about 700,000 population, is the only center of note in southeast Africa. Situated on a fine harbor and connected by rail to the interior, it is the principal port of South Africa. Sydney, with 2.6 million people, is the main city of Australia. It also has a splendid harbor, excellent rail facilities, and easy access to the great coal basin. Newcastle, emulating its English namesake, is the leading coal port and heavy industrial center of Australia.

twelve

LONG SUMMER HUMID CONTINENTALS

The continuous poleward progress in these regional studies has finally eliminated the tropical modifier. Continentality, denoting land control with high temperature ranges and marked seasonal differences, now assumes a foreground position. Furthermore, the book of climate now contains more varied chapters —only the summer section has tropical reminiscences.

The Long Summer Humid Continental regions encompass extensive areas of choice agricultural land. Almost everywhere the farmer is a potent force in the economic life. Grains govern the crop patterns, but remarkable differences in methods and standards are present from region to region. The farmers of the agricultural interior of the United States have the highest living standards—commercial farming is the rule. The state controls the destinies of the peasant in the communist countries, yet the usual oriental practices persist in the Far East.

Agriculture is "big business" in the United States; however, an abundance of raw materials, energy resources, skilled labor, transportation, and markets have established a firm base for industry. The world's greatest industrial concentration, the American Manufacturing Belt, lies almost wholly within the region. Factories, employing workers by the thousands, produce nearly every conceivable manufactured article that might be desired by humankind.

Location

Long Summer Humid Continental regions occupy interior and eastern margin locations in the northern hemisphere continents poleward of the subtropics and extending to about latitude 45° N. The interior borders are marked by the 20-inch *isohyets,* or lines connecting points of equal rainfall, and the northern and southern borders are formed respectively by the 150- and 200-day average frost-free season lines. Long Summer Humid Continental regions occur only in North America, Europe, and Asia (see Figure 12.1).

The North American region includes the tier of states extending from central Nebraska

LONG SUMMER HUMID CONTINENTALS   257

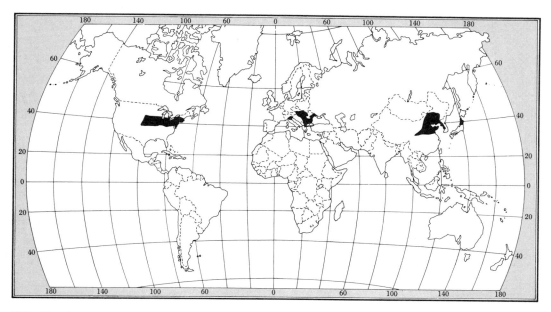

FIG. 12.1. The Long Summer Humid Continentals.

FIG. 12.2. The Long Summer Humid Continentals of the United States.

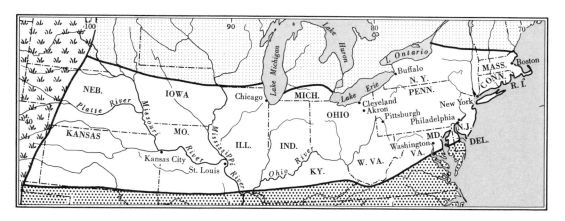

and Kansas on the west to the Atlantic coast (Figure 12.2). The European region, interrupted by mountains and seas, includes the Po Plain of Italy, the Danube Basin of Yugoslavia, Hungary, Bulgaria, and Romania, and the western portion of Ukraine in the Soviet Union (Figure 12.3). The Asian region consists of northeast China, including southern Manchuria, northwest Korea, and northern Honshu (Figure 12.4).

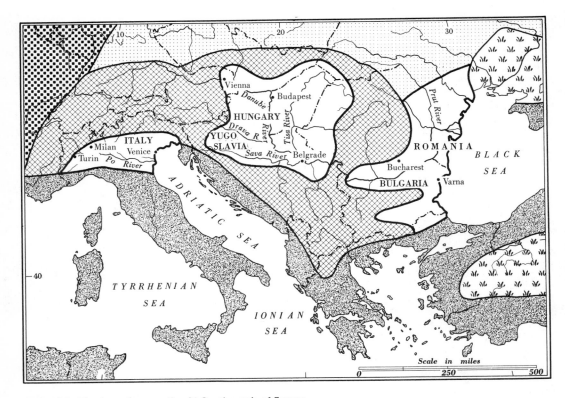

FIG. 12.3. The Long Summer Humid Continentals of Europe.

FIG. 12.4. The Long Summer Humid Continentals of Asia.

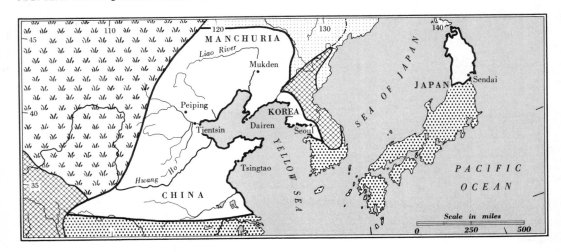

Physical Environment

CLIMATE

Cold winters are a distinctive feature of all humid continental climates—a season that distinguishes them from the tropical and subtropical regions. Annual range of temperature is wide, since the summer heat resembles that of the Rainy Tropics. The year is divided into four clearly marked seasons. Daily weather is highly variable and seasons differ considerably from year to year. Variability is due chiefly to the position of the humid continental regions in the middle latitudes, which are the scenes of conflict between polar and tropical air masses (see page 43).

Temperatures. Summer temperatures are high. July averages about 75° with daily maxima exceeding those of the Rainy Tropics. Summer days are sultry, and there is little relief from the heat at night. Winters are cold, with January averaging about 25° (Figure 12.5). Occasional cold waves send thermometer readings to zero and below. Location affects winter temperatures; stations deep in the interior or poleward are subject to colder winters than those located on coasts or equatorward margins.

In North America, the Great Lakes have an influence on temperatures. Cold air passing over the water bodies is tempered by the stored heat, which retards fall frosts on the eastern lake shores. The spring season is also late since the cold lake waters chill the eastward-moving warm air. Annual ranges for the region are about 50°. Growing seasons range from 150 days poleward to 200 days on the southern margin.

Precipitation. Precipitation occurs in the long summer humid continentals throughout the year with the maximum usually in the summer season. Summer rains result from frontal activities and from numerous convectional storms that are frequently associated with thunder and lightning. Winter precipitation is frontal (cyclonic), frequently in the form of snow although part falls as a cold, disagreeable rain.

FIG. 12.5. Typical climatic graphs of Long Summer Humid Continental stations.

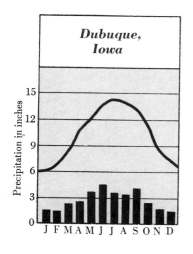
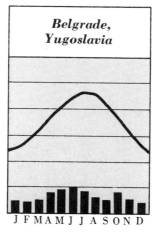
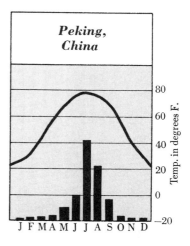

Total annual precipitation varies greatly in these regions. In North America, the eastern portion receives 35 to 45 inches. Amounts decrease westward, with 20 inches marking the subhumid limits of the region in the west. The Danube Basin and much of North China average less than 30 inches a year. Fluctuations from year to year are considerable. Snow falls 20 to 30 days a year, and amounts vary from about 10 inches on southern margins to as much as 40 inches in the north. Snow cover persists from a few weeks to several months, and the visible evidence thus gives the impression that winter precipitation is greater than summer; but from 5 to 15 inches of snow are required to melt down to one inch of water. A snow cover prevents deep freezing of the earth, acting as a blanket, keeping the heat in and the cold out.

Tornadoes develop during the spring and summer in the Mississippi Valley; the states of Kansas and Iowa average fourteen to sixteen a year. Great loss of life and property often result from these violent storms (Figure 12.6).

SURFACE FEATURES

A large percentage of the land is of low relief. Plains are the most distinctive features of every region; but there is considerable diversity of surface, and in no case are highlands lacking.

United States. The largest and most important plain occurs in the United States. Here the physiographic province known as

FIG. 12.6. Tornado. Notice that the tornado hangs from the large cumulo-nimbus cloud like a funnel. The movements from the funnel are erratic—rising, falling, turning, and swinging in various directions. When the end of the funnel touches the earth, there is near-complete destruction. (National Oceanic and Atmospheric Administration.)

FIG. 12.7. A cross-section of the United States region.

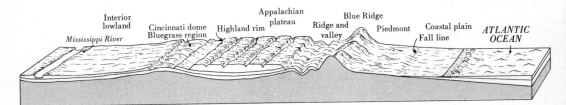

FIG. 12.8. The plains of Kansas. This view shows the slightly undulating topography of the western portion of the United States region. The fields have been terraced and contour listed in the fall. The snow collected in the furrows has melted, and the water is staying on the contours, gradually to soak into the ground and build up reserve moisture for next summer's crop. (United States Department of Agriculture, Soil Conservation Service.)

the Interior Lowlands occupies the western two-thirds of the region; the Appalachian Highlands separate these lowlands from the narrow Middle Atlantic Coastal Plain (see Figure 12.7). On the south, smooth terrain merges into hilly country such as the Ozark Plateau, but the smooth to gently-rolling surface continues westward through the Great Plains. The Corn Belt, which extends from eastern Nebraska and northeastern Kansas to central Ohio, is topographically one of the finest agricultural lands in the world (Figure 12.8). It owes at least part of its uniformity to favorable glaciation during the Pleistocene Period. The ice sheets invaded most of the area several times, smoothing off the hillocks and filling in low sections.

Three broad divisions of the Appalachians are recognized: the dissected plateau on the west, a series of linear ridges and valleys in the center, and the higher Blue Ridge and Great Smoky Mountains on the east. The most significant area of lowland in the Appalachians is the Great Valley, a composite lowland between the westernmost ranges of the older Appalachians and the easternmost ranges of the new Appalachians. It is made up, from north to south, of the Kittatinny, Lebanon, Cumberland, Hagerstown, Shenandoah, Tennessee, and Coosa valleys. (Figure 12.9).

FIG. 12.9. Farmland on the west edge of the Appalachian Plateau. Here the land is more rolling and strip cropping is practiced to prevent erosion. Such practices restrict soil loss to less than half that incurred by up-and-down slope cultivation. (United States Department of Agriculture, Soil Conservation Service.)

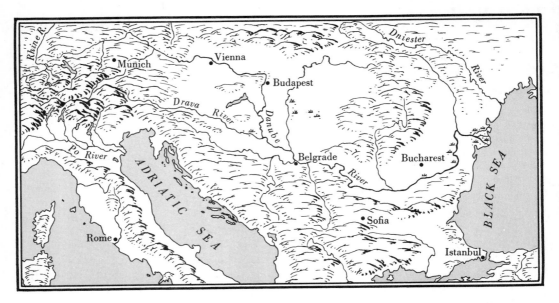

FIG. 12.10. A physiographic diagram of south central Europe.

Europe. Mountains divide the European region into three distinct lowlands: the Po Plain, the Danube Basin, and the Lower Danube Lowland, including the southwestern Ukraine. The Po Plain of northern Italy, formerly an extension of the Adriatic Sea, was created by sediments deposited by streams from the Alps and Apennine Mountains, and even today is being pushed seaward at a rapid rate.

Eastward from Austria, the ranges of the Alpine Mountains divide; the Carpathians, or northern branch, swing to the east then south in a broad arc to the Danube River; the southern branch extends through Yugoslavia and southern Bulgaria. The Danube Basin is thus encircled (see Figure 12.10).

Asia. Two lowlands—the North China or Yellow Plain and the southern half of the Manchurian Plain—form the core of the Asian region. The former is the largest plain and one of the most clearly defined regions of China, with highlands or sea borders everywhere but on the southeast, where it merges into the Yangtze Plain. Like the Po Plain, it has been built by river deposition, largely by the Hwang Ho or Yellow River. The same floods which originally built this lowland are now major problems. The Hwang Ho flows through vast streaches of dry land and enters upon the plain with a large load of sediments; reduced gradient and velocity diminish the carrying power of the stream, and it deposits silt, raising its bed in places above the level of the plain. Natural levees keep the river within its channel except during flood seasons. To minimize danger of inundation of fertile farmland, the Chinese have build retaining dikes, but even these are periodically breached, allowing flood waters to devastate broad areas. In contrast to this large depositional plain, the Manchurian Plain, featuring rolling topography and extensive terraces, was formed largely by erosion. Here the Liao River and its tributaries have worn down the land.

In addition to these plains, the Asian region includes: the Shantung Peninsula of North China, which is about equally divided between plains and uplands; the southern portion of the east Manchurian uplands; northwestern Korea, which is mainly hilly but has small river plains near the coast; and northern Honshu, which is largely mountainous with small alluvial coastal plains.

NATURAL VEGETATION

Hardwood forests consisting of maple, oak, chestnut, hickory, and other deciduous species characterized the original cover in the more humid areas of the United States region, which at one time contained the finest stand of hardwoods in the world (Figure 12.11). Farms have replaced the forests, and only a few remnants appear in woodlots and on the more rugged slope lands. Conifers grow on the sandy coastal areas or in the cooler and higher elevations. On the subhumid margins, forests are replaced by grasses. Prairie grasses once carpeted the western portions, but much of the prairie land has now been turned under by the plow. Fields of grain cover these former grazing lands.

The Italian region, the Po Valley, has no forest cover on the floor of the plains, but conifers ascend the higher alpine slopes.

FIG. 12.11. A stand of white oak on the Appalachian Plateau. In the American Midwest, forests have given way to a higher land use—agriculture—over broad areas of smooth topography. (United States Forest Service.)

Stately rows of Lombardy poplars have been planted along roads in the main part of the valley. Meadowland is common in the central portion of the plain, and marshes fringe the low coastline. The Danube Basin, chiefly in the western part, has dense forests on many of its surrounding slopes and often on the hills that rise from the basin floor. Deciduous species, usually beech, are found on the lower slopes, replaced at higher levels by conifers. Marshes are characteristic in the poorly drained areas, especially on the delta of the Danube. The boundary between forest and grass in the Danube Basin appears in eastern Romania, where trees disappear and short-grass steppe vegetation becomes dominant and continues into the Ukraine.

The quest for cropland and the need for building materials and fuel have consumed almost the entire forest cover of North China. Even the hills are bare. In some areas the land has been tilled for so many centuries that there is now some doubt as to the original nature of the vegetation. The only trees are found along roads and surrounding farmsteads, where poplars and willows have often been planted. Short-grass vegetation is the dominant natural cover in the more arid west. The Manchurian area, more recently settled, still contains forests on the mountain slopes. Deciduous species grow near the margin of the plain and are replaced by conifers at higher elevations. Similar forests were characteristic of the Korean area, but most accessible forests have been removed. The forest cover on the lower slopes of northern Honshu consists of broadleaf deciduous species, distinctive from the broadleaf evergreens in the warmer southern part of the island. Conifers replace birch, beech, poplars, and other hardwoods in the colder, higher elevations.

SOILS

The soils are generally more fertile than those of the Humid Subtropics, resulting from less precipitation, grass vegetation, and lower temperatures, with the soil often frozen during winter. Leaching is less intense and of shorter duration; organic content is usually higher. Three principal soil groups are recognized.

The *chernozems* occupy the drier margins in the United States, eastern Romania, and the Ukraine. This rich black soil has been subjected to only slight weathering and leaching and is therefore high in mineral nutrients. The tall-grass vegetation contributed an abundance of organic matter as a result of the death and decay of roots, stems, and leaves. The main limitation for these inherently productive soils is the relatively low precipitation, which restricts crops to grass and small grain types.

Prairie soils. These have developed in the grasslands of higher precipitation that border the chernozems. They are only moderately leached, dark in color, and rich in organic matter. They are excellent agricultural soils and form the heart of the American Corn Belt. Even though slightly lower in native fertility, they are more adaptable than the chernozems, due to the higher precipitation they receive.

Gray-brown podzolic soils. Predominate in the more humid portions of the regions, these soils develop under heavier precipitation and deciduous forest vegetation; they contain somewhat fewer plant nutrients than the other groups. The more abundant moisture causes greater leaching, and the forests contribute less organic matter. But they are easily cultivated, respond to fertilization, and well managed, are highly productive.

Special attention should be given to the soils of North China or the Yellow Plain. Much of this alluvial lowland, as well as the hills to the west, is mantled with wind-blown material, called *loess,* that has been carried by the strong winter monsoon winds from the drylands of the interior. The admixture of loess and alluvium produces a highly productive soil. Its characteristic yellow color is the basis for the names Yellow Plain, Yellow River, and Yellow Sea.

Occupance in the Long Summer Humid Continentals

The Long Summer Humid Continental regions rate high as a home for man. Farming, mining, industry, and related commercial activity, in varied combinations, form a broad base for supporting large populations. The United States region, with its many great cities and diversity of activities, is the most densely populated portion of the nation. In Europe, the Po Plain contains about 40 percent of the Italian people; the Danube Basin is the economic heart of the countries it comprises; and the Ukraine is part of the more densely settled section of the USSR. Although the Asian portion is overshadowed by the Humid Subtropics to the south, it nevertheless is populated to near its agricultural capacity and has the greatest numbers of all these regions.

Cultures differ markedly; there is especially great contrast between the North American and Asian regions. The former has a science and machine civilization, epitomized by a highly developed commercial agriculture, with farm families having high standards of living and education, and by the most important and advanced industrial concentrations in the world (Fig. 12.12.). The region is spanned east-west and north-south by the most dense transportation net in the world. The average American thinks nothing of traveling many miles to work or play. Subsistence farming has, until very recently,

FIG. 12.12. Agricultural regions of the United States. (Adapted from material from the Bureau of Agricultural Economics, United States Department of Agriculture.)

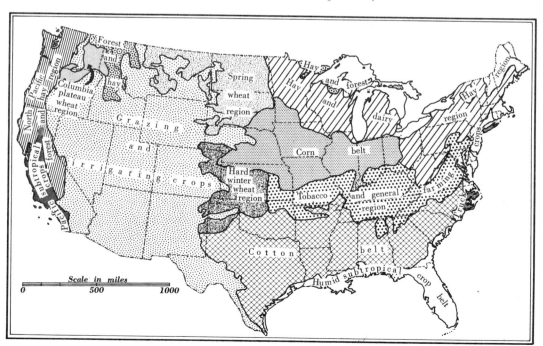

typified activity in the Orient, conducted in a framework of low buying power and traditional methods and ways of thinking. Although the Asians are still largely in a man-power stage, changes are underway. The communist regimes in China and North Korea have been attempting to modernize and diversify the economy through planning and government direction of development activity. Only Japan and the Manchurian Plain approach adequate transportation facilities; much of China cannot be reached by either rail or all-weather roads. Lack of modern surface transportation is a major economic and political weakness in China today.

The north American region also illustrates the tertiary sector of economic activity. As increasing efficiency has been achieved in the primary activities (agriculture, fishing, forestry, and mining) as well as in the secondary activities (manufacturing) the tertiary activities have become increasingly important. In 1970 the 81-million-person civilian labor force of the United States was employed as follows: manufacturing, 24.8 percent; wholesale and retail trade, 18.1 percent; government, 15.1 percent; services, 13.9 percent; transportation and public utilities, 5.5 percent, finance and real estate, 4.4 percent; agriculture, 4.3 percent; contract construction, 4.0 percent; other, 6.4 percent; and unemployed, 3.5 percent. Within the total structure of trade, spanning the scope from local to international, numerous people are employed in the tasks associated with commodity concentration, movement, and distribution—or in buying, transporting, and selling. Likewise, within the structure of services many are employed in a host of construction, repair, and maintenance activities and in a great variety of unskilled, skilled, and professional occupations, ranging from window cleaners to medical doctors. Included also are increasing numbers providing for the wants and needs of recreationists and vacationers for food, housing, equipment, and so on.

Whereas some aspects of trade and service are associated with essentially all the forms and stages, they are especially significant and representative dependents of commercial economies. More precisely, they are found in greatest concentrations as reciprocal elements in the more highly complex and productive commercial-industrial economies. Trade, or exchange, as noted earlier, is related to commercial orientation. Service activities flourish where concentrations of populations with high standards of living or high buying power exist; they have both promoted and responded to the growth of urban places. Thus trade and service activities are especially representative of the technically and economically advanced countries of the world. They loom particularly large in the economy complexes and employment patterns of Anglo-America, Europe-USSR, Japan, and in western countries and urban places elsewhere.

AGRICULTURE

In an overall view, and from the standpoint of numbers of participants, agriculture is the leading activity. Only in United States and Italian regions does farming employ significantly fewer people than manufacturing. The farming scene differs materially from one region to another, and even among portions of the same region. In general, grains tend to dominate; but the emphasis differs, and other crops, as well as livestock, are often significant. The greatest contrasts are evident in the techniques and production tools and in the organization and sizes of farms.

United States. Commercial agriculture characterizes the United States region. On the basis of dominant farm enterprises and systems, the land is usually broken up into the several agricultural regions shown in Figure 12.12. All the Corn Belt, the eastern half of the Winter Wheat Region, the northern fringes of the Tobacco and General Farming Region and the Middle Atlantic Truck-Cropping Region, and a portion of the Hay and Dairy Region are included in the Long Summer Humid Continental Region. This diversity in agriculture results in part from differences in the details of surface, soil, and climate, and in part from market possibilities.

Corn Belt farming is one of the most distinctive systems in the world. Smooth to rolling topography, excellent soil, and a warm, moist growing season, in combination with a progressive farm population and proximity to a high-standard-of-living market, have led to the development of a high order of agriculture in which science and machinery play major roles (Figure 12.13). Farms average 200 to 400 acres, and both farms and fields are generally rectangular in pattern. On individual farms 70 to 90 percent of the land is planted in crops. Feed and forage plants dominate the crops, with corn the leader. Oats, although declining, still fit well into the farming scheme: Planting comes earlier than corn in the spring, and value is also derived from its service as a nurse-crop for hay. Winter wheat is grown widely and is especially significant in the eastern part of the region—serving to protect rolling land against soil-erosion during the relatively wet winters. The importance of the soybean as a cash crop has increased greatly in recent years, responding to consumer market shift from animal fats to vegetable oils. The land base is well suited and the crop fits into the farming system. Today soybeans rank second only to corn in crop acreage in the Corn Belt and third after corn and wheat in the nation. Most farms are integrated feed crop–livestock operations; details of specialization however, vary with land quality, size of holding, markets, farmer preference, etc. Hogs and beef cattle are kept on most farms, and animal fattening is an essential part of the farmer's program, with most of the corn and oats and all the hay used as feed. Beef cow-calf sidelines are common on farms containing some rough topography. Sale of animals returns about two-thirds of the farm income. In the areas of smoothest topography there are concentrations of cash-grain farms, but

FIG. 12.13. A Corn Belt farm harvest scene. Farmers in this great agricultural region commonly require $20,000 to $30,000 worth of machinery for the efficient operation of their farms. The farmstead shown here is fairly typical of the entire Corn Belt. (Allis-Chalmers.)

FIG. 12.14. Self-propelled combines operating in the Winter Wheat Belt. Machines such as these have cut down the man-hours required for wheat production in this region from 50 hours per acre 100 years ago to less than two hours today. Custom harvest is common in the American wheat regions. Operators begin on the High Plains of Texas in June and work their way northward with the latitudinal progression of wheat ripening, ending in the Prairie Provinces of Canada in the fall. When it is recognized that machines such as these cost $10,000 to $15,000, and harvest season on an individual farm is only two to four weeks, one can understand that machines such as these are expensive. (Standard Oil Company of New Jersey.)

FIG. 12.15. A view across the Connecticut Valley. This is a significant cigar-wrapper-producing area—the long buildings are tobacco barns. (Standard Oil Company of New Jersey.)

even these keep some livestock, especially hogs. Dairying is important in the northern fringe and around many of the cities.

Southwest of the Corn Belt, lesser precipitation has encouraged the widespread production of winter wheat. Farms are larger, ranging from a half-section in the east to one or two sections or more in the west. Operations are completely mechanized (Figure 12.14). Grain sorghums are frequently raised in rotation with wheat. Beef-cattle grazing is also important in the Flint Hills area. In the Winter Wheat Belt, as well as the western fringe of the Corn Belt, many farmers in recent years have been developing individual irrigation systems based upon deep wells to protect against the vagaries of precipitation and increase the volume and value of production. Rising farming costs here as elsewhere in the nation have forced farmers to increase output or discontinue farming. As implied elsewhere, the numbers of farms and farmers in the United States have decreased significantly in recent decades; there were less than half as many in 1970 as in 1935. There are two ways a farmer can increase output: increase the size of holdings or intensify to get greater volume and/or value per acre. American farmers have used both approaches. In the national picture, farmers today are obtaining more production from less cropland than they employed three decades ago. This has been possible through greater capital inputs: more and better machines, improved seeds, and agricultural chemicals.

The Tobacco and General Farming Region, often termed the "middle country," is a transitional zone with elements of both southern and northern farming common. More rugged topography restricts arable land to about one-third of the area. In recent years considerable portions of former cropland have been converted to improved pastures. Corn, hay, and winter wheat are the leading cropland uses, but specialty crops are significant in certain well-defined districts. The Lexington or Bluegrass Basin of Kentucky is a leading tobacco center, ranking second in output only to the Virginia-Carolina area mentioned in the Humid Subtropics. Fruits and vegetables are local specialties in many areas, and for the region as a whole, animal industries return at least 50 percent of the farm income.

The Middle Atlantic Truck-Cropping Region is confined between the Piedmont and the Atlantic Ocean. The land is level to gently rolling and is usually under 100 feet in elevation. Wheat, corn, and hay are grown on a large part of the land under cultivation, but vegetable production is more important. This region is one of the world's foremost producers of commercial vegetables, even though these crops occupy but a fraction of the area. The mild, marine-modified climate with a long growing season is suited to many varieties, including sweet potatoes, sweet corn, cucumbers, cabbage, peas, tomatoes, spinach, and so forth. Two, three, or even four crops, depending on maturing season, can be grown yearly on the same land. As soon as a harvest is completed, the land is worked, fertilized, and a new crop is planted. Under such intensive production, farms are small in contrast to other portions of the United States region. The Middle Atlantic Truck-Cropping Region benefits from good transportation and proximity to large city markets. In recent years canning and freezing have developed extensively. Dairying and poultry are important enterprises in the northern segment of the region; in this portion urban sprawl is usurping a considerable amount of productive farmland. The process of urbanization is also a factor in intensive use of farm lands. It drives the values up and sets positive and negative forces to work on the farmer; for example, on the one hand, it increases the market value of land encouraging him to sell and, on the other, it increases his land taxes requiring that he intensify if he is to remain in business.

The southern part of the Hay and Dairy

TABLE 12.1
Agricultural comparisons for 1967

	United States	1967		USSR
PEOPLE / It takes about a third of the Soviet work force to produce the nation's food and fiber. Only 6.6 percent of our U.S. labor force is employed in agriculture.	197.1	National population	Millions	235.5
	74.3	Annual average employment	Millions	112.2
	4.9	Annual average employment in agriculture	Millions	39.5
	6.6	Farm share of total work force (annual average)	Percent	35.2
	1.6	Workers per farm	Number	418 Collective
				618 State
FARMS / About 97 percent of Soviet farmland is in huge state-owned or controlled complexes with hundreds of workers each. By comparison, U.S. farms are small. Most are operated by the farm owner and his family, with perhaps one or two hired workers.	3,146,000	Number of farms	Numbers	36,200 Collective
				12,783 State
	360	Land area per farm	Acres	30,077 Collective
				118,765 State
	95	Sown area per farm	Acres	7,031 Collective
				17,050 State
	225	Land area per worker	Acres	72 Collective
				192 State
	59	Sown area per worker	Acres	17 Collective
				28 State
INPUTS / The present Soviet regime is trying to step up agricultural efficiency by raising the level of such inputs as fertilizer and pesticides and by expanding irrigation and drainage facilities. It is also providing farmers more incentives—such as financial concessions—and is encouraging greater use of livestock products in the Soviet diet.	97	Fertilizer (plant nutrients) per sown acre	Pounds	33
	4,820	Tractors	Thousands	1,739
	3,125	Trucks	Thousands	1,058
	870	Grain combines	Thousands	553
	29.0	Electricity consumption (farm)	Billion kilowatt	25.8
OUTPUT / The Soviets have recently narrowed the agriculture gap in many aspects, but many deficiencies still exist. U.S. farmers, using less labor and land, and more capital, continue to achieve the greater output and generally higher crop yields. Both countries have adequate calories available per person. But the Soviet diet is heavy in cereals and potatoes, while the U.S. diet is high in vegetables, fruits, and livestock products.(25)	176	Four feed grains	Million short tons	49*
	51	Four feed grains	Million short tons	86*
	7,458	Cotton	Thousand bales	9,325
	976,060	Soybeans	Thousand bushels	20,209*
	112	Sunflowerseed	Thousand short tons	6,701*
	1,267,911	Tobacco	Thousand pounds	579,195*
	21,010	Beef and veal	Million pounds	9,557*
	12,550	Pork	Million pounds	7,440*
	650	Mutton lamb, and goat	Million pounds	1,764*
	8,108	Poultry	Million pounds	1,764*
	119,294	Milk cows	Million pounds	155,035*
	70.2	Eggs	Billion	33.9
				*USDA estimate

Region extends into the Appalachian Uplands. A high percentage of land in slope, poor soils, and a cool, moist climate discourages widespread crop agriculture here but favors the growth of good quality grass for pasture and hay. These factors, plus the nearby city markets, have stimulated the development of dairying. In the more isolated sections of this region and in the adjoining parts of the Tobacco and General Farming Region, agriculture is backward—probably the most backward in this nation. Corn, grown on surprisingly steep slopes, and livestock have been the mainstays in the farmer's subsistence economy. Conditions, except for rising populations and depletion of soils, remained little changed from the time of settlement shortly after the American Revolution until recent years, when improved transportation eased contact with the outside. This has made possible the development of coal, timber, and recreation resources and has allowed these isolated people to observe a more modern way of life. Yet, by national standards, Appalachia is classed as a "poverty area." The Finger Lakes District, south of Lake Ontario, on the northern margin of the Appalachian Plateau, is distinctive. In addition to dairying, the truckfarming activity here is comparable to that of the Middle Atlantic Coastal Plain. Grapes and fruit production are also important.

Europe. Grains dominate in the European Region, although contrasts are found in cropping systems and crop associations. The Po Plain, the principal agricultural area of Italy, is of particular interest because wheat, corn, and rice, the world's principal grain crops, are all grown in significant quantities, with wheat the leader. Hemp, flax, and sugar beets are notable, and many mediterranean-type crops were introduced long ago; there is wide distribution of vegetable gardens and vineyards, the latter found especially on sunny slopes. The small sericulture industry of Italy is centered in the alpine valleys on the northern margin of the plain. In the areas of poor drainage or on uncultivatable hillsides, cattle raising, and particularly dairying, has developed.

The countries of the Danube lowlands are primarily agricultural. Wheat is the leading crop; corn is usually a poor second, except in Romania. Barley and oats are important, and in some sections sugar beets, grapes, and tobacco are noteworthy. Cattle are found on most farms; sheep are numerous on the steppe; and horse breeding is traditional, particularly in Hungary. Since World War II the Danubian countries have become Soviet satellites, and farming systems are being patterned after those now practiced in the Soviet Union.

Wheat has long been the crop of the Ukraine. Barley and corn also are produced. Agriculture has been revolutionized since the communist regime took control. Two types of farms are now found in the Soviet Union, the collective farms formed by the amalgamation of former peasant holdings into large operating units and the state farms organized from former estates or in virgin areas where agriculture had not been practiced. In 1967 there were 36,200 collective farms and 12,783 state farms. The collective farms averaged 32,470 acres with an average sown cropland area of 7,156 acres; the state farms averaged 118,765 acres with 17,050 acres of sown cropland. Workers per collective farm averaged 418, and workers on the state farms averaged 618 persons. This organization has allowed greater government control and the implementation of science and technology. In Table 12.1 agricultural activities in the United States and the Soviet Union are compared.

Asia. The small-scale, intensive, hand-labor farming of the Asian region has contrasted sharply with the large-scale, mechanized farming of the United States. Farms on the North China Plain, which before communization averaged five acres and supported an average of six persons, have been organized into large units. Soil is fertile and productive, but relatively low, and unreliable precipitation restricts production of the heavy-yielding rice to the few areas where irrigation is possible. Moreover, multiple cropping can be practiced only with short season annuals, namely vegetables. Wheat, more suitable to subhumid conditions, is the distinctive crop but is not as dominant as rice in South China. Millet, corn, and the grain sorghum *kaoliang,* are widespread. Soybeans and many vegetables are grown throughout, adding variety to family diet and distributing the use of labor. In some cropping patterns it is possible to get three crops in a two-year cycle, a spring planted crop, followed by a fall planted crop, and then an early summer planted crop.

Only on the Manchurian Plain is there surplus for export. Northern Honshu merits special mention—it is a transitional zone between the agriculture of traditional Japan, where farming is set along subtropical lines, and that of Hokkaido or frontier Japan, which is more recently established and has many characteristics reminiscent of Europe and North America. In northern Honshu, farms are somewhat larger than in the south, averaging 3.5 to 4 acres; rice is still a major crop but, owing to severe weather, the fields lie fallow in the winter. The northern boundaries of many subtropical crops are reached in the southern margin of the Honshu region. Sweet potatoes and tea are grown only in the southern margins; latitude 38°N marks the northern limits of bamboo, and 39° or 40°N, the extent of significant sericulture. Other crops are added to the farming pattern, including apples, white potatoes, millet, and buckwheat.

FISHING

The waters off the Atlantic Coast offer a transition between fish species occupying warmer seawaters to the south and those inhabiting the colder waters of the North Atlantic. The fishing industry of the United States region, therefore, has characteristics similar to both areas. Shellfish, chiefly oysters and clams, are important seafoods harvested in the sheltered coastal bays. Large numbers of menhaden are netted and processed into fish meal for broiler producers and fertilizer for the truck-farmers on the sandy coastal plain and elsewhere. A variety of species such as herring, shad, and mackerel are a part of the commercial catch for the fresh-fish markets. Inland fisheries are found in the rivers of the Mississippi Valley, where catfish, carp, and buffalo fish comprise much of the commercial catch. The fisheries associated with the Great Lakes will be discussed in the following chapter.

Fishing in the European region is confined chiefly to the coastal waters of the Black Sea and the lower courses of the Danube and Dnieper Rivers. Several species of fish, including herring, salmon, carp, and perch, are caught, but the most valuable is the sturgeon, prized for its roe—caviar in the luxury trade.

Some fishing is practiced in the waters of the Yellow Sea, but the activity is far less important than in the waters south of Shanghai. The coastline south of Shanghai is indented with small harbors, whereas the coast to the north is more regular and provides less protection for small fishing boats. The inhabitants of North China have looked more to the land than to the sea for their food. Northern Honshu lies near the boundary between northern and southern Japanese fisheries, and species from both grounds are harvested. Many small Japanese fishing villages line the coast—the occupants often do part time farming.

FOREST INDUSTRIES

The forest-clad slopes of many of the mountains in the Long Summer Humid

Continental regions have stimulated lumbering industries. A few hardwood mills operate in the rougher upland areas of the Appalachians and Ozarks of North America where sound land management favors forest production. This region is, however, overshadowed by the Pacific Forests and the Southern Forests. Logging is most often a winter activity in highland borders of the Danube Basin, when the trees are easily transported over the snow. North China's bare mountain slopes afford no opportunities for lumbering and hardly enough material for fuel. Chinese housewives burn roots, straw, and the stalks of kaoliang, the giant sorghums, for cooking purposes. The mountain slopes in eastern Manchuria, unlike North China, still contain valuable forest resources of hardwoods and conifers, especially the Korean pine. Manchuria's lumbering activity will be discussed in the chapter dealing with Short Summer Humid Continental regions. Japanese lumbermen in northern Honshu follow a forestry program similar to that practiced in the southern part of the island—small-scale lumbering with the accent on complete use and careful forest management.

MINING

Minerals, particularly fuels, are major resources in several of the regions. Figure 11.35, page 247 graphically indicates the wide extent of bituminous coal in the United States region; here lie some of the greatest producing fields in the world. The great coal

FIG. 12.16. A continuous mechanical coal miner. This underground mechanical marvel reduces coal mining to one continuous operation. Cutting and loading simultaneously (eliminating drilling and blasting), the continuous miner is served only by a shuttle car that transports the mined coal directly to waiting mine cars or rubber conveyor belts. Use of this machine greatly speeds production. (Bituminous Coal Institute.)

deposits of the United States provide cargo for railroads, and are extremely significant in the foundation of the nation's industrial capacity. There is also considerable petroleum production within the United States region. Lead and zinc mines are located near the southern margin of the region in Missouri. Shifts to efficient machinery and technological innovations have steadily reduced the labor requirements in mining. The United States mining industry in 1940 employed 1,109,860 and in 1969 employed 628,000. The number in coal mining dropped from 652,000 to 136,000 during the same period.

Romania has the best mineral endowments of the European region; petroleum and coal are the most important products. Hungary ranks with France and Yugoslavia in bauxite reserves, but production has been relatively small. In Asia there are coal deposits in southern Manchuria and on the Shantung Peninsula; iron occurs near the coal in the Manchurian sector. Accessibility has led to the development of these reserves as major sources of supply.

MANUFACTURING

In a previous discussion it was established that manufacturing develops in response to certain basic requirements, including raw materials, power, labor, markets, transportation, and capital. It tends to locate at points which provide optimum access to these ingredients.

The major part of the American Manufacturing Belt lies within the boundaries of the American Long Summer Humid Continental region. A combination of favorable factors has made this belt the largest, most important, and diverse industrial area of the world. The most significant requirements of basic industry, iron ore and coking coal, are easily assembled here. Iron ore is obtained from the Great Lakes region, from Canada via the St. Lawrence Seaway, and from foreign sources through Atlantic ports. The Appalachian coal fields supply vast resources for power, and stores of petroleum are also nearby. Transportation facilities are excellent. The Great Lakes, linked by rivers and

FIG. 12.17. Coal transport on the Ohio River. Waterways allowing bulk transport lessen the cost of coal. (American Waterways Operators.)

FIG. 12.18. Unloading a Great Lakes iron ore boat. These great ore boats, averaging 600 feet in length, bring iron ore to coal. The huge unloading machine scoops 20 tons of iron ore in about three and one-half hours. The Great Lakes are the most significant waterway system in the world; they are of inestimable value to the American steel industry. (American Iron Steel Institute.)

canals, provide a 1,700-mile west-east waterway for the shipment of bulky raw materials to processing centers and markets. The St. Lawrence Seaway is a water outlet to the northeast Atlantic. Rivers of the Mississippi system are utilized for barge traffic, and the Illinois waterway provides connection with the Great Lakes (Figures 12.17 and 12.18). Numerous harbors serve the Atlantic seaboard, which is linked to the Trans-Appalachian area by the Hudson River and the New York State Barge Canal. The densest net of railways, highways, and airways in the United States provides rapid transportation. Petroleum products are brought to the region by a network of pipelines. The heaviest population of North America is concentrated within the belt, supplying skilled and unskilled labor, as well as large consuming markets.

The American Manufacturing Belt is not an area of continuous industrialization; it consists of a series of subregions (see Figure 12.19). Each industrial concentration is based on special advantages of assembly, raw material proximity, labor skills, markets, early start, and other factors. The list of commodities produced is large—only a few of the major activities will be used to characterize each subregion.[1]

The St. Louis district (A) is the most westerly subregion. Meat packing, brewing, shoe manufacture, petroleum refining, bauxite processing, and steel production are some of its higher diversified activities. The Chicago-Milwaukee district (B), bordering the south and southwest shores of Lake Michigan,[2] is outstanding for iron and steel plants, petroleum refineries, meat packers, and breweries. Clothing, agricultural implements, mining machinery, and furniture are other

[1] The following subregion classification is based upon C. L. White, E. J. Foscue, and T. L. McKnight, *Regional Geography of Anglo-America*, 3d ed. (Englewood Cliffs, N. J.: Prentice-Hall, Inc., 1965), p. 34.

[2] The northern fringes of this district are actually in the Short Summer Humid Continental region.

275

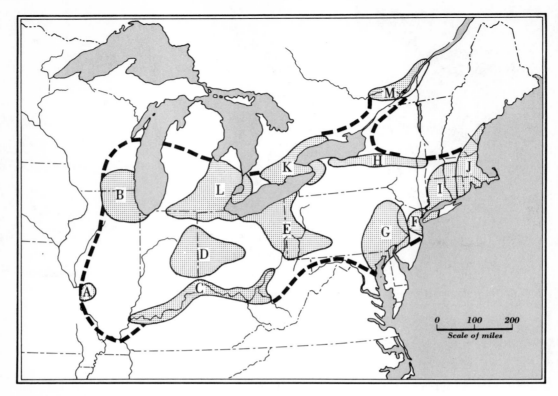

FIG. 12.19. The subregions of the American Manufacturing Belt. (A) St. Louis district; (B) Chicago-Milwaukee district: (C) Middle Ohio Valley district; (D) Inland Ohio-Indiana district; (E) Pittsburgh-Cleveland district; (F) New York Metropolitan district; (G) Philadelphia-Baltimore district; (H) Central New York district; (I) Southern New England; (J) Southeastern New England; (K) Niagara Frontier; (L) Southern Michigan Automotive district; (M) Middle St. Lawrence district. (Adapted from White, Foscue, and McKnight, *Regional Geography of Anglo-America*, 3d ed. [Englewood Cliffs, N.J.: Prentice-Hall, 1965].)

major commodities. The Middle Ohio Valley district (C) stretches 500 miles along the Ohio River. It specializes in electrical equipment, machine tools, auto parts, aircraft engines, liquors, tobacco, and household appliances. The highly diversified Inland Ohio-Indiana district (D), located between the eastern coal fields and western farmlands, concentrates on medium and light industry. It is the machine-tool center of the world and also manufactures electrical goods, cash registers, clothing, soap, and **agricultural** machinery. The Pittsburgh-Cleveland district (E) leads in heavy iron and steel and has the leading role in the rubber industry as well as producing electrical equipment, automobile parts, machine tools, and glass and clay products. The New York Metropolitan district (F) is extremely diversified, lacking mainly iron and steel production but concentrating on garments, chemicals, printing and publishing, and petroleum refining. Light and heavy industry characterize the Philadelphia-Baltimore district (G), which stresses textiles and chemicals, food processing, shipbuilding, and petroleum refining. This district is also noted for its tidewater steel plants, which utilize imported iron ore. The

Central New York district (H) stretching from Albany, New York, to Lake Ontario, contains a series of specialized industrial cities. Based on secondary raw material, products range from locomotives to cameras, electrical goods and clothing to optical equipment and chemicals. The Southwestern New England district (I) is concentrated chiefly in the Connecticut Valley. Manufacturing here is highly diversified, stressing products with high values and small bulk, including machine tools, hardware, firearms, clocks, watches, electrical goods, high-grade paper, and a host of others. The Southeastern New England district (J), paralleling the seaboard, specializes in diversified light industry including small metal goods, such as hardware, clocks and jewelry, woolen goods, cotton textiles, especially fine fabrics, and shoes. Other smaller districts can be recognized, including Omaha and Kansas City, especially notable for agricultural processing; the latter, however, also produces steel, furniture, cement, river boats, and airplanes.

Outside of the United States, the Po Valley of Italy contains the largest industrial area in these regions. Here, coal and iron are lacking; therefore heavy industry is minor and manufacturing in general stresses quality. Consumer goods dominate. Light industries are favored by hydroelectricity, skilled labor, a certain amount of raw materials, and a long tradition. Cotton, rayon, wool, silk textile milling, and clothing manufacturing form the single most important class. Food processing also is well represented. Other industries include manufacture of mechanical equipment, including automobiles, and light metal products. Chief centers are Milan and Turin.

Modern manufacturing is still in its infancy in the Danubian lands; handicraft and food processing are most typical. As the result of communist-inspired struggles to establish broader industrial bases, some development has occurred recently: for example, new iron and steel works have been established in each of the countries: in Hungary, near Budapest; in Romania, near Galati on the Danube; in Yugoslavia, one in each of

FIG. 12.20. Cuyahoga River ore dock and blast furnaces. This scene is in the environs of Cleveland, one of the major iron and steel centers of the nation. Here Great Lakes iron ore meets coal railed from the Appalachian Fields. (American Iron and Steel Institute.)

FIG. 12.21. Greater New York, one of the most densely populated areas of the world. Advantages for commerce have been the principal reasons for the growth of this urban area. Portions of the port facilities are visible in this scene. Lower Manhattan is recognized by its concentration of tall buildings. (The Port of New York Authority.)

the four major provinces; and in Bulgaria, near Sofia. National production capacities range from about 1.5 million tons in Bulgaria to nearly 5 million tons in Romania. Cement, chemical, and machinery plants are also now operating in the major centers of development.

During the period of Japanese control (1931–1945), a significant industrial area came into being in southern Manchuria, between Mukden and Dairen. Local coal, oil shale, iron, limestone, soybeans for oil, and nearness to the coast were favorable factors. The Japanese created a salient industrial structure based upon iron and steel, particularly for the production of railway equipment, with other manufactures including cement, chemicals, and soybean products. Tientsin and Tsingtao, as a result of their functions as ports and commercial centers, also have some industrial development, mainly based upon the processing of agricultural products, with cotton textiles being of particular importance. These areas have been given considerable attention by the communist leaders in recent years. It is probable that even greater attention has been given to new developments in the western part of the region which is less vulnerable.

TOURISM

The Appalachian Mountains, the Atlantic seacoast, and the Great Lakes are resort areas for the millions of people who inhabit the American region. The sandy beaches of the Atlantic have numerous summer resort centers, led by Atlantic City. Mountains are utilized for both summer and winter sports. New York City is a leading tourist attraction in the United States, and many other cities of the American Manufacturing Belt also draw thousands of visitors.

Tourism and recreational activities in the European region are closely tied to the urban centers. Historic Venice, Vienna, Milan, Budapest, and Belgrade are but a few of the colorful cities that attract local and world travelers.

URBAN CENTERS

United States. Industry has led to the development of the largest urban concentrations in the world in the American Manufacturing Belt. More than twoscore cities exceed 100,000 in size, and many more fall in the 50,000 to 100,000 class. New York, on the east coast at the terminus of the easiest route through the Appalachians, stands above all others (Figure 12.21). About eight million people live within the city limits, and that many more within the metropolitan area. It has the nation's best harbor, and is the largest port and commercial center, as well as being a leader in manufacturing and finance, and the center of styles and entertainment. Here is the core of Megalopolis, the eastern seaboard urbanized area, extending from Baltimore and Washington to Boston.

Chicago, second city of the United States, has also developed as the result of the combined benefits of commerce and industry. Its strategic situation at the south end of Lake Michigan assured the focus of land transportation routes. Philadelphia is the third city of the region. It is a great seaport and a major manufacturing center. Baltimore, Cleveland, Kansas City, St. Louis, Washington, Boston, Pittsburgh, and Cincinnati all are in the one- to two-million class, and a dozen more cities have populations exceeding 250,000. (Baltimore and Washington are close to the boundary between the Long Summer Humid Continental Region and the Humid Subtropic Region.) When metropolitan areas are included, all are found to be considerably larger, much of the recent growth having occurred in suburban margins. Manufacturing is a principal activity in all these areas except Washington, whose domi-

FIG. 12.22. Venice, city of canals. Venice has no automobile traffic. Waterways are the avenue of both commodity and pedestrian movement. Dating from the 5th century, Venice, for nearly thirteen hundred years, was the powerful capital of a rich trading area—the focal point of trade with the Eastern Mediterranean. In the more recent period, its situation on the outer margin of the Po Plain and its site on poorly drained coastal lagoons have resulted in a relative loss in trade advantages to other Italian cities such as Milan, Genoa, etc., which are in more favorable positions. Today, Venice is a notable tourist attraction and also maintains specialty industries such as glass manufacturing. (Italian State Tourist Office.)

FIG. 12.23. Budapest, Hungary. The Danube divides the city. On the west, the direction of this view, are the administrative, cultural, and intellectual institutions of the city; industrial and commercial activities are centered on the east. Budapest is a leading European flour-milling center. The city is the hub of a transportation net that provides good access to the farming lands of the Danube Basin. (Legation of the Hungarian People's Republic.)

nant functions are associated with the federal government.

Europe. The Po Plain is the outstanding urbanized area of the European region. At least one dozen cities exceed the 100,000 population mark. Milan is the metropolis of the plain and the largest city of Italy with about 3.2 million people. Its situation with respect to transportation routes has promoted a thriving commerce, and in recent years Milan has become that country's leading industrial center. Turin, on the slopes of the Piedmont at the west end of the plain, is also an important industrial and commercial center. Venice, at the head of the Adriatic Sea, is historically the most famous of the Po Plain cities, having reached its greatest heights as a commercial center during the Middle Ages, when trade between Europe and the Orient flowed through the Mediterranean Sea. Today its fame is based upon its picturesque canals, churches, palaces, and the romance of its historic past; it has become one of the leading tourist attractions in Europe (Figure 12.22).

Three large cities dominate the Danube countries—Vienna, Austria, with a population of nearly 2 million, Budapest, Hungary, around 2 million (Figure 12.23), and Bucharest, Romania, 1.6 million, Belgrade, major city of Yugoslavia, with about 750,000 people, is also in the basin. Each tends to dominate urban life within its respective country.

Asia. The Asian Long Summer Humid Continental region has many villages but few cities. Tientsin and Peking are the leading population nodes—the former has about 4 million and the latter, nearly 8 million. Tientsin is the major commercial and industrial center of the North China Plain; Peking owes its growth to its services as the capital of the Chinese Empire and the present communist government. Mukden, with a population of over 2.4 million, is the chief city of Manchuria. Its location on the main railway at the junction of lines to Peking and Korea has favored its growth. Tsingtao, on the south side of the Shantung Peninsula, has an excellent harbor and a population of over 1 million. Dairen, at the southeast tip of the Kwangtung Peninsula, has a fine harbor and an important function as a port.

thirteen

SHORT SUMMER HUMID CONTINENTALS

The Short Summer Humid Continental regions are the poleward frontiers of relatively continuous settlement; beyond lie the winter-dominated lands. In many ways these regions resemble their Long Summer neighbors. Climate differs only in intensity—summers are shorter and cooler, winters are longer and colder. Agriculture remains the dominant form of land use, and grains, the leading crops; corn for grain disappears and spring wheat replaces winter wheat. Hardier cereals such as rye, barley, and oats, and root crops, especially potatoes, assume important roles. Conditions favor dairying; the North American region contains America's Dairy Belt. Industry is also important; the northern margin of the American Manufacturing Belt, including the heart of Canadian industrialization, is in the North American Short Summer Humid Continental region, and there are a number of important manufacturing districts in Eurasian regions. In addition, two world fisheries are off the eastern coasts of North America and Asia.

Location

The Short Summer Humid Continental regions, found only in the Northern Hemisphere, lie poleward of the Long Summer regions (see Figure 13.1). The North American region extends from the Atlantic Ocean through northern New England and the Maritime Provinces of Canada westward through the Great Lakes to the vicinity of the 100th meridian in the United States; but it reaches the 120th meridian in Alberta, Canada, as the result of progressively lower heat and a lower evaporation rate, which allows an increasingly higher precipitation efficiency (Figure 13.2). The Eurasian region stretches from south Sweden and central Germany through Poland to include a major share of the Russian lowland between latitudes 50° and 60°. Beyond the Urals, a narrow belt straddles latitude 55° through the West Siberian Lowlands and terminates near the northern range of the Altai Mountains. Much of the USSR region is bordered by steppe on the south (Figure 13.3). The third region is in east Asia, occupying

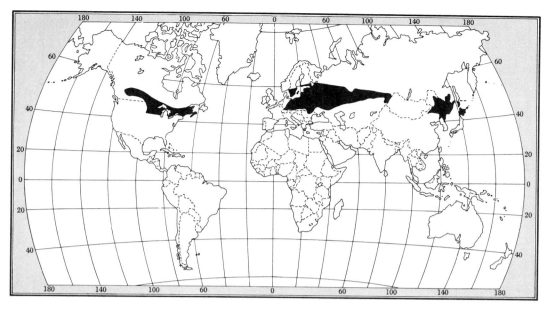

FIG. 13.1. The Short Summer Humid Continentals.

FIG. 13.2. (lower top) The Short Summer Humid Continentals of North America.

FIG. 13.3. (bottom) The Short Summer Humid Continentals of western Eurasia.

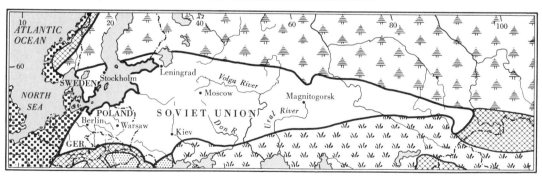

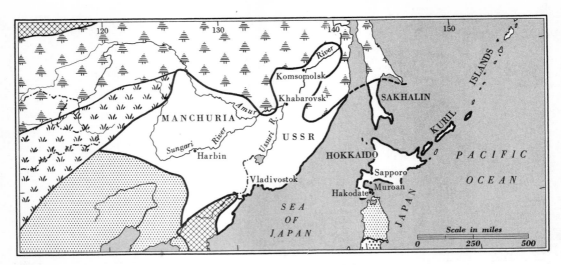

FIG. 13.4. The Short Summer Humid Continentals of eastern Asia.

northern Manchuria, northeast Korea, the neighboring portion of the Soviet Union, and the entire Japanese island of Hokkaido (Figure 13.4).

Physical Environment

CLIMATE

Location in the higher latitudes and greater continentality account for climatic differences that distinguish these regions from the Long Summer regions. Cold winters, short warm summers, large annual ranges of temperature, moderate precipitation, and lower vegetative growth potentials characterize the Short Summer Humid Continentals.

Temperatures. Winters are long and severe, with January averages ranging as much as 30 degrees lower than those in the Long Summer Humid Continental regions. Daily temperature fluctuations of 30 to 40 degrees are common. Winters also differ from year to year. Summers are short, but a few months are warm and the daytime maxima frequently reach 90° and above when skies are clear. July averages are generally only 5 to 10 degrees cooler than the Long Summer Humid Continental regions. Although summers are short, the 16 to 18 hours of daylight help to compensate for the brevity of the season; nevertheless, plant growth potentials are less than in the regions to the south. Annual ranges are high. Winnipeg, Canada, for example, has a yearly average range of about 70° with an average monthly low of −3° and a monthly average high of 67° (Figure 13.5). Frost-free seasons are short, ranging from 90 to 150 days.

Precipitation. Summer is the season of maximum precipitation. Yearly totals vary from 20 to 40 inches in typical places, but northern interior areas often receive less than 20 inches. Winter precipitation is chiefly in the form of snow, and 60 to 80 days with snowfall are common. The total fall may amount to as much as 50 inches in the more humid portions. Cold weather maintains snow cover for four months or more.

Strong winds accompanied by zero weather, called blizzards in North America and *burans* in the USSR, occasionally occur in the interior portions of the regions. Powerful winds whirl snow particles high into the air, and visibility is then practically zero. The combination of freezing temperatures, wind, and blinding snow is extremely hazardous to man and livestock.

Location and local differences. The oceanic influence on the leeward sides of continents is generally not strong enough to cause the development of a definite climatic type; however, there are some modifying effects of the sea on the eastern margins. Summers are cooler and winters warmer than in the interior; precipitation tends to be more evenly distributed.[1] The St. Lawrence estuary in the North American region carries maritime influences some distance inland, and the Great Lakes also produce modifications.

Cool currents—the Oyashio off Asia and the Labrador off North America—have a cooling effect when northeast winds blow onto the continent. These, however, are not the prevailing winds. Thick set fogs occur when moist warm air comes in contact with the chilled air above the cold currents; the neighboring littorals are frequently shrouded in fog, making coastwise navigation dangerous.

SURFACE FEATURES

North America. The major portion of the North American region is contained in the Interior Lowland Physiographic Province, a vast expanse of relatively smooth land. In the west this lowland merges into the Great Plains, which are somewhat higher but of monotonously slight relief. A fringe of the Laurentian Uplands, an area characterized by gentle hills with a few low, rounded

[1] These differences are sometimes used as the criteria for the recognition of a modified or New England type of climate on the eastern peripheries of North America and Asia in these latitudes.

FIG. 13.5. Typical climatic graphs of Short Summer Humid Continental stations.

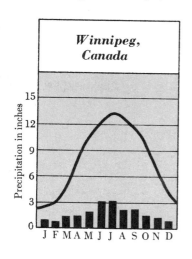
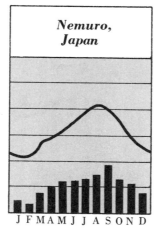
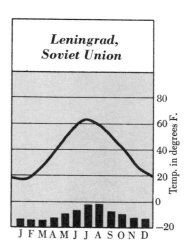

mountains and literally thousands of lakes and extensive swamps, is included in the north (Figure 13.6). In the east there is a predominance of hills and low mountains known as the Northeastern Uplands. These subdued uplands long have been exposed to the forces of weathering and erosion, and in general summits range between 1,500 and 3,000 feet. Only two areas of low relief are noteworthy: the Aroostook Valley of Maine (Figure 13.7) and the Annapolis-Cornwallis Valley of Nova Scotia. There are, however, numerous small stream valleys. Coastal plains are usually lacking, with slopes meeting the sea in an irregular coastline.

Most of the region was affected by continental glaciation during the Ice Age, with extensive modifications in topography, soils, and drainage. Through much of the Laurentian Upland and the Northeastern Uplands the results were adverse—in many areas the ice sheet was an eroding agent, leaving bare rock surfaces exposed. In others deposition was the principal activity. In some cases the load of material being carried by the ice sheet was dropped as a *ground moraine* consisting of various sized stones intermingled with clay, sand, and gravel, collectively known as *glacial drift*. At places where the margins of the ice were more or less stationary, with a balance between advance and melt, ridges of glacial drift accumulated as *marginal moraines*. Much of the glacial load was carried beyond the margins of the glacier by streams of meltwaters. Like all streams, these sorted their material according to weight, depositing the heaviest first and the lightest last. In some cases these glaciofluvial deposits were arranged in floodplain form in existing valleys and are known as *valley trains;* in

FIG. 13.6. Landscape in the Laurentian uplands. The streams of this region have a great power potential. This view shows the Chelsea Power plants on the Gatineau River, Quebec. (Royal Canadian Air Force.)

FIG. 13.7. The Aroostook Valley, Maine. The village is Limestone, Maine, an important shipping point for potatoes. Cool summers and a short frost-free season limit the range of crops in this area; however, both climate and soils are well adapted to potatoes, the principal cash crop. Forage production has been increasing, resulting in gains in dairying. (United States Department of Agriculture, Soil Conservation Service.)

others they were spread in fan-like deposits about the margins of the ice to form *outwash plains*.

Other glacial features include eskers, kames, kettles, and drumlins (see Figure 13.8). *Eskers* are long, narrow ridges, commonly sinuous, composed of stratified drift. They range in height from a few feet to 50 or more, in length from a fraction of a mile to over 100 if gaps are included, and in breadth from 20–40 to 300–500 feet. *Kames* are low, steep-sided hills with a knoll-like form, consisting of stratified drift. *Kettles* are small depressions that occur in drift, usually stratified. Few exceed a mile in greatest diameter, and most are no more than 20 to 30 feet deep. *Drumlins* are low elliptical hills of glacial drift, streamlined in form. They average nearly a mile in length, 1,200 to 1,800 feet in width, and 60 to 100 feet in height. Usually drumlins occur in groups or "fields" running into the hundreds.

Glaciation in the Interior Lowlands portion of the region was more beneficial, with resulting conditions similar to those described for the Corn Belt. Several subdivisions, however, should be recognized. The Great Lakes district is mantled with recent glacial deposits

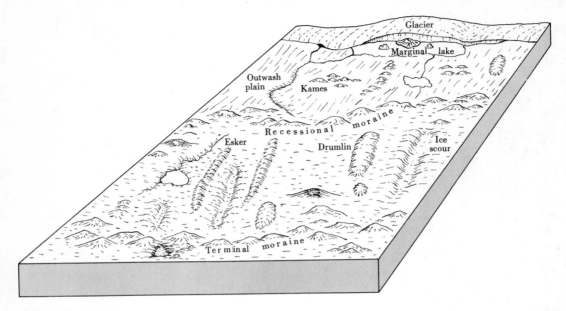

FIG. 13.8. Glacial features.

in which there are thousands of lakes, ponds, and swamps. Bordering the Great Lakes, there are *lacustine plains* that have resulted from deposition of sediments into water and their eventual exposure by shrinking of the lakes. The Driftless Area, lying largely in southwestern Wisconsin, is particularly distinctive since it was not covered with drift.

The northwestern portion of the lowlands is also mantled with recent glacial deposits and has numerous lakes and ponds. There were many more lakes and ponds immediately after the shrinkage of the continental ice sheet than now; some drained naturally and others were drained by man and converted to farmland. The most famous, former glacial Lake Agassiz, is evidenced by a large lacustrine plain and the present-day Lake Winnipeg. The tremendous prehistoric lake was formed during the waning stages of the Ice Age; as the ice retreated poleward, northward-flowing meltwater was blocked and covered the extensive and fertile plain that is today drained by the Red River of the North. When the ice retreated into Hudson Bay, the present-day drainage by way of the Nelson River was established.

Europe and the USSR. The European region is part of the North European Plain. In the entire expanse there are few elevations that rise to 1,000 feet. The most notable exceptions are the Central Russian Uplands west and south of Moscow, which in general rise in low hills of 600 to 1,000 feet elevation and do not break the continuity of the plain except where steep slopes face the west bank of rivers. The Ural Mountains are the only real break in the topography and, although they separate the western plains from the flat Siberian Plain, are not a significant barrier.

Like the North American region, most of the plain was affected by continental glaciation, which disrupted drainage and covered the surface with glacial and glaciofluvial deposits. Three north-south zones are

SHORT SUMMER HUMID CONTINENTALS

apparent: the *ground moraine*, which dominates in southern Scandinavia; the Baltic *terminal moraine*, or *end moraine*, which attains elevations up to 900 feet, forming a long zone of sandy hills paralleling the Baltic Sea from the Jutland Peninsula of Denmark into the Soviet Union; and a great *outwash plain*, built by the meltwaters pouring from the Baltic Moraine, extending across Germany and Poland into the Soviet Union (Figure 13.9). Much of the outwash plain is poorly drained, with large swampy areas and soils that are frequently sandy. In contrast, the southern fringe of the North European Plain is mantled with fertile loess, presumably brought by winds that blew across the material deposited by the glacial meltwaters.

East Asia. The east Asian region is dominated by uplands. On the mainland notable lowlands do occur in the Sungari Plain of north Manchuria and the adjacent Ussuri Valley and Middle Amur Lowland of the USSR, but elsewhere mountains often descend to the sea with little or no coastal plain. There are three sizable lowlands in Hokkaido, of which the Ishikari Plain, extending north-south through the west-central part of the island, is the most significant. The others are the Tokachi Plain on the southeast coast and the Nemuro Plain on the east coast.

FIG. 13.9. North Germany's "mountains" are by the sea and with their lovely chain of lakes by Plön (shown here), they form between the Baltic ports of Kiel and Lübeck (German Federal Republic) the so-called "Holstein Switzerland." Only a few miles away is the "Iron Curtain," the zonal frontier which divides Germany from Germany. On the other side of this frontier, the landscape merges into the Mecklenburg Lake Plateau, which together with the Schleswig-Holstein sea coast was a popular resort for holidaymakers before Germany was divided. (German Information Center.)

NATURAL VEGETATION

Forests are the natural cover in the wetter portions of the North American region. Pine, spruce, hemlock, and fir dominate the evergreen species, and a mixture of birch, aspen, beech, poplar, maple, and other broadleaf deciduous trees occur on the warmer southern margins. The foliage of the hardwoods during the fall season adds brilliant splashes of color to the landscapes. The forests have been cut in many of the areas, especially around the Great Lakes and in the Northeastern Uplands. Second growth and rotting stumps now cover thousands of acres of forest land that at one time appeared inexhaustible and as impenetrable as the forests of the humid tropics. Grasses gradually supplant the forest in the drier western interior.

Forests constitute the typical cover of the western portion of the Eurasian region, where the land is either too rough or too infertile for agriculture. Conifers are most common in the higher elevations but are also scattered throughout the sandy areas of the glaciated plains. Heath and moor are usual in the poorly drained lands. A belt of mixed forest stretches into the USSR, forming a distinctive vegetative zone between the colder coniferous forest lands in the north and the drier grasslands in the south. Marshes occur along the flat east banks of Siberian rivers that are flooded during the spring thaw.

The mountains in the Far Eastern section of the Asiatic region are covered with stands of conifers made up of spruce, fir, larch, and Korean pine. Broadleafs grow at the lower elevations. Broadleaf forests compose the vegetative mantle of southern Hokkaido; there is a gradation into coniferous forests northward.

FIG. 13.10. A stand of 95-year-old conifers in the Chippewa National Forest of Minnesota. The trees shown in this view are red pine. (United States Forest Service.)

SOILS

Soils show transitional characteristics paralleling the change in climate and vegetation. Gray-brown soils like those of the more humid phase of the Long Summer Continental region prevail in the warmer mixed-forest margins. Poleward with lowering temperatures and a change to coniferous forest, *podzol* soils occur. This group, as its Russian name indicates, is ash-gray in color, acid, and low in soluble salts and organic matter. The forest cover takes little and returns little to the soil, but the surface drainage water continually carries away nutrients. More fertile soils are found in the interior; paralleling the lowering precipitation and the transition to grass vegetation there is a gradation from gray-brown or podzol soils through prairie soils to chernozems.

Effects of glaciation have in places interfered with the natural soil-building processes. Poor drainage, frequently indicated by swamps and marshes, has restricted normal development in a number of areas. Excessive drainage occurs in stony ground moraine and coarse, sandy glaciofluvial deposits—particularly in the European region. Adverse effects also result from scouring, which has produced bare rock surfaces in some northern, hardrock areas. On the other hand, the lacustrine deposits, such as those of old glacial Lake Agassiz and the borders of the Great Lakes, are excellent agricultural soils. The wind-deposited loess also is usually high-quality soil.

FIG. 13.11. Balsam fir stand with an overstory of aspen. Many acres of this type are found in northern Minnesota. (United States Forest Service.)

Occupance in the Short Summer Humid Continentals

The Short Summer Humid Continental regions mark the northern limits of the world's major zones of relatively dense and continuous population. Within each region, population density tends to thin out both poleward and toward the drier interior. The nature of the topography also affects distribution, with rough upland sections discouraging settlement. The greatest concentrations are found along the southern margins of the North European Plain, where fertile soil and mineral deposits have favored agriculture and industry, and in the eastern portion of the North American region in part of the manufacturing belt, including many cities and the industrial heart of Canada. Interestingly, however, there is a distinct contrast between the northern portion of the east coast urban belt and rural New England. Rather pioneer landscapes are found within a few miles of the major urban center margins; poor, shallow soils and sloping surfaces are restrictive to crop farming and second growth forest and woodland prevail. The lowlands of northern Manchuria and the Japanese island, Hokkaido, are the most densely populated purely agricultural areas. Unlike the regions to the south, the Manchurian sector and Hokkaido are the oriental frontier lands and are not as crowded as the regions to the south.

The agricultural possibilities are not as promising here as in the Humid Subtropic and Long Summer regions, but when fish, minerals, and forests are added to the base, the foundations for supporting comparatively large numbers of people are strong in many portions.

AGRICULTURE

Farming is the most widespread activity. Dairying and small-grain and root-crop production comprise the major enterprises best suited to the climatic and soil conditions. The

FIG. 13.12. A farmstead in the Prairie Provinces of Canada surrounded by a sea of wheat. (National Film Board of Canada.)

emphasis varies considerably in response to differences in environment and culture.

North America. Two distinct agricultural regions are found in the North American area. The Spring Wheat Region occupies the western, lower-rainfall portion, and the Hay and Dairy Region, the more moist eastern portion. In the Spring Wheat Region commercial agriculture prevails, and the bulk of the production is consumed elsewhere. Farms are fully mechanized and are commonly 400 to several thousand acres or more in size. Rural population is sparse and dispersed. Long, cold winters, with a relatively slight snow cover to offer protection to a crop, require that wheat be planted in the spring rather than the fall. Wheat overshadows all other crops and has first choice of the land; barley, hay, and flax for seed are the common rotation crops. The growing season is too short and the nights too cool to mature corn for grain; however, some corn is grown for silage use on the southeastern margin. On the eastern margin the Red River Valley has become an area of more general farming, with some dairying, potatoes, and sugar beets as well as wheat and other grains.

To the east, cooler, moister climate, poorer soil, and rougher surface restrict profitable grain production but favor forage crops and pasture. The large concentration of urban centers provides markets for milk products. These factors have induced a large portion of the farmers to become dairymen; the area is known as America's Dairy Belt. Farming here is an intensive, year-round activity. Farm units are smaller, usually 160 to 320 acres, and are devoted to raising the feed and pasture for an average of 30 to 60 dairy cows. The number of dairy farms, as farms in general in the United States, is decreasing whereas the number of cows per unit is increasing and so is the milk production efficiency per animal. In 1950 the nation's dairy cows numbering 23.8 million produced 116.6 billion pounds of milk; by 1969 the number had been reduced to 14.1 million but the production was essentially the same, 116.2 billion pounds.

The typical dairy farm west of Lake Erie has fields of hay, oats, and corn, a silo in which the corn is preserved as silage, and a large barn to house the stock and store winter hay. New England dairymen tend to have a higher portion of land in pasture than those in the Lake States.

The nature of the salable dairy products depends upon location with respect to markets and cost of transportation. Near cities, fluid milk is sold, but manufactured products such as butter, cheese, and condensed milk are more important in the areas distant from cities. Pricing systems and bulk fluid milk transport, however, have altered this traditional pattern to some extent.

Within the dairy belt, city markets of the manufacturing belt have encouraged some vegetable and fruit production in a number of favored areas bordering the Great Lakes and in New England. Aroostook County, Maine, is one of the major potato-producing areas of the United States; the Cornwallis-Annapolis Valley of Nova Scotia accounts for

FIG. 13.13. Wheat elevators in Saskatchewan. Such grain storage facilities are common sight throughout the wheat lands. (National Film Board of Canada.)

FIG. 13.14. Dairy farms in Wisconsin. (Wisconsin Department of Agriculture.)

FIG. 13.15. A farming community on the north shore of the St. Lawrence River. Cool climate, adverse slopes, and podzol soil restrict farming opportunities and make dairying one of the most profitable land uses. (National Film board of Canada.)

FIG. 13.16a and b. Harvest time in the south of Sweden. The typical Swedish farm contains 25 to 50 acres. (Oliver H. Heintzelman.)

an important share of Canada's commercial apple crop; and the western border of peninsular Michigan is a noteworthy fruit district important for cherries, apples, and peaches. The Minnesota River Valley is representative of a number of sites where cool and mild season vegetables are produced for processing; Le Sueur is the home of Jolly Green Giant. The region also accounts for much of the cranberry and blueberry production of the United States.

Europe and Western USSR. The mixed-crop livestock region of the Northern European Plain ranks among the great agriculture regions of the world. It differs significantly from the Corn Belt in crop and animal associations, scale of farms, organization, and purposes as the result of differences in land base, culture history, numbers of farmers, and political framework. Over much of the region the summer climate is not sufficiently warm for corn to mature as grain; over much, limitations of soils and cool, moist summers favor rye over wheat, and these same conditions, plus possibilities of relatively high yields, favor widespread production of food and feed root crops. Potatoes and sugar beets are outstanding in the land use pattern. Spring wheat is important in the southern margin of European Russia, as well as in croplands newly opened in the 1950s east of the Urals (these are mainly in the Dry Continental Region). Although the animal enterprises vary somewhat, the general average is an emphasis on dual-purpose cattle; these provide home milk supplies and meat for sale. Except in the Baltic states of the USSR where hogs are of commercial importance, these animals are primarily reared for home needs.

Traditionally, farms have been small through the entire region and, in the main, privately owned and operated. In the communist areas, however, reorganization has been under way with resulting increase in unit size and state control. Other differences from the Corn Belt include greater dependence upon hand labor and draft power, fragmentation in some areas, and the greater attention given to producing a significant share of home needs—thus greater emphasis on household gardens, and so on. Some portions of the region are well known for specialties. These are exemplified by the concentration of market gardening around many of the cities.

Eastern USSR and Asia. In northern Manchuria, spring wheat and soybeans are the main crops. Farms and fields are somewhat larger than those of the North China Plain. Yields per person are also higher, and a surplus is available. Development in the Amur and Ussuri Valleys has been recent and includes both agriculture and other activities. Since the early stages of World War II strategic reasons have prompted the Soviet Union to concentrate major effort on this Far Eastern appendage over 5,000 miles from the heart of the nation. Only one railroad, the Trans-Siberian, provides land contact with the west. Large-scale, mechanized state farms have been established for the production of wheat, rye, oats, barley, and sugar beets. Some of the poorly drained, low-lying areas are devoted to hay crops and livestock rearing, including herds both of dairy and beef cattle.

The agriculture of Hokkaido differs markedly from that of Humid Subtropical Japan. The former presents a pioneer landscape—farms average 10 or 12 acres in size and are usually in single, rectangular plots. Many of the crops of the south are not grown; instead more adaptable species are planted. Beans have the greatest distribution and rank first. White potatoes and sugar beets are prominent cash crops. Rice, specially developed for more restricting climatic conditions, is the most widespread grain, but it does not dominate land utilization as is the case elsewhere in Japan. A considerable acreage is given over to forage crops for dairy cows and for horses, which are used for draft purposes. Hokkaido has the most

FIG. 13.17. Autumn plowing on a Polish state farm. (Polish Embassy.)

important dairy industry of Japan. During the early stages of settlement, the Japanese government brought agricultural experts from New England to Hokkaido to help plan the development of the island. The influence is noted in the rectangular survey patterns and in the New England style of some of the houses, barns, and silos, as well as the crops. Despite early planning, many elements of old Japan are present in the form of conventional southern house types, crops, and methods. The blending that results has given Hokkaido a cultural landscape distinctive in the Orient.

FISHING

One of the major fishing grounds of the world is in the coastal waters off northeast North America. Physical conditions here are extremely favorable for the growth of an abundance of *plankton,* minute plant and animal organisms, the basis of fish food. A series of submarine elevations called banks are located on the wide, shallow continental shelf that extends from southern New England to Newfoundland. Three banks are most important: the Georges Bank east of Cape Cod, the Sable Island Bank southeast of Nova Scotia, and the largest, the Grand Banks, southeast of Newfoundland (Figure 13.18). The shallow water of the banks, 300 feet or less, allows an abundance of light to penetrate to the ocean floor, and this stimulates the growth of plankton. The depths are aerated and nutrients brought to the surface by contrasting currents—the cold Labrador Current and the warm water from the Gulf Stream—and by tides and general storminess which renew the supply of oxygen necessary for life. The indented coastline provides many excellent harbors for the numerous fishing vessels, and the smooth bank bottoms favor the use of otter trawls or dragnets for capturing *demersal*, or bottom-feeding, varieties. Cod, haddock, and flounder are among the important commercial species, taken chiefly by use of otter trawls. Large quantities of herring and mackerel constitute the catch of *pelagic* species, or surface feeders. In addition this area has a valuable shellfish industry based on lobsters, clams, oysters, and scallops. The Great Lakes contain the most important United States and Canadian freshwater fisheries. Blue pike and lake herring make

FIG. 13.18. Fishing operation of the Grand Banks off Newfoundland. The *purse seine* in use here is long net with a draw rope at the bottom. In use, one end is held firm, and the other is pulled by a small boat to encircle a school of fish; the bottom is then closed and the net pulled aboard the fishing boat. (National Film Board of Canada.)

FIG. 13.19. Ladder fishing. When fishermen from the tiny village of Minudie, Nova Scotia, clear their nets, they drive a team of horses and a wagon three miles from shore. They take a ladder along to gather in the fish. The tremendous tides of the Cumberland Basin, which have a range of 22 to 28 feet, make this operation possible. At times the shad, caught by their gills, stick stubbornly. (Imperial Oil)

FIG. 13.20. Finnish fisherman marketing brined herring in Helsinki. (Finnish National Travel Office.)

up the largest portion of the commercial catch, but other species, particularly whitefish, perch, and carp, are also taken. The fisheries of the Great Lakes are now severely threatened by pollution from littoral urban centers and industries and from streams that drain into them. Many of the Canadian lakes, such as Winnipeg, Winnipegosis, and Manitoba, have important winter commercial fisheries.

The Baltic Sea fisheries are so overshadowed by the rich fishing grounds of the North Sea that their importance is often minimized; however, the presence of cod, mackerel, and especially small herring, as well as oyster beds, along the west coast have stimulated a fishing economy among the Baltic shore countries. The highly indented coastline of Sweden and Finland furnishes numerous harbors for fishing craft. Finnish fishermen are particularly active in the Baltic and the Gulf of Finland and also make important catches in their numerous streams and lakes. The major Finnish fishing grounds are in the coastal waters lying between Helsinki and Hangö (Figure 13.20).

The shallow water over the continental shelf contiguous to eastern Siberia and Hokkaido is another major world fishing ground. Conditions favoring a fishing industry are similar to those found off northeast North America. Species include herring, cod, mackerel, and salmon; crab is the most important shellfish. Approximately one-fifth of the total Japanese catch comes from waters near Hokkaido. The Soviet Union has given major attention to the development of fisheries on her Far East border during the last three decades. The fishing industry is organized very much like its counterpart, the agricultural industry.

FOREST INDUSTRIES

Lumbering in the United States had its beginning during the colonial era along the northeastern seaboard, especially in the present state of Maine. After the peak of lumber production was reached here, the loggers moved westward to the forests of the Great Lakes states. By the early decades of the present century, the saw timber was cut. Today, harvests are from stands of second growth, which are under improving management for sustained yield. Even so, the

cool, relatively short frostless season along with the relatively poor soils of these forest lands results in comparatively slow growth. The emphasis is on pulpwood. Lumbering in the southern margins and more accessible areas in the Canadian portion of the region has had a similar history, and the best saw timber has been cut. The smaller trees and the forests avoided earlier now furnish raw material for pulp mills. Logging in both countries is chiefly a winter season activity—the snow cover and frozen ground facilitate log transportation. The lakes and rivers are utilized for moving logs to the mills in the spring and summer (Figure 13.21). Pulp and paper mills, situated on water sites, are scattered throughout the Great Lakes area and the northern New England states. Canadian pulp and paper mills are found from the lower St. Lawrence River westward to the province of Manitoba (Figure 13.22).

The tapping of sugar maple trees for the manufacture of maple sugar and syrup is unique in the North American region. The province of Quebec accounts for most of the Canadian supply, and New York and Vermont produce two-thirds of the United States' total. And the demand of city-dwellers for Christmas trees has stimulated a small but profitable forest industry.

There is little contrast between the components of the forest industries in the Eurasian and North American regions; however, the forests of the former are more

FIG. 13.21. (top) Floating newsprint. This log harvest is floating down the Gatineau River to the mills, where it will be processed into newsprint. The importance of the many waterways in this area, for both transportation and storage, is clearly shown here. (Malak, Ottawa.)

FIG. 13.22. (bottom) Canadian Paper Company, Cornwall, Ontario. Water for transport, hydroenergy, and the production processes are attractions at this site. (Malak, Ottawa.)

carefully managed; the waste is nil, with even the smallest branches gleaned for fuel. Trees are felled mainly during the winter, natural waterways are significant in log movement, and pulp and paper mills utilize much of the softwood cut, especially in Sweden, Finland, and the western USSR. The emphasis on pulp and paper is not as strong in Eurasia as in North America, since the forest must also furnish logs for lumber and large quantities of wood for winter fuel.

MINING

North America. Some of the most significant iron ore deposits in the world occur in the Lake Superior District, which includes parts of northern Minnesota, Wisconsin, Michigan, and southern Ontario. Eight distinct ore bodies, known as iron ranges, are present. The Mesabi Range of Minnesota is outstanding—it has been the most productive iron ore deposit yet known. Mining here has been favored by the richness and size of the deposit, its accessibility, and the ease of working. The overburden of glacial drift is easily removed. The ore, mined with power shovels, is loaded directly into railway cars for haulage down to the lake ports of Duluth and Superior (Figure 13.23). In the other ranges a portion of the ore is mined from underground seams. This district has a fortunate relationship to the American manufacturing belt, with the Great Lakes providing the highway for the low-cost transport of bulky iron ore to the margins of the coal fields in huge, specially designed ore boats. The major portion of the steel industry of the United States depends upon the Lake Superior district for its ore. The rich deposits, drawn upon for more than half a century, are being depleted, but there are enormous reserves of lower-grade ores that will be used increasingly in the future. Processes have been developed for the utilization of taconite, a siliceous iron-bearing material containing 25 to 35 percent iron—thus making available reserves comparable to the alltime output of the Mesabi Range.

Tremendous lignite, or brown coal, and sub-bituminous reserves are found in the Prairie Provinces of Canada and the western edge of the region in the United States. A huge petroleum field in Alberta has major developments centered in the Edmonton-Leduc area. Highest quality Canadian coals are mined in Nova Scotia. The world's largest nickel deposit and production occur in the Sudbury area of Ontario; platinum and copper are also mined there. Large deposits of asbestos are worked near Quebec. Building stones have long constituted a noteworthy product of New England. The abun-

FIG. 13.23. Open-pit iron mining in the Lake Superior District. The huge equipment in use from mines to mills, the great reserves of iron, the ease of mining, the downhill rail grade to lake ports, the ore boats, and the loading and unloading equipment are responsible in a large measure for America's industrial strength. (American Iron and Steel Institute.)

FIG. 13.24. Rock-crusher north of Lake Ontario. Throughout the glaciated areas, road building and concrete construction are facilitated by an abundance of rock. (International Harvester Company of Canada.)

dance of high-quality granite, marble, and slate close to large urban centers has made this area a leader in the stone industry. Glacial sands and gravels are abundant for construction (Figure 13.24).

Eurasia. Upper Silesia is an important coal-producing area in Central Europe and also contains small iron deposits, giving rise to a significant industrial development. Large deposits of sub-bituminous coal and lignite occur in the Moscow Basin, and there is also lignite near Leningrad. One of the large iron ore districts of the Soviet Union is found in the Southern Urals. The greatest coal reserves of the USSR occur near the Tom River, a tributary of the Ob, in an area known as the Kuznetz Basin. An oil-producing field is located between the Volga River and the western flanks of the Urals, especially on either side of the Kama River; the large reserves there have prompted the name "Second Baku," after the great oil field near Baku on the western shores of the Caspian Sea. In the Far East, one of the two largest coal deposits of Japan is located on the Ishikari Plain of Hokkaido. One of the principal iron deposits of Korea occurs near the northeast border.

MANUFACTURING

Several subdivisions of the American manufacturing belt are located in the North American region (see Figure 12.20). The Niagara Frontier district in western New York and adjacent Ontario contains the industrial centers of Hamilton, Toronto, Buffalo,[2] and Niagara. Iron and steel, flour milling, and electrochemical and metallurgical industries are dominant. The Southern Michigan Automotive district is centered at Detroit-Windsor, with a ring of tributary cities in the lower Michigan peninsula and northern Indiana and Ohio. About 90 percent of the automobiles and automobile parts of North America are produced in this area. The Middle St. Lawrence district, with its manufacturing cities of Ottawa, Montreal, and Three Rivers, is the eastern Canadian manufacturing center. An abundance of water power and timber has stimulated aluminum production and pulp and paper industries—Three Rivers is one of the world's largest paper-manufacturing centers. Other manufactures include flour, iron and steel, clothing, and textiles. In addition to the subregions of the manufacturing belt, there are numerous centers that specialize in processing such raw materials from their hinterlands as wheat, fish, wood, and dairy products.

In response to coal deposits, as well as other factors favoring industrial location, several significant manufacturing districts are

---

[2] Buffalo is in the previous region, but it is mentioned here because of its association with this industrial subregion.

distributed throughout the European and Asian regions. Sweden is the industrial leader among the Scandinavian countries. Although the nation is rich in iron ore, heavy industry is not particularly characteristic, since coal is lacking. Products are diverse, ranging from expensive glassware to automobiles and the fine steel cutlery, bearings, and other articles manufactured at Eskilstuna and nearby centers; billions of matches are produced at Jönköping. Germany has a number of large and small industrial concentrations whose products are infinite. Prior to World War II, Berlin was the largest manufacturing city in Germany. Saxony, with centers at Leipzig, Zwickau, Chemnitz, and Dresden, is a significant central German industrial area. Chemicals, textiles, optical goods, and porcelains are the characteristic specialties. Bavaria has a number of specialized manufacturing centers such as Nürnberg and Stuttgart. The Silesian Coal Basin is the industrial heart of Poland; in addition to one of the great European coal reserves, iron and other metal resources are nearby. Manufacturing accents are on iron and steel, but chemicals and textiles are significant. The adjacent section of Czechoslovakia, together with the Bohemian Basin, contains the concentrations of that country's industry. Ostrava is the leading iron and steel center; the Skoda munition works are at Pilsen. Foods, beer, shoes, and fine glass are notable Czechoslovakian products.

The USSR has its share in the industrial developments of this region. Leningrad is a focus of shipbuilding, machinery, wood products, chemicals, and electrical equipment. The Central Industrial District around Moscow is one of the great Russian industrialized areas. Products range from textiles through autos, agricultural machinery, precision equipment, and a score of others. A dozen major industrial cities are situated in the Urals, dominated by the steel and machine centers at Magnitogorsk, Sverdlovsk, and Chelyabinsk. Relatively recent is the industrial rise of the Kuznetz Basin. Factory growth in the past quarter-century has made the district the fourth largest in the country, with important centers at Novokuznetsk, Leninsk-Kuznetskiy, Kemerovo, and Novosibirsk. Komsomolsk and Khabarovsk, on the lower Amur, and Vladivostok on the coast are the industrial centers in the Far East, specializing in steel and ships, with some general manufacturing. Hokkaido has a steel plant at Muroran where ore and scrap are imported to use with coal from the Ishikari deposits.

TOURISM

Tourism is well established as one of the major economies of these regions. Charming rural landscapes, historic attractions, mountains, streams, lakes, and a picturesque coastline provide recreational outlets for the densely populated areas of the eastern United States and Canada. Scenic beauty and sports opportunities attract thousands of tourists, vacationists, skiers, hunters, and fishermen.

FIG. 13.25. In the European manufacturing region the artisan still has an important role in industry. This picture shows German glassblowers at work. (German Tourist Information Office.)

Summer resorts and homes are scattered along the coastline and many of the lake shores. Tourism is less developed in the Eurasian regions, although Sweden, Finland, and Germany draw many visitors from Western Europe as well as the Americas.

URBAN CENTERS

A surprising number of large commercial and industrial cities are found within the Short Summer Humid Continental regions. Principal concentrations in the American region occur in (1) the Toronto to Windsor sector of Ontario, the leading industrial district of Canada, (2) the St. Lawrence Valley between Lake Ontario and Quebec, and (3) the Detroit area of Michigan. The population of Detroit, the automobile capital of the world, with its metropolitan district is about 4.25 million. Montreal, formerly head of navigation on the St. Lawrence, has nearly 2.6 million people and is the leading city of Canada. Cities such as Milwaukee and Buffalo benefit from splendid situations with respect to transportation routes and derive considerable importance from commerce as well as industrial activity. Ottawa is predominantly a political center, serving as the capital of Canada (Figure 13.26). The western cities have grown largely as the result of favorable transportation and are

FIG. 13.26. Ottawa, capital city of Canada. The British influence is apparent here. One hundred miles to the east, Montreal displays both British and French influences, while Quebec, about 150 miles downstream from the latter, is clearly the capital of French Canada. (Royal Canadian Air Force.)

bulk-breaking, assembly, and distribution points: for example, the east-west railway lines, as well as the main north-south lines of central Canada, pass through Winnipeg. The twin cities, Minneapolis–St. Paul, have a somewhat similar situation that led to flour milling and meat packing; more recent diversification has added farm implements and a variety of general manufactures. Edmonton, oil center, capital, and major commercial city of Alberta, benefits from its position on the main line of the Canadian National Railroad. Calgary, on the Canadian Pacific Railroad, serves the urban needs of the southern farming and ranching district of Alberta and is also known as an oil center.

There are many cities in the North European Plain. The two outstanding centers are Moscow, with about 9.2 million people, and Berlin, with more than 3 million. Both cities benefit from transportation, commerce, and industry in addition to their political functions. Berlin is unique among the major cities of the world, divided between two governments. About two-thirds of the population is in West Germany and the remainder in East Germany. At least twenty cities in Germany exceed 100,000 population, most deriving their significance from industrial activity and transportation. Nearly as many are found in Poland with Warsaw, the capital and largest city, exceeding one million. Leningrad, with over 4.4 million people, is the second largest city of the Soviet Union. There are a number of industrial centers within the Soviet area; especially important are those of Northern Ukraine, Volga Valley, Southern Urals, and Kuznetz Basin.

The main city of the Far East region is Harbin, Manchuria, at the point where the Chinese Eastern Railway crosses the Sungari River and makes junction with the rail line southward. The present population is about 1.5 million. Khabarovsk and Vladivostok are the principal cities in the Soviet Union

FIG. 13.27. A city within the metropolis describes the recently completed Europe Center in Berlin. In the immediate vicinity of the Gedächteniskirche (Commemorative Church) on the Kurfürstendamm, in the smallest possible space, a city was built of glass, steel, aluminium, and cement, towering upward to the sky. The Europe Center, designed by the Berlin architect Karl-Heinz Pepper, took eighteen months to build. It is to be a meeting place for shoppers, businessmen, visitors to Berlin, gourmets, film fans, cabaret enthusiasts, night revellers, and ice skaters. The shop windows alone total more than a mile, and the Mercedes star on top of this 275-foot-high skyscraper has a diameter of 33 feet. (German Information Center.)

portion. Khabarovsk has a favorable situation on the banks of the Amur River just below the confluence of the Ussuri River, where the Trans-Siberian railroad bends southward to Vladivostok, which is the Soviet Far Eastern port. More than one-quarter of the population of Hokkaido are urban dwellers, half of whom live in Sapporo, the capital, and Hakodate, the main seaport.

fourteen

MARINE WEST COASTS

Marine West Coast regions are aptly named; in no other lands is the sea such an active force in creating climatic homogeneity. The cyclonic westerlies flowing over vast ocean bodies import a constant supply of marine air that brings ample precipitation, keeping summers cool and winters mild. Moderate temperatures and liberal rainfall are reflected in nature's landscapes. Greens dominate the color scheme, interrupted only briefly by the yellows and browns of late summer and the bright hues of fall.

Whereas climate provides a salient element of regional similarity, other natural endowments, cultural features, and stages of development have created great contrasts from region to region. Western Europe, the "heritage land" of most of the peoples of the United States, Canada, Australia, and New Zealand, represents maturity. Here energetic men, utilizing their ingenuity, rich resource base, and geographic position, have created a most significant segment of the cultural and industrial world; Western Civilization and the Industrial Revolution epitomize their accomplishments. Through colonial empires and world wide trading activities, the influence of the Western Europe nations spread throughout the world.

The other Marine West Coast regions, less endowed, count their white settlement in years, not in centuries. In contrast to Western Europe, their occupancy is immature. The Pacific Northwest of North America shows the most promise, based on its land, water, and forest resource and intimate relationships with two advanced nations. Isolation from the great world communities is the major handicap of areas in the southern hemisphere.

Location

The Marine West Coast regions are located poleward of the Dry Summer Subtropics on the western margins of the continents, in the heart of the cyclonic westerlies (see Figure 14.1). The greatest region is in Western Europe, extending from northwest Spain to northern Norway. Included in the bulk are the British Isles, western France,

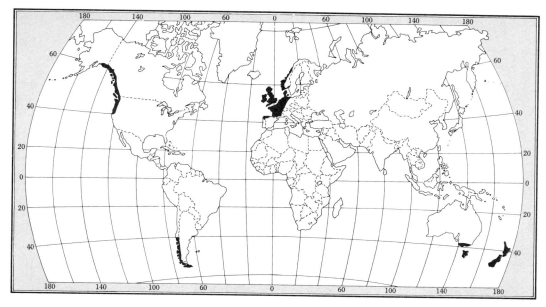

FIG. 14.1. Marine West Coasts.

western Germany, and Low Countries, Denmark, and the southern fringe of Iceland (Figure 14.2). The North American region, known as the Pacific Northwest, includes northwestern California, the Pacific slopes of Oregon, Washington, and British Columbia, and the southern fringe of Alaska (Figure 14.3). The South American region is confined to the southern one-third of Chile (Figure 14.4). Australia, with a limited area influenced measurably by the cyclonic westerlies, has only a small region that includes most of Victoria and all of the neighboring island of Tasmania. New Zealand, 1,200 miles southeast of Australia, is an isolated island marine region (Figure 14.5).

Physical Environment

CLIMATE

Marine West Coast regions owe their climate chiefly to continent-margin locations in the path of the cyclonic westerlies. The eastward streams of air are modified by passage over the sea; therefore the atmosphere over the land is ruled by marine air masses. Temperatures are mild, for the latitude and annual extremes are small. Clouds and rain are typical for many months of the year, but considerable sunshine occurs during summer and early fall, especially on subtropical margins.

The modifying effect of the comparatively warm ocean on the incoming air masses explains why marine climates extend beyond 60° N latitude in Norway and Alaska. The

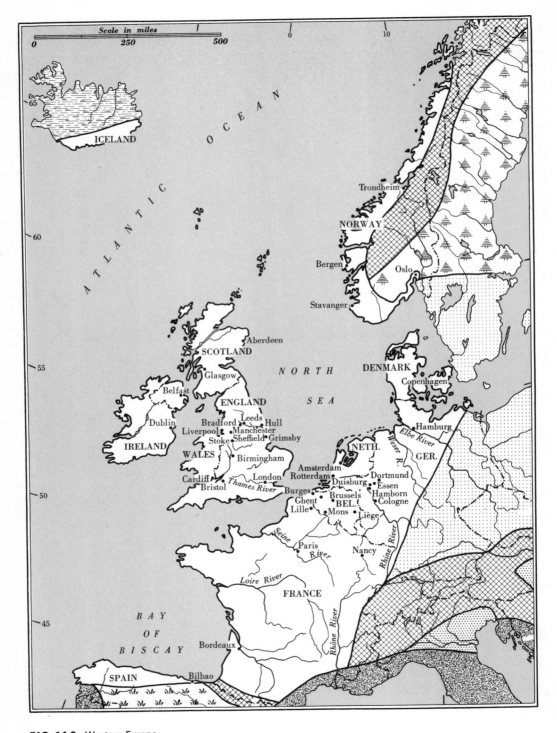

FIG. 14.2. Western Europe.

MARINE WEST COASTS 309

shoestring shapes of the Marine West Coast portions of Norway and Alaska, as well as that of the region in Chile, are due to mountain ranges that closely parallel the coast and restrict marine modification inland.

Temperatures. Summers are cool with average July temperatures usually about 65°.

Portland, Oregon, has a July average of 67.2° and London, England, 62.7° (Figure 14.6). Weeks of hot weather are rare, but maxima occasionally reach 90° and above. Summer nights are seldom too warm for sleeping, and blankets are usually needed for comfort. Winters are mild, with January averages in the neighborhood of 40°. Seattle,

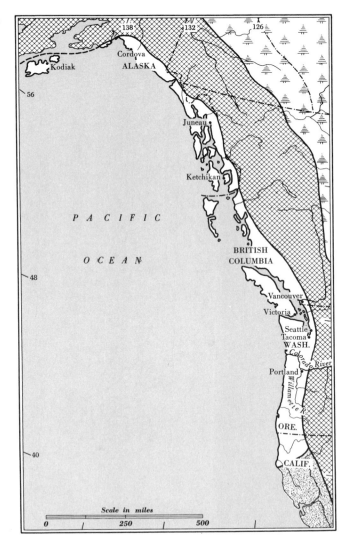

FIG. 14.3. The Pacific Northwest.

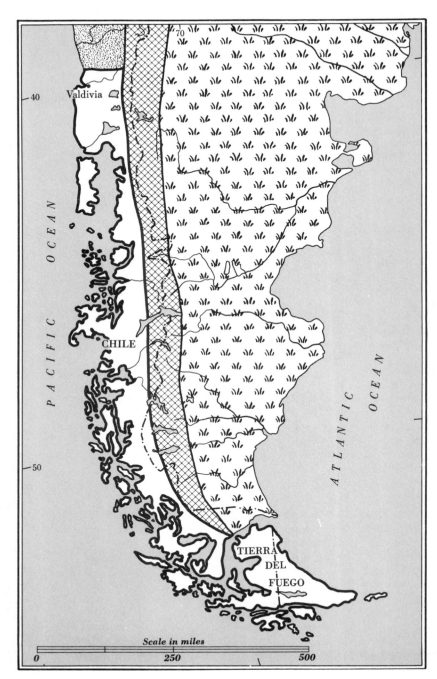

FIG. 14.4.
Southern Chile.

FIG. 14.5. Australia–New Zealand.

Washington, records a January average of 39.8° and Dublin, Ireland, 42.1°. Two factors account for winter mildness: the stored heat in the offshore ocean bodies and warm ocean drifts (with the exception of Chile), which bring tropical heat from equatorial waters. Thus the incoming air is warmed, causing winter temperatures to average about 25 degrees higher than inland areas of Europe in similar latitudes. W. G. Kendrew writes, "The air over the ocean west of Norway is more than 50 degrees warmer than the average for the latitude, the greatest anomaly of temperature known."[1] Vancouver, on the coast of British Columbia, has a January average of 35.6°, 38 degrees warmer than the −3° recorded at Winnipeg, Manitoba, in the interior, but less than one degree higher in latitude. Although winter tempera-

[1] W. G. Kindrew, *Climates of the Continents* (New York: Oxford University Press, 1961).

FIG. 14.6. Typical climatic graphs of Marine West Coast stations.

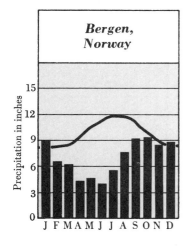
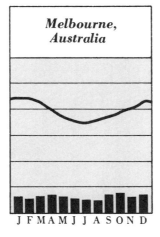
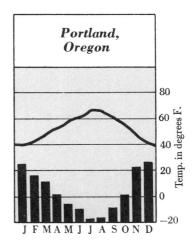

tures are mild, one may feel uncomfortable due to the high relative humidity. Sensible temperatures at 40° are chilly, raw, and penetrating, whereas in the interior of the continent 40° would be considered fairly comfortable. There are few spells of cold weather, especially in regions where mountains block out cold continental air; but in western Europe, where high protecting mountain ranges are lacking, the area is subject to cold waves from the Eurasian interior. Winter diurnal ranges are low, because clouds and water vapor retard nocturnal cooling. The atmosphere operates on the same principle as the panes of glass on a greenhouse, and the process is sometimes called the "greenhouse effect." Incoming shortwave radiation passes through the atmosphere, but longwave earth radiation is largely absorbed and only a small part is lost directly into space. The earth's longwave radiation absorbed by the atmosphere is reradiated, and therefore earth heat is retained in the lower atmosphere. Yearly ranges of temperature are from 15 to 25 degrees. Length of growing seasons is from about 150 days near poleward margins to as high as 300 days equatorward.

Precipitation. Mild, extended rain and cloudy weather are truly Marine West Coast characteristics. Averages are difficult to state, since amounts vary with relief, exposure, and distance inland (see Figure 14.7). Windward slopes near coasts receive from 80 to over 100 inches a year, whereas leeward locations commonly receive from 30 to 40 inches: for example, 200 inches of rain falls annually on a small windward area near the south end of Vancouver Island, but the city of Victoria on the lee side of the mountains a few miles to the east has only 30 inches. Where coastal mountains are lacking, as in Western Europe, moderate precipitation is carried well inland. Precipitation occurs during all seasons, but many areas tend to have winter maxima. Cyclonic disturbances are most active at this time, and severe storms often lash the seacoast.

Toward the eastern margin of the Marine region in Europe, the continental influence becomes stronger, and there is a tendency for precipitation to become more concentrated in the warmer half of the year. Precipitation is chiefly cyclonic in origin, falling in long, continuous drizzles. Thunderstorms are infrequent, ranging from three to ten a year. On the lowlands of the equatorward margins there are only a few days with snow, which seldom remains on the ground for long. With increasing latitude the snowfall and persistency of cover on lowlands increase, and highlands throughout experience heavy snow. There is a large number of rainy days; Seattle, Washington, has 151 days with rain, and Valencia, on the

FIG. 14.7. Topography and precipitation, Western United States.

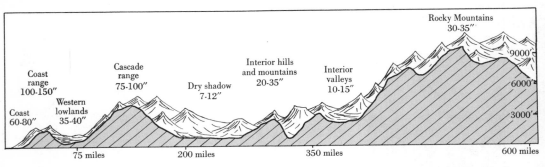

west coast of Ireland, has 252. The intensity is low, as illustrated by Seattle with a total precipitation of only 32 inches. There is much cloudy weather, which gives these regions the distinction of being the cloudiest on earth. Fogs are prevalent, especially along the coast; Great Britain averages 50 foggy days a year. Average relative humidities are high, and on the littorals the air is nearly always damp. During summer, however, humidities are occasionally so low in the Pacific Northwest (30 percent and under) that logging operations are stopped to prevent forest fires.

Precipitation has a high degree of reliability; drought is infrequent. But subtropic margins, particularly in North America, have fairly dry summers reminiscent of mediterranean regions. Sprinkler irrigation during the summer has become a common sight in the Puget Sound-Willamette trough; even in the coastal valleys of Oregon, where yearly averages are about 90 inches, supplemental irrigation is practical (Figure 14.8).

SURFACE FEATURES

A maximum of rough topography and highly irregular coastlines are characteristic features of most of these regions. The coasts of Norway, Scotland, British Columbia, Alaska, southern Chile, and southern New Zealand are similar. All are fringed with islands and deeply indented with long, steep-walled, narrow arms of the sea known as fjords. These features were produced by the submergence of ice-scoured mountain valleys that fronted on the sea. Such coastlines have numerous harbors and beautiful scenery, but leave little room for man. The limitations are indicated by the fact that less than 4 percent of Norway is arable (see Figure 14.9).

Most Marine West Coast regions are favored with an abundance of water resources resulting from the combination of mountain watersheds in their backgrounds and the

FIG. 14.8. Sprinkler irrigation in the Nestucca River Valley of western Oregon. This area receives approximately 90 inches of precipitation annually, but despite a high degree of cloudiness, only about three inches fall during July and August. Dairymen have found supplemental irrigation of pastures profitable during these months. (United States Department of Agriculture, Soil Conservation Service.)

persistent precipitation. Hydroelectric power sites are available in many areas, and the obtaining of industrial water supplies is relatively easy. In addition, streams such as the Rhine, Elbe, and Thames in Europe, and

FIG. 14.9. Hardanger Fjord, Norway. Human distribution and activity in this country are intimately related to the fjords, the most spectacular of the landform features, which provide attractive scenery, a lure for tourists, protected transport arteries, and fisheries. Settlements and farms are found on the limited level lands scattered here and there, often miles apart, along the margins and heads. Each fjord tends to develop distinctive cultural characteristics, reflected in dress, dances, folksongs, and so on. The Hardanger Fjord is 105 miles long, and its waters are reported to be as much as 4,000 feet deep. The farming community in the background is typical of larger areas of level land. (Oliver H. Heintzelman.)

the Columbia in the United States, are excellent transport arteries.

Western Europe is the most favorable of the regions. The North European Plain begins at the Pyrenees Mountains and swings northeastward through the Aquitaine and Paris Basins of France and continues through the Low Countries into Germany. Throughout the greater part of the plain, elevations seldom exceed a few hundred feet, and in the Netherlands about 35 percent of the country is actually below the mean level of the sea and is protected by sand dunes and dikes.

The Puget-Willamette Lowland in the Pacific Northwest is the only other major area of low relief. Even here the Puget Sound Lowland, as a result of glaciation, has an irregular surface; glacial and glaciofluvial deposits are widespread and often consist of infertile sand and gravels, the finer materials having been washed free. The Willamette Valley in Oregon is more favorable, floored with alluvial materials carried down from the bordering mountains by the Willamette River and its tributaries; only in the middle and northern portion do hills break the continuity of the flat or undulating surface (see Figure 14.10).

Southern Chile is virtually without level land. The forested slopes of the rain-drenched Andes descend sharply to the sea. Glacial erosion and deposition in the past created numerous lake basins, which are kept brimming full by the abundant precipitation. In contrast to middle and northern Chile, which is without a good natural harbor, the south has a highly indented coastline; how-

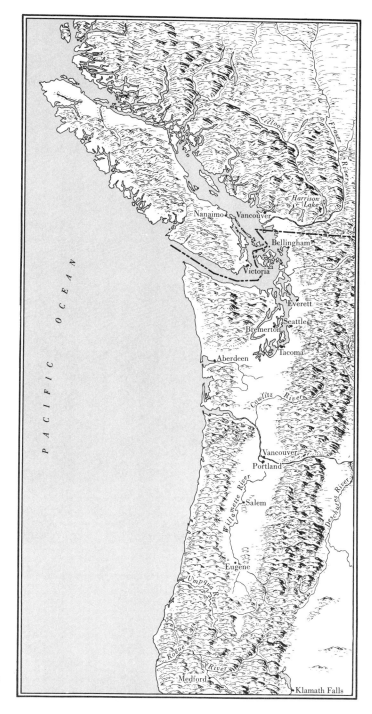

FIG. 14.10. Physiographic diagram of the Pacific Northwest.

FIG. 14.11. (top) The Netherlands, lowest of the Low Countries. If if were not for the rampart of coastal sand dunes and dikes, much of this nation would be under water. This scene shows a Friesland canal south of Sneek. Barge transportation is an important means of bulk movement in Western Europe. (Standard Oil Company of New Jersey.)

FIG. 14.12. (bottom) Hedgerows enclosing farmlands in Somerset, England. These are similar to the bocage of France. (British Information Services.)

ever, the limited development of the back country provides little commerce to stimulate use of the excellent harbors.

In Victoria, Australia, the continent's mountain divide bends westward and forms the northern boundary of the Marine West Coast region. The coastal strip is somewhat broader than that of the Humid Subtropic region in the east and is altered slightly in character. The Great Valley of Victoria, separated from the sea by low hills, is the major lowland. The Great Valley is divided into two parts by Port Phillip Bay. Tasmania, essentially an outlier of the Eastern Highlands, is a mass of mountains and hills with small fertile valleys. New Zealand is largely a mountainous country; less than one-quarter of its surface is below the 650-foot contour. The two large islands are divided by a main mountain axis extending from Fjordland in the south to East Cape in North Island. On either side of the highlands, small areas of level land occur. These are very limited in extent and are not continuous.

NATURAL VEGETATION

Forests, favored by the temperate rainy climate, comprise the typical natural vegetation. Uniformity of cover, however, is lacking, and each region has differences in species whose qualities for use vary. Similarities occur mainly in densities of stands and sizes of trees. The luxuriant growth of many forest areas makes them comparable to the selva of the Rainy Tropics.

The forests of Western Europe fell prey to the cultivator and were cleared years ago. Oak and beech were the principal species; conifers were scarce and confined chiefly to the higher elevations. Areas with poor soils and drainage were covered by marshes and moors, many of which remain today; in fact, human modification has increased the area in moor. With the exception of Norway, many of the forests existing in Western Europe have been planted. Some areas, too rugged or with soils too poor for agriculture,

FIG. 14.13. Emmeloord on the northeast polder of the Netherlands. Less than five decades ago this land was on the bottom of the Zuider Zee. Through Dutch engineering skill and tenacity it has been reclaimed, and now, with a productive agriculutral base, it supports many farms ranging from 30 to 120 acres. The entire project was planned in detail and developed according to a program and schedule sponsored by the government. This picture was taken from the new water tower of Emmeloord, the planned master town of this polder. Note the tabletop flatness of the terrain. (Oliver H. Heintzelman.)

FIG. 14.14. Stand of virgin redwoods in Six Rivers area of northern California. These giant, moisture-loving trees grow relatively close to the coast within the fog belt. The man beside the middle tree gives a dramatic size comparison. (United States Forest Service.)

now maintain stands of commercial trees. Southwest France has a planted pine forest in the sandy Landes area along the Bay of Biscay. Lowland landscapes viewed from a distance in England and France appear forested due to the many English hedgerows and ornamental trees on estates, and in France, because of the trees and hedges planted for screens and farm and pasture boundaries by the peasants of the *bocage* country (see Figure 14.12).

The Pacific Northwest contains the most valuable coniferous forests in the world. Thick stands parallel the coast from northern California to Alaska. The redwood is the outstanding species in the hills and coastal valleys on the south margin, with a major concentration in the Eel River Valley of northern California. This tree is magnificent, reaching a height of 300 feet with a diameter exceeding 10 feet; a redwood forest is one of the most majestic sights on earth (Figure 14.14). Dense stands of Douglas fir dominate the conifers in Oregon, Washington, and southern British Columbia. Mature trees attain heights of 200 feet and more with diameters ranging from 6 to 9 feet. Douglas fir constitutes the most important lumber tree of the region (Figure 14.15). Other large conifers include the western hemlock, western cedar, and spruce. In the Alaskan panhandle, hemlock and Sitka spruce are the major species.

The Chilean region is clothed with rain forests from the coast to the tree line of the Andes. Unlike the Pacific Northwest, the forests are dominated by less valuable broadleaf evergreen southern beeches; however, conifers are found in the higher elevations. Luxuriant forests in the Australian region contain tall, straight eucalyptus reaching heights of 300 feet. Beautiful understories of tree ferns cover the forest floors. Prior to white settlement, approximately three-fourths of New Zealand was covered with broadleaf and coniferous forests. Scrub forests and grass grew on areas of less precipitation. Today nearly half of the forest cover has been cleared for crops and pasture. The remaining broadleafs are chiefly species of southern beech. Among the conifers, the Rimu or red "pine" is the principal timber tree. The Kauri, the largest tree of New Zealand, is prized for timber and a valuable resin called kauri gum. Here as in Australia, tree ferns form the understory in many of the forests. To improve the base for wood products industries, significant plantings of exotic species have been made in Australia and New Zealand. Monterey pine (*Pinus radiata*) is the most popular of these planted trees. Under favorable site conditions of climate and soils this softwood grows more rapidly than native species, and apparently produces better timber trees than in its native habitat in the central coastal area of California.

SOILS

The soils of the Marine West Coast regions do not fit as smoothly into the world classification scheme as those of other regions. The diversity of surface, variations in vegetation, glaciation, and alluvial deposition have produced many local differences. In general, most are leached of their soluble elements and are acid in reaction. Podzols prevail in the poleward areas with conifers, and gray-brown soils are common in those portions that have deciduous forests. In many of the mountainous districts severe glaciation left bare rock exposed at the surface; at the other extreme, youthful alluvial valley soils can be very fertile, with drainage the main agricultural limitation.

Occupance in the Marine West Coasts

The Marine West Coast regions, for their size and limitations of surface and soils, support surprisingly large numbers of people. Western Europe has one of the great population blocks of the world. The United Kingdom alone has more than 55 million people, and average densities of nearly 800 and 900 persons per square mile are found in Belgium and the Netherlands. The Pacific Northwest, in contrast, presently supports relatively few people; nonetheless, it is in a stage of rapid development, and rather large concentrations are forming in the Willamette-Puget Lowland, including the Frazer River

FIG. 14.15. Forests for the future. The young stand of Douglas fir in the foreground will be ready for thinning in about 35 years. This selective harvest will provide materials for poles and pulpwood. Approximately 50 years later, the stand will be ready for a complete harvest. (Weyerhaeuser Company.)

FIG. 14.16. The wealth of Denmark in a large measure is related to its livestock. This pasture scene with the herd of red Danish cattle is located in the eastern portion of Jutland. (Oliver H. Heintzelman.)

floodplain and delta of British Columbia. About one-fourth of the Australians live in the state of Victoria, and there are nearly 3 million people in New Zealand. South Chile is the only region with a small population; heavy and persistent rainfall, lack of level land, and remoteness have made it unattractive.

Primary production—agriculture, forestry, and fishing—dominates the activities of most regions, but Western Europe ranks with east-central North America as a manufacturing belt and preceded that area in industrial development. It also has a long-time envolvement in trade and service activities.

AGRICULTURE

Marine West Coast regions are not endowed with the best conditions for farming. Frequently surface, soils, and the moist, often cloudy maturing season place restrictions on the crops that can be grown. In Western Europe there is competition from other forms of land use, especially from industries and urban centers. Fortunately the major plains are located towards the equatorward margins, where summers are warm and not excessively wet. Grass, root crops, vegetables, and certain fruits do well. Of the small grains, oats are best adapted, although barley is grown extensively and wheat is found in the areas with relatively dry summers; in these areas the wheat plant grows and produces well, however, the grain is in the soft wheat class and is poorly suited for bread flour. Not all wheats are equally suitable for bread flour. Some of the proteins in wheat are in the form of gluten which makes the dough elastic. This allows the trapping of the gas bubbles formed during fermentation; these bubbles are solidified during baking to give the light, porous loaf. This introduces an interesting point: the optimum climate for the growth of the wheat plant—ample soil moisture, cool weather during the formative period, and mild, gradually warming temperatures during the growth and maturing period—allow starches to accumulate faster and in greater quantities than proteins. Toward the dry margins of tolerance the protein accumulation is greater, hence the bulk of the commercial hard wheat production for bread flour is in the subhumid and semiarid regions.

The long, cool, frost-free season and the mild winters favor pasture production, allowing a long grazing season, and animal industries are usually major enterprises.

Western Europe. The European region leads in agricultural development. Farming here is an old activity, but its nature has

FIG. 14.17. Agriculture in Brittany. This region, long known for its fishing, is also engaged in significant agricultural activity. Here, in carefully constructed and tended beds, oysters are being grown for the Paris market. (Oliver H. Heintzelman.)

undergone basic changes in the modern era, especially since World War II with the greater development of technology and the labor demands of increasing industrialization and services. Today it is usually secondary to industry. Barely 3 percent of the laborforce of the United Kingdom is engaged in agriculture; about 5 percent of the Belgians are farmers. Even in Denmark, well known as an agricultural country, less than 18 percent of the labor force are engaged in farming pursuits. No country approaches total selfsufficiency.

The majority of countries have tended to specialize in or stress those enterprises for which their environment is best suited and for which markets are readily available; all seek to avoid major imports. Most are, in fact, importers, but several are net exporters in terms of value, notably Denmark and the Netherlands. These two countries typify the best in Western European practices. Denmark, for example, with science, industrious people, standardization of high-quality products, and elimination of the middleman by cooperative buying and selling organizations

FIG. 14.18. A French farmstead in Normandie. Cheese, the significant product of the region, is mainly marketed in Paris.

FIG. 14.19. Hothouses near Brussels, Belgium. In the crowded Low Countries, hothouses make possible the production of out-of-season vegetables and fruits. (Belgium Tourist Office.)

has made agriculture profitable. The Danes have geared their farming activities to the market needs of the neighboring industrial countries, specializing in dairy products, particularly butter, lean bacon hogs that utilize much of the available skimmed milk, and poultry, mainly for eggs. Farms, averaging 60 acres, are intensively managed and devoted to raising pasture, feed grains, and forage crops. Dairying is the chief phase of agriculture throughout the European region.

Dutch agriculture is characterized by relatively small farms and intensive operations. Some 90 percent of the holdings are less than 50 acres, and about 50 percent of those are under 10 acres. The prosperity of the country's agriculture is due more to the ingenuity of its engineers and farmers than natural conditions. Intensive reclamation and drainage projects and high rates of fertilization underlie high yields of horticultural, root, and cereal crops. Of the 6.3 million acres in the productive farm base of this tiny country, about 59 percent is pasture and meadow which support a significant dairy industry; swine and poultry are, also, important. The success of Dutch agriculture is indicated by the fact that the country produces an exportable surplus; production is now about one-fifth greater than domestic consumption. Exports are regulated mainly under the Common Agricultural Policy of the European Community, of which the Netherlands has been a member since its inception.

Beef cattle and sheep are also raised in large numbers, especially in the United Kingdom, where primogeniture inheritance practices have kept farm holdings somewhat larger than elsewhere in Europe.

Generally throughout the European region, at least one-half of the land is given over to pasture and hay. Many fine breeds, now important in other parts of the world, were developed in the area bordering the North Sea. Crops other than oats, barley, and root crops are important in the drier areas. For example, wheat is grown extensively only in eastern England and the Paris Basin of France.

The Pacific Northwest. Agriculture in the North American regions is confined largely to the Puget-Willamette Lowland and small plains scattered along the coast. General farming on units of several hundred acres is common, but enterprises are similar to the European region. Dairying and poultry production are widespread but are especially important around Puget Sound and in the coastal valleys. Tillamook County, Oregon, has gained a national reputation for quality cheddar cheese. The Willamette Valley is the outstanding agricultural area; favorable gentle surface, fertile soils, and warm dry summers allow the production of a wide range of agricultural commodities, including grains, hay, grass and forage seeds, a variety of fruits, walnuts and filberts, other horticultural crops, poultry, and livestock (Figure 14.20). Development of supplemental irrigation by taking water from shallow wells or surface sources with small pumps, and using portable aluminum pipes and sprinklers, has increased the possibilities greatly; improved pastures and a significant growth in the acreage of processing crops, such as sweet corn, beans, carrots, and beets, has resulted.

Southern hemisphere. The southern hemisphere's basic limitations for commercial agriculture are small local populations and great distances to major consuming centers. Agriculture is further restricted in south Chile by excessive rainfall and ruggedness of terrain; here a scant population is engaged in sheepraising and lumbering; however, the northern part is the chief potato and dairy area of the country. Agriculture is more important in Victoria, Australia, and New Zealand. In both, the pasturing of animals is the leading phase of land utilization; sheep raising and dairying are the main farm enterprises. Both are admirable adjustments to environmental conditions, providing at the same time exportable products—butter, lamb, wool, and cheese comprise the greater part of the items entering foreign trade. Agricultural commodities comprise 70 percent of the value of exports of Australia and 90 percent of New Zealand.

FISHING

The Marine West Coast region of Western Europe possesses one of the world's major fisheries. Rich fishing grounds extend from northern Norway to the Bay of Biscay, but the concentration of the activity is in the waters of the North Sea. Halfway between

FIG. 14.20. A Willamette Valley landscape south of Salem. Complete utilization of the fertile valley awaits further population growth and expanded markets for agricultural commodities. (Oregon State Highway Commission.)

FIG. 14.21. Stabanger, Norway. This city is one of Norway's most notable fishing ports. A large portion of its population is supported in canneries, which process herring and sardines—products that flow out to world markets. This scene shows a part of the harbor. Note also that the waterfront is the marketplace for both fish and farm produce. (Dorothy A. Heintzelman.)

England and Denmark are the Dogger Banks, one of the most productive fishing banks in the world. A combination of factors has favored the development: Plankton is abundant on the shallow continental shelf; smooth bottoms favor the use of otter trawls; aeration of water and surfacing of nutrients are provided by tides and currents; rivers bring in large amounts of fish food; warm ocean drifts keep seas open through the winter; drowned river mouths and fjords serve as excellent ports; and a high per capita fish-consuming population provides excellent markets. Cod, haddock, herring, and mackerel constitute the principal commercial species taken.

All the countries of Western Europe are engaged in fishing, but the British Isles and Norway are the leaders. Numerous fishing villages border the east coast of Great Britain; Aberdeen, Hull, and Grimsby are the major centers. About 6 percent of the Norwegian adult male population is engaged in commercial fishing; many divide their fishing occupation with farming or lumbering. Bergen, Trondheim, and Stavanger are Norway's most important fishing ports (Figure 14.21).

Important commercial fisheries are located along the North Pacific Coast of North America from California to Alaska. Salmon are the most valuable species and comprise the major portion of the catch. These *anadromous* fish, which live in salt water and return to fresh water to spawn, are caught in rivers

FIG. 14.22. Ketchikan, Alaska. Timber and fishing provide the economy of Ketchikan. Most of the southeastern portion of the state is incorporated in Tongass National Forest and, under the supervision of the U.S. Forestry Department, the dense crop of trees is harvested for pulp and lumber. (Ketchikan Chamber of Commerce.)

FIG. 14.23. Harvest scene in the Douglas fir region of Washington state. Harvesting methods are based on the growth habits of the Douglas fir tree. Seedlings of this tree will not thrive in the deep shade of older and larger trees. For this reason, a system of logging called forest area selection is used. Selected blocks or strips of timber are cut and intermittent islands of trees are left unharvested to serve as a seed source. In this way forests are being replaced faster than they are being harvested. (Weyerhaeuser Company.)

FIG. 14.24. Loading Douglas fir logs at the spar pole landing for movement out of the woods. This is the high lead system used in areas of rugged terrain. Donkey engines and a system of pulleys and cables snake logs from the woods to the landing. (Weyerhaeuser Company.)

or near river mouths. The Columbia River is the major salmon stream of the region. Halibut and tuna are other significant fish. An oyster industry has been developed in some of the sheltered bays of Oregon and Washington with Willapa Bay of Washington outstanding. Many of the coastal cities are fishing ports; Seattle on Puget Sound is the leader.

Fishing in the southern hemisphere regions is a minor activity in comparison to the northern hemisphere regions; however, both Australia and New Zealand have become increasingly aware of the sea's potential, and some fishing activities are being accelerated.

FOREST INDUSTRIES

The forest contributes to the economy of all the regions. The Pacific Northwest region, however, leads in reserves, production, value, and quality of saw timber; the extensive forests constitute its greatest source of raw material.

The forest industry of the Pacific Northwest is firmly based upon the dense stands of large conifers—Douglas fir, redwood, western hemlock, western cedar, and Sitka spruce. The Douglas fir is the most prevalent and most valuable tree (Figure 14.23). Its high-quality wood is excellent for both lumber and plywood. Western hemlock, the second-ranking tree, and Sitka spruce are used mainly for pulp for paper and rayon stock; and western red cedar is used for shingles. Logging operations are highly mechanized. Specially designed, diesel-powered crawler tractors do much of the heavy work in the woods, but the high lead system employing spar poles, cables, and donkey engines is needed to assemble logs at landings in areas of rugged terrain (Figure 14.24). From these landings, the logs are transported by truck or rail to sawmills or to rivers for further movement. Power saws are used for most of the tree felling, a task that was formerly accomplished by man-operated crosscut saws. Logging is usually possible throughout the year, but the woods are sometimes closed by winter snow or by low summer humidities when there is danger of fire.

Logging in Western Europe is most significant in Norway. Large stands of Scotch pine and Norway spruce are concentrated near Trondheim and in the southeast. The trees are small in comparison to those in other regions and are cut mainly for the pulp and paper industries. Many Norwegian farmers become loggers during winter.

Logging in Chile is a small industry considering the large forest area. Several factors have discouraged this activity: the rough and rain-drenched terrain makes logging operations difficult; many of the species are inferior for lumber and thus there are low yields of valuable trees per acre; furthermore, Chile is remote from good markets.

Logging is an active industry in New Zealand—in fact, timber demand has been such as to encourage development of forest plantations of fast-growing, nonindigenous species, as well as to import lumber from North America and Australia. Rimu, or red "pine," and Monterey pine (an introduced species) constitute about 75 percent of the cut. Both are used chiefly for building purposes. There is some activity in Australia, including Tasmania, based on eucalypti.

MINING

Minerals have played major roles in the industrial growth of Western Europe. This region is favored with both coal and iron, but all the industrial countries of the region do not have sufficient quantities of these necessities and must trade with neighbors (see Figure 14.25).

Great Britain was the first major coal-mining nation, and the large reserves of this fuel have been a major factor in her industrial growth. Coal has provided not only the

FIG. 14.25. Coal and iron in Western Europe.

basic power source but also a valuable item of exchange that has been hauled as ballast in British ships to many lands, to trade for raw materials and food. The large and excellent coal deposits of the Ruhr area have had an important role in Germany's great industrial development. Coal also occurs in the Sambre Valley of France, and the seams continue into Belgium and the Netherlands (Figure 14.26). The major iron ore deposits of Western Europe are in the Lorraine District of northeast France and Luxembourg, and lesser deposits are found in Germany, Great Britain and in northern Spain, at Bilbao; the ores of the latter are some of the best in Europe. West Germany and France rank third and fourth as world potash producers, with centers around Strassfurt and Mulhouse.

Other Marine West Coast regions are largely without minerals. A few small coal deposits occur in the Pacific Northwest, but their low quality has restricted major use. A small deposit of nickel is being worked at Riddle, Oregon, noteworthy principally because it is the only primary nickel mine in the United States. New Zealand has several small coal fields and several lignite deposits; there are also a number of lignite fields in Victoria, Australia.

MANUFACTURING

Western Europe. Western Europe is almost synonymous with manufacturing. Here modern industry had its genesis, and today the region contains one of the largest and most diversified industrial concentrations in the world. Factories employ from 25 to over 35 percent of its working population. Despite sharp competition from more recently developed areas, Western Europe continues to produce an important share of the world's goods.

The broad Western European manufacturing belt, extending from northern Ireland to western Germany, is endowed with a number of favorable factors that have stimulated and consolidated industry. Numerous coal fields supply power, and a variety of raw materials is available. There is an abundance of highly skilled labor with a rich background of experience dating back long before the Industrial Revolution. Transportation facilities are excellent, with networks of railroads, highways, rivers, and canals; great ports are situated on the river estuaries and a "Marine Broadway" is at the front door. Domestic markets are large and outlets well established abroad.

Manufacturing does not dominate throughout the belt; rather there are many major and minor districts, usually located on or near coal fields and in ports and transport centers. Eight such districts are recognized in the British Isles: (1) Northern Ireland, the most western district, specializing in ships and linen textiles, is centered at Belfast. (2) The Scottish Lowland is the industrial heart of Scotland with centers, dominated by Glasgow, on the Clyde River. Industries are varied, but shipbuilding is outstanding. (3) The northeast England district consists of a number of iron and steel and shipbuilding cities along the lower courses of the rivers Tyne and Tees. Shipbuilding, significant here as it is in the former districts, exemplifies Britain's dependency upon commerce. Bordering the coal fields on the east

FIG. 14.26. Coal-mining scene near Charleroi in the Sambre-Meuse valley of Belgium. The buildings at the pit heads and the slag piles are as typical of Western European industrial areas as the factory smokestacks. (Belgium Tourist Office.)

side of the Pennine Mountains is (4) the Yorkshire subregion. The emphasis is on woolen textiles at Bradford and Leeds; however, iron and steel are also important, especially at Sheffield, long known for its high-grade cutlery. (5) Lancashire, the cotton textile district, is west of the Pennines. Manchester, dominating the district, is contiguous to a number of smaller cities having textile specialization. (6) The Midland or the "Black Country" of England, south of the Pennines, manufactures nearly every type of metal goods. Birmingham is the center. Coventry is noted for automobiles and bicycles. Shoes, hosiery, textiles, rubber goods, and electrical supplies provide diversification; large potteries are located at and near Stoke-on-Trent. (7) South Wales and Bristol constitute a subregion on the southwest coast. South Wales refines metal and tin-plates steel. Bristol's industries are based chiefly on products of colonial trade, such as tobacco, chocolate, and sugar. (8) Metropolitan London, the final British district, is the hub of the country's transportation and the great entrepôt of the world. Heavy industrial plants include shipyards, machine shops, chemical works, and oil refineries; nearly every type of light industry also occurs.

Three significant subregions stand out in France. Paris, the largest and most important, has long been associated with luxury items such as style-setting clothes, jewelry, perfume, and other quality goods; however, the manufacture of engines, automobiles, and machinery, the refining of oil, and printing and publishing are also important. Many firms have located in Paris to capitalize on the city's reputation for quality production. The Northern Industrial subregion, containing the richest coal field of France, is the metallurgical and textile center and also produces glass and pottery and refines sugar. Lille, chiefly a textile city, is the major center. The Eastern France subregion, located around Nancy in the Moselle Valley, is a metallurgical district based on the Lorraine ore deposits. To the northeast is the Saar Basin, important for coal production and iron and steel.

The Low Countries are significant industrial nations despite their small size. Central Belgium contains the Sambre-Meuse district, where coal production has fostered a line of industrial cities extending through the valley from Mons to Liége. The emphasis is upon metallurgy, chemicals, and pottery. Verviers, east of Liége, is Belgium's woolen textile and glass center. A second district, the Low Countries Manufacturing subregion, extends from the Flanders Plain of Belgium into the Netherlands. Textiles lead in the Belgian section, focused at Brussels, Ghent, and Bruges; Breda and Tilburg are the Dutch textile cities. Southeast of these centers is

FIG. 14.27. A skilled craftsman at a potter's wheel in Limoges, France—a world-famous pottery center. (French Embassy Press and Information Division.)

Eindhoven, the great electronics center. Much of the Netherlands' industry centers around the port of Rotterdam and is based on tropical raw materials such as sugar, copra, and rubber; however, chemicals, food processing, textiles, and others are important, too.

Germany's most important industrial area is located in the Ruhr district, on western Europe's largest coal field. This is the greatest single concentration of heavy industry and coal mining in Europe, as well as in the world. Here iron and steel are basic, and a cluster of industrial cities, such as Essen, Dortmund, and Duisburg, manufactures every conceivable item of heavy equipment. Chemical industries are also noteworthy.

These subregions by no means complete the list of industrial concentrations—only the most outstanding have been highlighted. Hundreds of smaller areas are scattered throughout western Europe. Furthermore, only the typical products of the districts are listed—a complete inventory would be almost endless.

The Pacific Northwest. Industry in the Pacific Northwest is based chiefly on the raw materials of the forest, farm, and sea, hydroelectric power and abundant water, and a reservoir of skilled labor. Wood processing —in lumber mills, plywood plants, paper and pulp mills, and furniture factories—is the major industry. Lumber mills are scattered through the entire length of the area and are the most typical manufacturing plants. Some lumber mills have an enormous production, sawing as much as one million board-feet a day. Many are multiple-purpose plants producing lumber, pulp and paper, plywood, and often other products; such utilization of the saw log eliminates waste and is in the interest of conservation as well as economic efficiency (see Figure 14.29). Paper mills, located in response to large supplies of water and raw material, produce several grades of paper, ranging from paperboard to quality magazine paper.

The canning and freezing of fruits and

FIG. 14.28. Nitrate plant in east Norway. Cheap hydroelectric power is the advantage for this industry.

vegetables is a significant industry, with plants distributed throughout the Puget-Willamette Lowland; Salem, Oregon, is outstanding; salmon canneries located on tidewater are scattered from Alaska to Astoria, Oregon. Low-cost hydroelectric power has attracted aluminum plants to the Lower Columbia River area (Longview, Vancouver, Troutdale) and to Tacoma. At Ruston, near Tacoma, is one of the world's largest combined copper smelters and refineries. Shipbuilding, the aircraft industry, wool textiles, and the manufacture of work and sports clothes are also important. The Puget Sound Lowland is the major area of industrial concentration and includes the cities of Van-

FIG. 14.29. Integrated forest-products mill at Springfield, Oregon. Here finished lumber, plywood, pulp, container board, and Pres-to-logs are manufactured. Former waste products are almost completely utilized. (Weyerhaeuser Company.)

couver, in British Columbia, and Bellingham, Seattle, Bremerton, and Tacoma, in Washington. The Portland-Vancouver metropolitan area on the Willamette and Columbia Rivers is the second largest district. The significance of low-cost electricity in the location of the aluminum industry is well illustrated by the Kitimat project on the primeval coast of British Columbia. Dependent upon bauxite from Jamaica and markets far beyond the site, it was developed to utilize untapped water resources. Production began in 1954, and long-range plans, when completed, will make the facility the largest producer in the world.

Southern hemisphere. Industry is minor by comparison in the southern hemisphere regions. All are handicapped by their remoteness from markets. Forest-based industries are dominant in Chile; however, a modern steel complex is located at Huachipato near Concepcion based, in part, on local coal deposits. Much of New Zealand and Australian industry is based on agriculture and forestry. Several huge pulp and paper mills are opening to use New Zealand's new Monterey pine (*Pinus radiata*) forests. Melbourne is the principal center of all the southern regions and does manufacture a variety of articles.

TOURISM

Marine regions are endowed with a wealth of resources that foster tourism. Western Europe's assets include not only beaches, charming landscapes, historic sites, shrines, churches, and cities, but also a dense population. Cultural and historical attractions vie with nature's contributions. All the countries have active promotional programs to entice the foreigner as well as the local inhabitant. Norway advertises "the Midnight Sun" and fjords. Colorful posters emphasize "Come to Britain." The "pull of Paris" is traditional.

Nature is the dominant tourist attraction in the Pacific Northwest. The ocean, lakes, streams, mountains, and forests, combined with cool summers, draw visitors from every part of the United States and Canada. Although the Pacific Northwest does not possess a large population, excellent transportation facilities, services, and accomodations have favored tourist movement into the area.

Furthermore, there is considerable interregional travel. Tourism, the third-ranking industry in the region's economy, is growing rapidly.

Remoteness and small populations are the major handicaps of the southern hemisphere regions for tourism. Activity is largely limited to local participation. Superb alpine scenery

FIG. 14.30. (top) Sidewalk cafe in Paris—Left Bank. Throughout continental Western Europe this custom provides a relaxing pastime. This cafe is located on the Rue St. Germain near the University. Here Frenchmen and tourists leisurely enjoy refreshments and talk with friends, read newspapers, or simply sit and watch the passing parade. Traffic noise, congestion, and exhaust fumes are minimizing the pleasures of cafe sitting. (Conrad L. Heintzelman.)

FIG. 14.31. (bottom) Tourists in the Rhineland. This celebrated valley has many attractions for the European traveler—picturesque towns and cities, castles, and steep slopes devoted to intensive viticulture. (German Tourist Information Office.)

and numerous glacial lakes have stimulated a resort trade in the lake district near Puerto Montt, Chile. New Zealand also possesses natural wonders such as Fjordland, mountains, and hot springs. Air transport with reasonable fares has improved the accessibility of these southern Hemisphere attractions in recent years.

URBAN CENTERS

Industry and transportation functions have accounted for the development of many concentrations of cities in the Western Europe area. In recent years the urban growth has been at an accelerated rate, accompanying growth in the trade and service sector of economic activities. The urban character of a high portion of the population is indicated by the fact that about two-thirds of the British are found in cities or urban clusters with over 50,000 persons. Greater London, commercial, governmental, industrial, and financial center, has more than 11 million people. The population of Birmingham is nearly 3 million, and of Glasgow and Liverpool is nearly 2 million. Manchester, Sheffield, and Leeds have well over one-half million. There are at least fifty cities with

FIG. 14.32. Portland, Oregon. The panorama of the city looking eastward to Mt. Hood. The Willamette River bisects the city and provides a navigable waterway for ocean-going vessels from the Columbia River. (Portland Chamber of Commerce.)

FIG. 14.33. Champs Elysées, Paris—one of the most famous boulevards in the world. Here are located the many smart clothing shops and some of the better theaters of the city. Notice the symmetry of the buildings and skyline. These are some of the characteristics for which Paris is noted; in recognition of their tourist attraction, city planners minimize change. (Conrad L. Heintzelman.)

populations ranging between 100,000 and 500,000.

Paris, with nearly 9 million people, dominates the cities of the mainland (Figure 14.33), but there are many industrial centers in north France, Belgium, Netherlands, and Western Germany. One of the major concentrations is found in the Ruhr District in the triangle formed by Duisburg-Hamborn, Cologne, and Dortmund. There are eight cities with populations nearing or exceeding 500,000. Commerce and associated activities have promoted the growth of a number of large port cities such as Hamburg, Germany, with more than 1.8 million persons, at the head of the Elbe River estuary; Antwerp, main port of Belgium, on the Scheldt River, with about 700,000 people; Rotterdam, Netherlands, with 1.1 million, an international transit harbor at the mouth of the Lek, a distributary of the Rhine; Amsterdam-Haarlem, Netherlands, economic center of the nation, with about 1.3 million people, and a port by virtue of a canal to the sea; Copenhagen, Denmark, with about 1.4 million, commanding the gateway between the North and Baltic Seas; and Oslo, population around 600,000, capital and main port of Norway, located on a well-protected fiord in the southeast of the country. Brussels, a little over one million population, differs from most of those mentioned—it developed neither as a port nor industrial center, but rather as the political capital of Belgium. Later it developed considerable general manufacturing and financial and cultural functions. Small sea-going vessels can reach the city by way of a canal to the Scheldt River.

Four ports dominate the urban centers of the Pacific Northwest. From north to south, with metropolitan populations, they are: Vancouver, British Columbia, population about one million; Seattle, Washington, 1.4

million; Tacoma, Washington, over 400,000; and Portland, Oregon, approximately one million. Each owes its growth to strategic situation with respect to water and land transportation. The first three have excellent harbors on Puget Sound, a deep penetration of the Pacific Ocean into the land, where routes through the Cascade Mountains join the north-south transportation lines. Portland, 90 miles inland, is situated on the Willamette River, a few miles upstream from its juncture with the Columbia. The latter, the great river of the region, allows ocean-going vessels to move inland to Portland and barge traffic to pass through the Cascades to the heart of the Columbia Basin; at the same time it provides a low level land route through the Cascades and Oregon Coast Range to the Pacific.

Melbourne, Australia, with about 2.3 million people, is the largest city of the southern hemisphere regions. It has a fine position at the head of Port Phillip Bay, centrally situated on the southern lowland. More than one-third of the people of New Zealand live in the four major cities. Auckland is the largest urban center of North Island. This commercial capital, located on a spacious harbor, serves the richest lowland. Wellington serves the southern lands of North Island. Christchurch on the Canterbury Plains and Dunedin in the Otago grazing area are the urban centers of South Island. The far south is without a major city.

fifteen

DRY CONTINENTALS

The Dry Continental regions are lands of variety. Great stretches of middle latitude grasslands and deserts occupy seemingly endless expanses of plains and plateaus. Climate from region to region differs greatly in yearly temperatures, due to variations in latitude, but scanty precipitation is the common denominator. Risk and failure haunt the croplands of semiarid portions, where farmers attempt practices common in the humid lands. Low precipitation raises the land-man ratio—holdings change in size from acres to sections or square miles and in name, from farm to ranch or station. Crops are restricted to drought-tolerant cereals, and these can be profitably produced only in the portions of highest precipitation and best soils. Grass is the chief resource in many areas, and there grazing is the basic activity. The pastoral economy varies in complexity from that of the wandering nomad of central Asia to large commercial livestock ranches with huge herds of beef and bands of sheep in other regions. Settlements are widely dispersed, having developed mainly near mineral deposits and in oases where dependable supplies of water stimulate irrigation and intensive agriculture. Transportation advantages have been the stimulus for some urban developments. Nevertheless, man and his works still are almost obscured by the vastness of space in all the Dry Continental regions.

Location

The Dry Continental regions have two characteristic types of locations: one, deep in the interior of the large continents, distant from the oceans, the principal sources of atmospheric moisture; the other, in the dry shadows on the lee side of high coastal mountains. These locations account for scant precipitation. Regions are found on all the continents, but the widest distribution is in the great landmasses of the northern hemisphere (see Figure 15.1).

The largest region is found in the heart of Eurasia, filling the vast space that extends from the western Ukraine and the Caspian Sea to the Loess Highlands of northwest China. Distance from the Atlantic

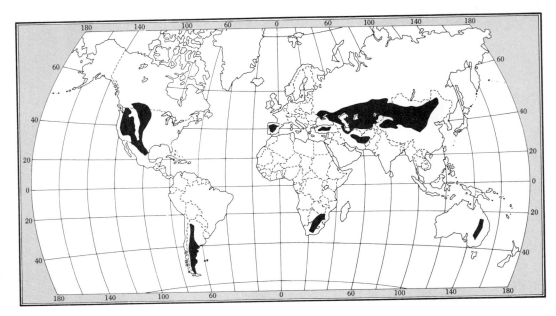

FIG. 15.1. The Dry Continentals.

FIG. 15.2. The Dry Continentals of Eurasia.

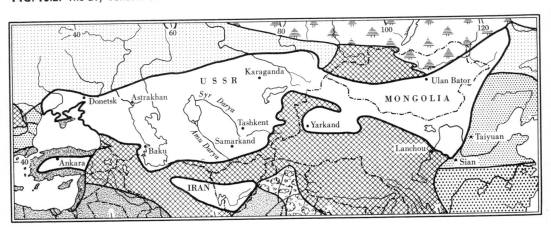

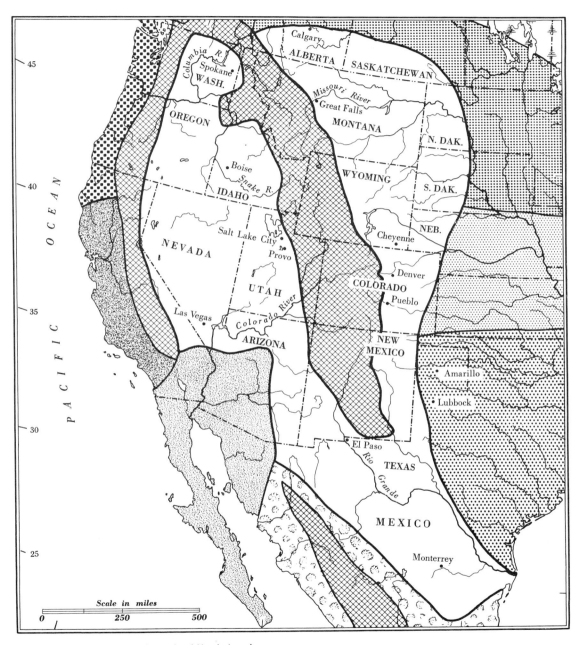

FIG. 15.3. The Dry Continentals of North America.

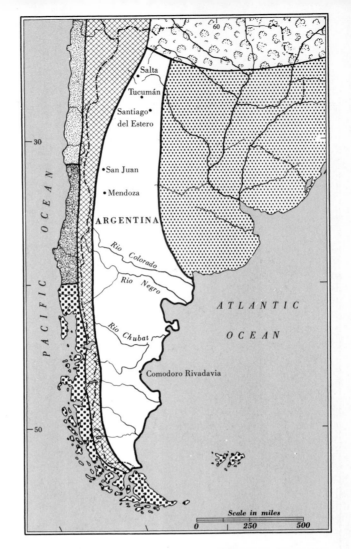

FIG. 15.4. (top) The Dry Continentals of South America.

FIG. 15.5. (bottom) The Dry Continentals of Australia and Africa.

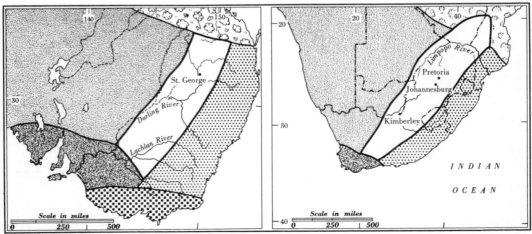

Ocean and the barrier of the earth's major highland system place the entire area out of the range of normal rain-bearing winds (Figure 15.2). Smaller Eurasian regions include the interior plateaus of Turkey and Spain—the Anatolian Plateau and the Spanish Meseta.

The Dry Continental regions of North America occupy a major portion of the middle latitudes between the 100th meridian on the east and the Sierra Nevada–Cascade Mountains on the west. The Rocky Mountains separate two divisions—the Great Plains on the east and the Intermontane Province on the west (Figure 15.3).

South America contains the only landmass in the southern hemisphere that extends far enough into the middle latitudes to develop extensive Dry Continental climate. A region lies in the lee of the Andes Mountains in Argentina. Owing to the narrowness of the continent, the entire area known as Patagonia and the western part of the pampas are in a dry shadow, and the region reaches the sea along its eastern border (Figure 15.4). Limited land extensions into the middle latitudes restrict the size of the Dry Continental regions in Africa and Australia. In Africa a region lies west of the Drakensbergs, and in Australia a small region occurs west of the Eastern Highlands in New South Wales (Figure 15.5).

Physical Environment

CLIMATE

The Dry Continental climates are lacking the homogeneity characteristic of the previous regions. Temperature contrasts are the major differences. The regions fringe mild, subtropical lands in nearly all cases. In Asia and the Americas, the great north-south extent of the regions through more than 20° latitude results in cold winters poleward. Narrow Patagonia faces the south Atlantic and is marine-modified. Only during the summer months do all the regions have temperature similarities; however, they all have extremes of heat and cold. Aridity and its environmental influences are the unifying factors.

Temperatures. There is no typical temperature for any one Dry Continental region as a whole. The mean July temperatures range from about 60° in the poleward margins to over 80° near the subtropics. Maxima exceeding 100° occur throughout all regions. Winter temperatures show even greater contrasts. The January means for stations near poleward margins range from below 30° to as low as −16°, at Urga, Mongolia. Blizzards and nonperiodic cold spells drive thermometer readings far below average. Stations in the northern Great Plains record lows of −40°. In contrast, winters are mild on the subtropical margins. In Africa and Australia coldest months are usually above 50°. Two stations in Patagonia indicate the influence of latitude on temperature: Santiago del Estero in the north has a January average of 83.1° and a June average of 55.9°; whereas Santa Cruz, 32 degrees farther south in latitude, has a January average of 58.6° and a June average of 35.2° (Figure 15.6).

Sunshine is abundant, especially during the summer. Clear skies promote both rapid daytime heating and nocturnal cooling, causing high diurnal ranges. Relative humidity is usually low, especially in summer. Frost-free periods range from all year in parts of Australia and South Africa to about ninety days in poleward areas of North America and Asia.

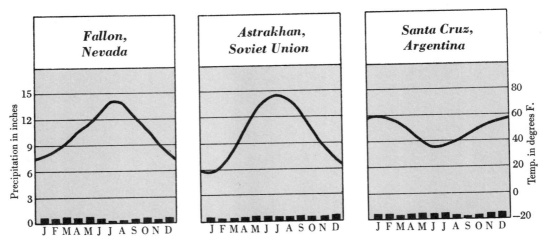

FIG. 15.6. Typical climatic graphs of Dry Continental stations.

Winter temperatures in the lands bordering the eastern foothills of the Rockies are modified by warm, dry *foehn* or Chinook winds. These winds develop when an air mass is forced up and over a mountain range. The air, becoming cooler as it rises, causes water vapor to condense, and the release of latent heat retards cooling. When the air mass descends on the lee side of the mountain, it is heated by compression. Consequently the descending air is relatively warm and dry, raising lee-side winter temperatures appreciably. This warm wind is sometimes called the "snow eater," as its warming effect rapidly melts the snow cover, exposing the grass. Temperatures have risen from below zero to 40° or 50° in a few hours. The Chinook is especially welcomed by cattlemen with stock on the range. Strong winds are common in the deserts during summer, filling the air with dust and sand; in the USSR the desert casts its spell beyond its boundary and dry, scorching winds play havoc with crops in the middle and lower Volga Valley.

Precipitation. Precipitation characteristics of low amount, irregularity, and undependability are common in all areas. On the basis of amount, two phases can be recognized: (1) the steppe or semiarid lands, on the humid margins such as are found in the Great Plains of North America and grasslands of western Asia, and (2) the desert, such as the Takla Makan and Gobi of Central Asia. Semiarid areas receive between 10 and 20 inches of precipitation. Evaporation rates, however, must be considered in measuring precipitation effectiveness. Where temperatures are high during the season of precipitation, evaporation rates are also high; therefore, 20 inches of rain on subtropical margins are much less effective than the same amount poleward. In fact, 20 inches of precipitation poleward often place an area within the humid limits—such is the case in the Prairie Provinces of Canada discussed in the Short Summer Humid Continentals. Middle latitude desert rainfall is almost the counterpart of Tropical Desert—10 inches and under.

Precipitation is concentrated chiefly in the summer half of the year, except on Dry Summer Subtropical margins where winter maxima are the rule. Elsewhere, winters are relatively dry, because of the presence of high pressure air masses; some snow falls, but the cover is usually light. Convectional showers, accompanied by thunder and lightning, are usual in summer, and destructive hail storms often occur. In the desert, some years are almost rainless and in others, sudden violent rainstorms drench the land. Irregularity is also typical of the semiarid phase, and often several wet years are followed by several years of drought. During the humid years farmers and stockmen prosper, but they suffer great economic loss during years of drought.

SURFACE FEATURES

Surface features are very similar to those of the tropical dry regions. The same types of weathering, degradation, and aggradation are at work; there is the same grass-shrub vegetation and the general absence of permanent streams.

Eurasia. Extensive plains, basins, and plateaus characterize the great Eurasian region. Three major landform divisions comprise the Soviet Union segment: the plain of the southern Ukraine, the Turan Lowland, which includes the plains between the Caspian Sea and the central Asiatic mountains, and the Kazakh Uplands, an ancient mountain mass worn down to rolling hills and plains that separate the Turan Lowland from the West Siberian Plain. Mountains divide the drylands of Inner Asia into several distinct units. The almost rainless Tarim Basin lies north of the Plateau of Tibet between the Altyn Tagh and the Tien Shan Mountains. The Dzungarian Basin, which forms a lowland corridor between Mongolia and Soviet Middle Asia, is located between the Tien Shan Mountains and the Altai ranges; both basins are in Sinkiang. The largest unit is the Gobi Desert of Mongolia, the world's most northern desert. Surrounded by mountains, this broad basin-like depression appears as a rather featureless plain with many areas of desert pavement and shallow hollows without exterior drainage.

The Loesslands, one of the most unusual areas in the world, form the transition from the Gobi Desert to the subhumid North China Plain. This region of hills, mountains, and valleys is mantled with wind-laid silt or loess. The fine, powdery loess, varying from a thin film on the steepest slopes to depths of 300 feet, has apparently been carried by the strong winter monsoons from the Ordos Desert, which lies to the west. Many people here live in caves cut into the loess walls of the valleys. Such dwellings are cool in summer and warm in winter, but disastrous when earthquakes occur, as they sometimes do.

North America. Two broad physiographic provinces may be recognized in North America: the Great Plains east of the Rocky Mountains, and the Intermontane Plateaus and Basins between the Rockies and the Sierra Nevada-Cascade Mountains. The former stretches northward from the Rio Grande to the Mackenzie River in Canada, a distance of 4,600 miles, with an average width of about 400 miles; however, as Figure 15.3 indicates, not all this area is included in the Dry Continental region. The general surface is remarkably smooth, but the elevation declines gently from 5,500 feet at the eastern base of the Rockies to 1,500 feet at the western border of the Central Lowlands. There are local variations in surface. For example, in northwestern Nebraska there is a hilly sand dune belt; in Wyoming and the Dakotas there are extensive areas of badland topography where minute dissection has produced exceedingly rough surfaces with very complex forms; there are also several mountain outliers, such as the Black Hills of South Dakota and Wyoming. Terraces are present in the major stream valleys, indicating several uplifts of the region, causing rejuvenation of downward erosion.

The Intermontane Province is much

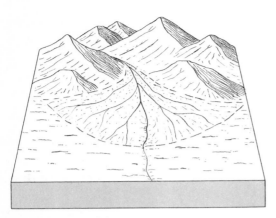

FIG. 15.7. An alluvial fan.

more complicated than the Great Plains. The Mexican Plateau occupies the southern portion; this large expanse of arid land is characterized by many isolated ranges rising 2,000 to 3,000 feet above the general surface. These mountains are usually asymmetrical fault blocks, formed by upward thrusts along zones of fracture in the earth's crust. Among the ranges there are alluvial fan-fringed basins of interior drainage with playa lakes (see Figures 15.7 and 15.8). These basins have been termed *bolsons,* from the Spanish word meaning purses. The Colorado Plateau occupies a major part of New Mexico, Arizona, and Utah. This great area of some 130,000 square miles is made of brilliantly-colored and nearly-horizontal sedimentary strata. The work of running water has carved fantastic landscapes, including many vertical-walled canyons, the most magnificent of which is Grand Canyon of the Colorado River. Northwest of this plateau is the Great Basin, an area of interior drainage and mountain and bolson topography. The characteristic features are numerous basins separated by block mountains which trend north-south. The ranges are commonly 50 to 75 miles long and rise 3,000 to 5,000 feet above the basin floor. The surface of the Great Basin grades almost imperceptibly into the Columbia Basins and Plateaus in the northern part of the region (Figure 15.9). The unifying features of the Columbia Basins and Plateaus are the extensively distributed surface of lava and the common drainage by the Columbia River system. Actually there is a variety of landforms. The Blue-Wallowa-Seven Devil Mountains roughly divide the province into the Columbia Basin in the north and the High Lava and Snake River

FIG. 15.8. A bolson and ranges.

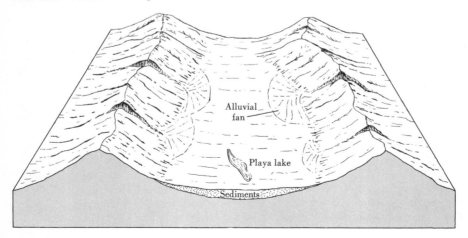

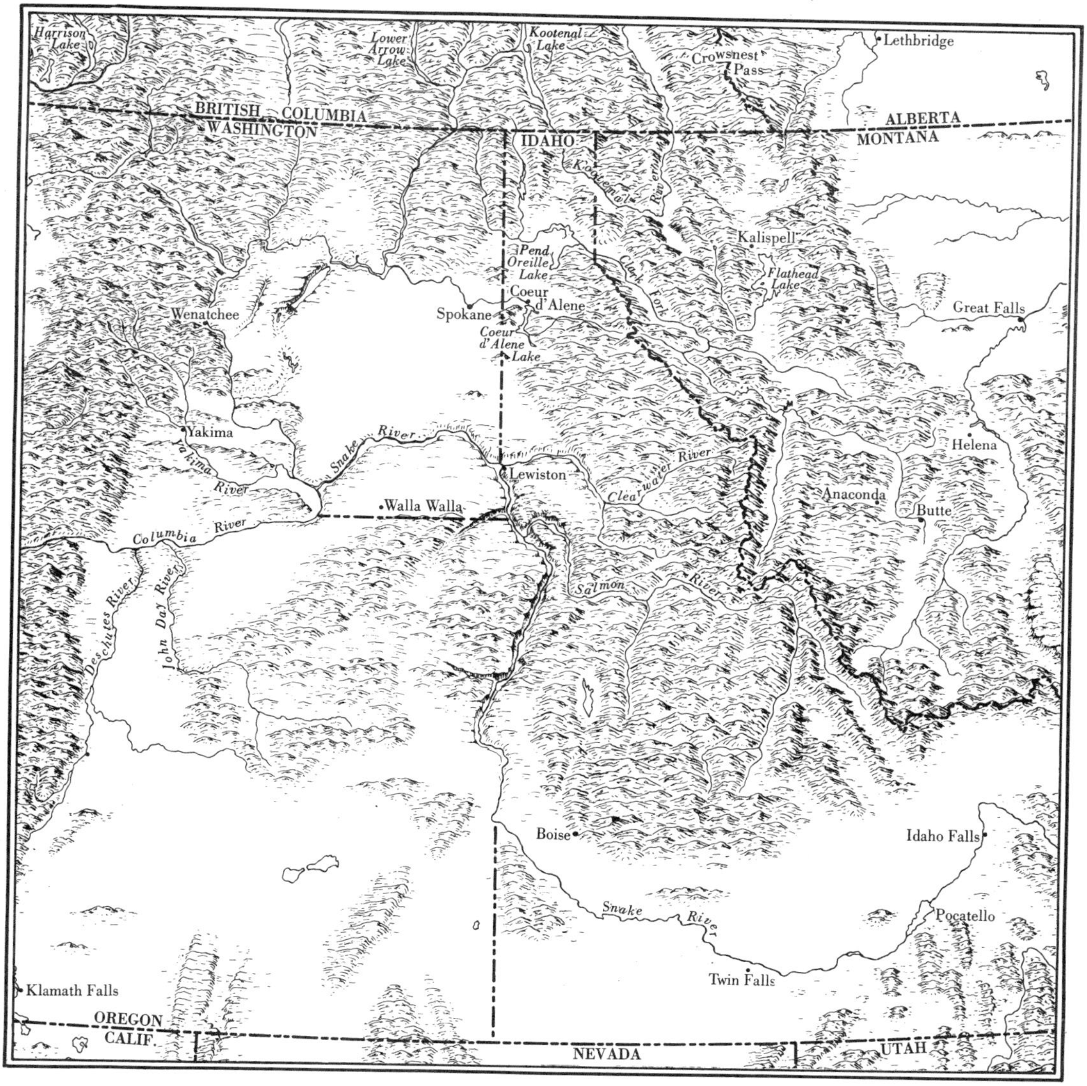

FIG. 15.9. Physiographic diagram of the Intermontane Province of the Pacific Northwest.

Plains on the south. The northern part of the Columbia Basin was considerably altered by glacial meltwaters that poured across it to rejoin the Columbia River and, in so doing, carved many channels, the most famous of which is Grand Coulee (Figure 15.10).

Southern hemisphere. In Argentina, the plain of the north part of the Dry Continental region gives way to plateau topography in

Patagonia, which extends southward from latitude 38° or 39°. This southern sector slopes eastward from an elevation of about 5,000 feet near the base of the Andes to a few hundred feet along the coast. The regions of Africa and Australia also occupy piedmont positions. The former, an area known as the High Veld, slopes westward from the Drakensbergs; the Australian region occupies a similar position to the west of the Eastern Highlands.

NATURAL VEGETATION

Low precipitation largely restricts the natural vegetation to grass, shrub, and desert xerophytes. Trees appear only on slopes of higher elevation, where precipitation is more abundant or where they grow along water courses.

Steppes. Grass, the vegetative cover in semiarid lands of nearly all the regions, varies in height and density of stand with precipitation and available soil moisture. The Great Plains of North America is a short-grass region, dominated by buffalo and grama grass in the north and central portions and by mesquite grass in the warmer south. Bunch grass is found west of the Rockies. The grasslands of South Africa are known as the *veld*. Short grasses form much of the cover in Eurasia and Australia. Nor-

FIG. 15.10. (left) An aerial view looking southwest over Grand Coulee Dam with the Third Powerplant Construction area in the foreground. This great man-made structure has a generation capacity of nearly two million kilowatts and is a key to the irrigation of 600,000 acres of arid land, now being developed in the central plains of the Columbia Basin. The water is lifted by giant pumps 280 feet from Lake Roosevelt into an equalizing reservoir that occupies the northern portion of the Grand Coulee, the great Ice Age meltwater channel of the Columbia River. From the southern end of this reservoir a system of canals and laterals is being constructed to carry water to the land. (United States Bureau of Reclamation.)

FIG. 15.11. (below) Grass is the dominant vegetation in areas with 10 to 20 inches of annual precipitation. When care is given to proper stocking and grazing management, there is provision for a permanent livestock adjustment. (United States Forest Service.)

FIG. 15.12. Subtropical vegetation in the Intermontane Province of the United States. In this Nevada scene are blackbush and the Joshua tree. Notice that the minute details of the landforms are clearly visible. Vegetation such as this provides a very low grazing capacity. (United States Forest Service.)

FIG. 15.13. House Rock Valley Buffalo Herd (maintained by the state of Arizona as a reminder of the once vast herds that roamed the plains) grazing under special use permit. (United States Forest Service.)

FIG. 15.14. Big sagebrush type, Colorado. (United States Forest Service.)

thern Patagonia is distinctive for its large number of low shrubs; farther south the country appears as a vast, treeless moor with grass tussocks.

Deserts. In response to lessening amounts of precipitation and/or increasing rates of evapotranspiration, grasses are gradually replaced by low shrubs on the drier margins with species differing latitudinally and to some extant from continent to continent. Sagebrush is the usual vegetation in the North American deserts, and yuccas and cacti appear on the subtropical margins. Other species include greasewood, creosote brush, mesquite, and shadscale. The scanty desert vegetation is composed of xerophytes and seedproducing annuals which utilize all the protective devices of nature to fight aridity and high rates of evaporation.

NATIVE ANIMAL LIFE

The grasslands are the habitat of a variety of herbivorous animals as well as carnivores, rodents, reptiles, and birds. Insect life is common in all regions; grasshoppers and locusts are the greatest pests, doing considerable damage to range and crops. The steppes of Asia are inhabited by antelopes, wild asses and horses, bactrian camels, and wolves. In North America there are wolves, coyotes, antelope, and smaller animals such as the jackrabbit, gopher, and prairie dog. The bison at one time was the dominant animal of the North American region; today only a few protected herds exist (Figure 15.13). Patagonia has many small rodents but few large animals. On the veld of South Africa, the animals are quite similar to those inhabiting the savanna to the north. Marsupials, typified by the kangaroo, are common in Australia, and the emu and the dingo, or wild dog, also range the Australian steppe. Rabbits, introduced during early settlement, are extremely numerous and constitute a problem for stockmen.

SOILS

The soils fall into the major division of zonal arid or pedocal—soils in which carbonates accumulate, usually in the B horizon.

FIG. 15.15. Buck pronghorn antelope. (United States Fish and Wildlife Service: E. P. Haddon.)

In response to lowering precipitation and decreasing vegetation cover, several groups are formed. The chernozems and the brown soils are found in the semiarid grasslands. Chernozems develop in the higher-rainfall margins under tall, deep-rooted grass. Precipitation is not great enough to produce leaching, and as a result this group is high in mineral plant food as well as organic matter supplied by the extensive root systems. These black earths are high in fertility and excellent agricultural soils; the major limitation is the low precipitation which restricts their normal crop adaptability. There are two large areas of chernozems: One stretches through the southern Ukraine and across the southern margin of the West Siberian Lowland; the other occupies the eastern margin of the Great Plains of the North American region. The brown soils have developed on the margins of the chernozems under slightly lower precipitation and shorter grass. The soluble salt accumulations are nearer to the surface, and the grass roots provide less organic matter, thus the lighter color. These soils are fertile but are strictly limited for crop agriculture, owing to deficiencies in rainfall, and are used predominantly for livestock grazing.

The soils of the Middle Latitude Deserts largely fall into the gray desert soil group, although parent material may result in other light colors. The content of organic matter is low. Lime and other soluble substances accumulate near or even on the surface. Usually concentrations are not sufficient to harm plants; however, alkali areas do occur where the surface accumulations of soluble salts are so great as to be injurious to plants.

The fertile loess merits special mention because the semiarid phase has the greatest distribution of this wind-laid soil of any of the regions of the world. The middle-latitude deserts, with little protective vegetation, are easy prey for the winds. Silt-laden winds from these arid lands, blowing across the bordering grassland, deposit their load to be held by the more complete vegetative cover. In addition to the Loess Highlands of China, extensive areas occur through the central Ukraine into the steppe lands of western Siberia, in the Palouse Hills of eastern Washington, and in the Missouri-Mississippi Valley. The Palouse Hills are unusual. Viewed from the air they are like a billowy sea. Despite a high percentage of slope land, deep, rich soils have encouraged almost complete utilization for cropping.

Occupance in the Dry Continentals

The Dry Continental regions are frontiers of modern settlement; but the occupancies of some are very old. The oases towns of the deserts of Inner Asia are ancient, and the neighboring steppes have been utilized for grazing by Mongol nomads for centuries. Major development in other regions is relatively recent, and in many cases still evolving. Much of the progress has been made in the twentieth century, resulting from improved transportation, mechanization of farming, better techniques, large-scale irrigation projects, and growth in markets. Even the Inner Asian region is also undergoing some modern development under Soviet direction.

Agriculture is the principal activity. Three phases are significant: dryland grain farming, irrigation crop production, and livestock grazing. The first and third are extensive types with dispersed settlements supporting relatively few people. The densest populations occur in the irrigated districts. The Loesslands of China support the largest population. Despite the semiarid character of the climate, there are more than forty million farmers. In contrast, extensive areas of the Gobi Desert are entirely without people, as is also true in many other portions of the arid regions. Mineral discoveries have added to the importance of several of the regions; in the USSR region, these have

FIG. 15.16. A native dwelling in the Orange Free State. Cultures differ greatly in the Dry Continentals of South Africa. This simple thatched-roof hut contrasts sharply with the modern farm home of the Europeans in the region. (South Africa, Government Information Office.)

FIG. 15.17. Modernization of agriculture is under way in Turkey. This method of cutting wheat is a vast improvement over the old hand sickle. Notice that this Caterpillar Diesel was manufactured in Peoria, Illinois. (Turkish Information Office.)

made possible the development of a major industrial area in the eastern Ukraine and in several scattered, lesser centers.

DRY FARMING IN THE SEMIARID LANDS

There is some crop production in all the major semiarid regions, and this phase of agriculture has been increasing in significance. The climate is best suited to grain and grass-type plants. Wheat is the chief cultivated crop, and, together with the bordering subhumid lands, these regions constitute the wheat grainaries of the world. The Chinese, Soviet Union, North American, and Australian regions are especially important, and parts of the plateaus of Turkey and Spain are also cultivated. The Soviet Union's Dry Continental region contains much of the better farmland of that nation.

One, two, three, or even more sections of land commonly constitute the farm units in the free enterprise countries. In communist countries units are much larger. Operations are fully mechanized, with tractor-powered equipment worked in gangs so that 20 to 60 or more feet of land can be cultivated or seeded in one swath around a field. In those areas where precipitation is between 10 and 15 inches, crop production usually requires a summer fallow rotation. In alternate summers the land is cultivated and kept free of weeds, but no crop is produced; the soil thus has an opportunity to store moisture and nitrogen for the following year. In the areas of 15 to 20 inches of reliable precipitation, a crop is often produced each year; exceptions, however, would be found in the areas of high evaporation. For example, a rough approximation is that 8 to 10 inches on moisture-retentive soils are required to grow and mature wheat along the northern border of the United States and, in response to increases in the evapotranspiration rate, another inch is required for each 150 miles movement southward. Several crops other than wheat are also grown in the semiarid regions. Some cotton is found in the southern Ukraine and in the southern part of the Great Plains. Grain sorghums and millet locally replace wheat as the principal grain or are sometimes used as a rotation, especially in areas with hot, dry winds. Large acreages of sunflowers are grown in the Ukraine.

The Loesslands of North China are the most distinctive semiarid region. Despite adversities of climate and slope, population densities rival those of other areas in the Orient. Methods are necessarily more intensive, but crops are similar to those of other semiarid regions. Millet, winter wheat, and kaoliang (a sorghum) lead the food crops, with cotton, tobacco, and opium the major cash crops. The ratio of population to cultivated land, nearly 1,200 per square mile, has made this one of the most serious famine areas in the world; a slightly drier than normal year results in lower crop yields for the farmer and insufficient food for all.

Farming can be treacherous business in any of the semiarid regions, because droughts appear to be part of the normal climatic cycle. A series of wet years may encourage extensive modifications of natural conditions that may lead to disaster. The United States Great Plains serve as an excellent example of man's struggle with nature in these regions. Following the War Between the States, a major wet period, coinciding with the building of transcontinental railroads and an increase in westward population movement, encouraged settlement on the Great Plains. The 160 acres allowed by the Homestead Act was far too small to support a family by grazing; settlers were required to put this land under the plow, regardless of its suitability. Not until 1909 was the

FIG. 15.18. A view of wheat fields in the north central Oregon plateau. Notice that the fields occupy smooth to rolling lands in the interstream areas. The slopes are avoided because of erosion danger and the difficulties of utilizing machinery. (United States Department of Agriculture, Soil Conservation Service.)

homestead tract enlarged to 320 acres; and although in 1916 grants of 640 acres were authorized, a provision was made that they be used solely for grazing. Thus the homesteader often had to plow when plowing was harmful to the land and was sometimes forbidden to plow when plowing might have been profitable and unharmful. Much of the land remained in grass because of these limitations. Large herds of cattle grazed on what was practically one great open "pasture," extending from Texas to Montana. Drought began in 1886 and continued until 1895, and the "cattle barons," faced with financial disaster, almost faded from the scene. Uncontrolled grazing on free range was not a viable form of occupance.

Farming during the nineteenth and early twentieth centuries expanded in wet years and contracted somewhat in dry years as farms were abandoned. On the whole, however, cultivated land expanded at the expense of the range. After 1910, the advances in mechanical equipment and the high prices paid for wheat became powerful influences. Large acreages were turned by the plow—including lands with a history of drought. If sufficient rain fell there was a harvest; if drought supervened there was not. Dry years characterized the climate of the plains in the early thirties, and severe drought conditions burned the plains in 1934. Scores of farmers and cattlemen lost their land and livestock. Just as overgrazing impaired the natural pastures, so wrong tillage methods injured what had been considered tillable land. Soil exposed to the winds was lost in great quantities, especially in the south. Only now is man beginning to learn that he must work with, rather than in opposition to, nature if he is to live permanently in this region. Progress in land classification, adjustment of use to capability, cultural practices, windbreaks, and other techniques are improving greatly his ability to adjust permanently to the region.

Similar problems have long plagued the Russians. Much of their agricultural area is periodically subject to drought. The Middle and Lower Volga Valley and adjacent areas constitute the principal portion with recurring dry years. In the period 1891 to 1937 there were only 11 years of good moisture supply; 22 had partial droughts and 15 had full-fledged droughts. More or less extensive droughts have been recorded in a significant portion of the years since 1937.

Finding ways to combat these calamities has been of special concern to the Soviet government. In 1948 they announced a fifteen-year program that included moisture-conserving practices, some local irrigation, and an ambitious program of tree planting. The most important phases of the latter scheme involved the planting of national forests in the southwest European region on the watershed divides and on river banks such as the Volga, Ural, and Don and the establishment of tree shelter belts on the collective and state farms. The forests would have a double purpose: first, to help retain the snow more uniformly spread on the ground—this would protect the fall-sown grain and become a significant source of moisture with the spring melt; second, to reduce the soil erosion and soil moisture evaporation produced by the winds off the Asiatic desert. The success of these projects cannot yet be predicted; much depends upon whether or not the trees can be grown—experience has shown that unless good care is given the young trees, especially during the critical first three years, they will usually die.

IRRIGATION AGRICULTURE IN THE DRYLANDS

Irrigation, the artificial control of soil moisture by application of water, was the earliest basis for crop agriculture in the middle-latitude drylands. Works for the collection and distribution of water were established in ancient times in the dry heart of Asia. Today in this vast region, people are permanently settled only at the small and scattered oases. Most of these developments are in the Tarim Basin of Sinkiang and are situated on the upper margins of the alluvial fans bordering the mountain. The ancient kanat system of underground tunnels is still the most widespread technique for water distribution.

The southern Turan oases, together with those on the north flanks of the Caucasus and in the lower Volga Valley, are extremely important to the economy of the Soviet Union. The longer frostless season in these areas makes possible the growth of crops that can not be produced in other sections of the country. Cotton is the chief crop, occupying two-thirds of the irrigated acreage in the Turan oases. Wheat and rice are the principal grains, and sugar beets and a variety of fruits are important. About 25 million acres are being supplied with water. Irrigation techniques vary from the small ancient kanats to large-scale projects of modern design developed by the Soviet regime. The bulk of the irrigated acreage, however, is contained in the modern schemes.

Irrigation agriculture in the North American region is widespread. Including the sections described in other regions, nearly 33 million acres are irrigated in the United States West. Some of the acreage is in large-scale projects developed by the United States Government. Other enterprise forms, however, account for a larger share of the acreage; these include cooperative and mutual systems, incorporated and unincorporated, water districts organized under state laws, and individual and partnership enterprises. The latter have been the major growth category in recent years. The projects are too numerous to mention, since almost all the permanent streams provide some irrigation water. Alfalfa, potatoes, sugar beets, and a host of horticultural crops constitute the main enterprises; cotton is a leading crop in the southern areas. Some projects have become famous for their specialties, but in most, agriculture tends to be diversified, and livestock production is usually of some importance. Alfalfa and by-products of other crops, such as sugar-beet tops and pulp, are used for feed.

During the last three decades the High Plains of Texas (in part included within the Cotton Belt) and the adjoining area of New Mexico have been the site of a spectacular development of irrigation based upon the tapping of groundwater through deep

FIG. 15.19. The raw desert before irrigation. This barren ground near Moses Lake, Washington, on the Columbia Basin Project, marks the site of the Moses Lake Development Farm on November 1, 1946. (United States Bureau of Reclamation.)

FIG. 15.20a and b. Columbia Basin Project, Washington—and aerial view of the Quincy Valley showing the development of the irrigated lands. The West Canal may be seen flowing through the valley, and the town of Quincy is in the background. Figure 15.20b shows the same area in 1952, before development. (United States Bureau of Reclamation: J. D. Roderick.)

wells. Cotton and grain sorghum are the chief crops (Figure 15.23). This development, however, has a questionable future because the water supply is limited, being based upon a stored stock that is receiving little recharge. There also have been noteworthy irrigation developments in some of the northern basins and valleys of the Mexican portion of the region. Cotton is the major enterprise in most of these areas.

In Argentina a number of oases have developed along the piedmont of the Andes, diverting water from mountain-fed streams. Since they extend through about 15 degrees of latitude, a variety of crops is produced, but there is some local specialization. Mendoza, with about one million acres under irrigation, is the largest, and together with the San Juan oasis to the north produces the bulk of Argentine wine grapes. Tucumán, farther north, is perhaps the most famous oasis; it is synonymous with sugar, producing more than three-quarters of the nation's supply on large plantations. Developments also have been made along the Rio Colorado, Negro, and other streams that traverse Patagonia. In this area, Argentina is constructing the major Chacón hydro complex to develop over a million killowatts of electric generation capacity, to control the river, and to

FIG. 15.21. Siphoning water from irrigation ditches to the fields. Plastic tubes convey water to the corrugations between the sugar-beet rows. This is the principal system for irrigating row crops. Where land can be leveled and forage or grain crops grown, low dike borders are commonly constructed and water is turned into the field at one end, to progress slowly as a sheet down a slightly graded slope. This latter system is known as flood irrigation. (United States Bureau of Reclamation.)

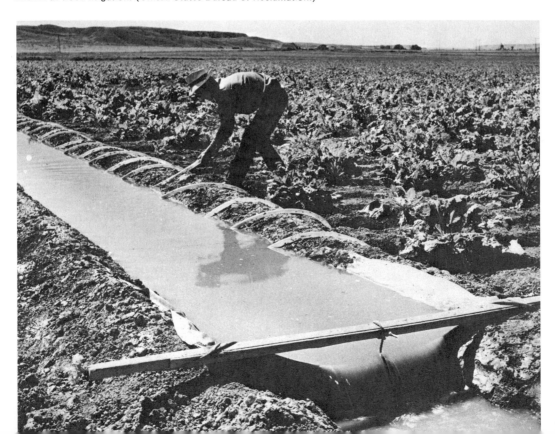

FIG. 15.22. Rosa Irrigation District, Yakima Project, Washington. This panoramic view looks east over the Rosa Irrigation District from the east end of Snipes Mountain. In the foreground are the Upland Winery Farms with cherries on the left and concord grapes on the right. Many varied crops are grown in this valley such as mint, apples, peaches, pears, hops, hay, and grain. (United States Bureau of Reclamation: J. D. Roderick.)

FIG. 15.23. Cotton near Lubbock on the High Plains of Texas. Since the last part of the nineteenth century, agriculture here has evolved from range livestock activity, to dry farming in the early part of this century, to extensive irrigation beginning in the 1930s and developing rapidly after 1950. In the late 1960s, about 50,000 wells were tapping the stored water supply to irrigate about 5 million acres, which accounted for approximately 10 percent of the nation's cotton crop and 35 percent of its grain sorghum. The aquifer has been cut off from the source area by the Pecos River, and most of the local runoff collects in playas (due to impervious subsoil) and is lost through evaporation. Without adequate recharge the water source will be exhausted in time. (Soil Conservation Service.)

FIG. 15.24. Loading alfalfa hay in Africa. Draft animals are still important in the farming scheme. (South Africa, Government Information Office.)

provide irrigation water for about 2 million acres of land in the Negro Valley.

GRAZING

The Dry Continental regions contain the most significant grazing lands of the world and are the centers of the commercial range livestock industry. Land use is dominated by grazing except where precipitation is sufficient for dry farming or where sufficient water is present for use in irrigation agriculture.

Pastoral nomadism in Mongolia stands in sharp contrast to the commercial livestock industries prevailing in other regions. On the steppe lands of Inner Asia, the Mongol nomads keep horses, sheep, goats, yaks, and camels. Animal specialties differ with nomad tribes and from area to area, but sheep are preferred. Animals furnish meat, milk, hides, fleece, fuel, are used for barter, and have

FIG. 15.25. A view across the grazing and dry-farming area of the Orange Free State. The Drakensbergs are in the background. (South Africa, Government Information Office.)

social as well as economic significance. The Mongols have no permanent homes but follow their animals along regular grazing routes. Living is much the same as in the days of Genghis Khan, but the Soviet government is doing much to settle these groups permanently.

State livestock farms have been established by the Soviets in the USSR steppe, operated by herdsmen who have a nomad ancestry. The collective farms also raise sheep, cattle, horses, goats, and camels. Many improvements have been made in the traditional system; pasture rotation is practiced, breeding stock has been improved, and provisions made for winter feeding. The area is noted for the fat-tailed Khirghiz sheep and for the karakul sheep which produce Persian lambskins.

The range livestock industry of the Great Plains and Intermontane Province of North America is based on beef cattle and sheep. Hereford is the main type of beef cattle, and sheep are Merino or dual-purpose crossbreeds. Cattle are most widely distributed, whereas sheep are chiefly concentrated in the cooler northern sections and the more rugged and arid areas. Sheep are able to utilize steeper slopes and have the ability to subsist on shorter grass than cattle; they can browse on shrubs and require less water. The greatest concentration of Angora goats in the United States is found on the Edwards Plateau of southwest Texas. Ranches are large, usually several thousand acres, with some exceeding 100,000 acres. Sizes increase with decreased precipitation and quantity and quality of forage, which necessitates more land per-animal-unit. Details of range quality influence the beef cattle specialization also. In the northern plains, under good management, the forage base is excellent through the frostless season and the grass cures well on the stem to provide natural hay when the ground is not covered with snow. Many livestock men here specialize in putting solid gains on mature animals which they buy for this purpose. In contrast the Southern Plains have a high evapotranspiration rate and poorer forage, but grazing can be year-long. In consequence more ranchers concentrate on cow-calf operations with the young beef animal sold as a stocker or feeder.

In the ranch country the human imprint upon the landscape is slight. The occasional fence, the watering facility, and the widely spaced ranch headquarters are the principal observable features. The isolated ranch sites are usually determined by the presence of water and transportation access. Buildings consist of the ranch home, bunkhouses for the ranch hands, barns, and corrals. Irrigated hayfields for winter supplemental feeding are often maintained near the ranch headquarters.

Spring and fall roundups are the busiest times for cattlemen; calves are branded and marketable cattle segregated from the herds at these times. The balance of the cattle is scattered over the range. Spring, the lambing and shearing period, is the busy season for the sheepmen. Sheep are maintained in bands, about 1,250 ewes plus lambs, and cared for by a lone sheepherder and his dogs. In the Intermontane Province and on the western edge of the Great Plains, livestock are often driven into the mountain pastures during the summer. Texas cattlemen shuttle livestock seasonally between humid and dry portions of the plains. Such movement of herds, practiced by grazers in all parts of the world, is known as *transhumance,* which denotes a seasonal movement of flocks or herds of domestic animals between two areas of different climatic conditions.

The Southern Hemisphere Dry Continental regions are especially noted for sheep. Patagonia is one of the great sheep-producers of the world, and grazing of these animals is the major utilization of the land. Large *estancias* are maintained by sheepmen of British ancestry. These holdings average 22,000 acres in size, but many are larger.

FIG. 15.26. A band of sheep in New South Wales. Australia normally has about 175 million sheep and is a principal producer of the world's finest wool. Sheep numbers fluctuate with precipitation cycles; during "wet" years, numbers increase. (Australian Department of Information.)

Isolation has made wool the specialty, and this product must be moved great distances by truck. Australia is the leading world wool-producer, and the principal sheep areas are located in the semiarid lands (Figure 15.26). Here huge ranches, known as stations, also specialize in the Merino breed for quality wool. Some stations keep herds of cattle. Australia's semiarid region is favored for grazing by mild winters, artesian water supply, and a scarcity of animal diseases. In both South Africa and Australia, drought often is a serious problem, causing a scarcity of water and grass. Predators, too, take a toll; in Australia an abundant rabbit population competes for forage, and dingos kill many animals; jackals and other wild beasts kill thousands of sheep in Africa.

FISHING

South America is the only continent having a Dry Continental region whose boundaries have an extent of ocean shoreline; the regions in the other continents are landlocked. Fishing is almost a nonexistent means of livelihood in these regions—with one major exception.

The Soviet Union has several interior seas, lakes, and lower river courses where important fisheries exist. These are the Black, Azov, Caspian, and Aral seas, and Lake Balkhash, and the Don, Volga, and Ural rivers, which drain into these water bodies. The catch includes carp, herring, salmon, and sturgeon. The shallow water around the north shore of the Caspian, where the Volga and Ural rivers empty large quantities of organic matter into the sea, is one of the Soviet Union's richest fishing grounds.

MINING

Mining has become an increasingly important activity in the Dry Continental regions in recent decades. The Soviet Union and American areas are especially important now, and the Loesslands of China are the locus of recent Chinese efforts. The latter region has large coal deposits but, owing to the difficulties of transportation and limited industrial development in pre-communist times, these reserves are only beginning to contribute to China's economy.

Large and diversified deposits in the Soviet region have made possible its rise as a leading mineral producer. The iron deposit at Krivoi Rog, immediately west of the bend of the Dnieper River, has been the leading producer for many decades. The best ores are nearly exhausted, but as in the case of the Mesabi Range of Minnesota there remain large reserves of lower-grade ores. The Donetz Coal Basin, long the outstanding producer, lies about 200 miles to the east. A manganese deposit is located at Nikopol farther down the Dnieper. The leading petroleum production is at Baku on the Apsheron Peninsula, which juts into the Caspian Sea from the west; other fields occur around the eastern and northeastern shores and to the north of the Caucasus Mountains. The principal metallurgical coal source for the Ural steel industry is the Karaganda deposits in the midst of the Kazakh Upland. Near the north shore of Lake Balkhash is a great copper mine, at Kounrad. Salts, common to most deserts, are taken in large quantities in several localities, but particularly from Kara-Bogas Gulf on the eastern side of the Caspian Sea.

Copper, petroleum, and salts are major mineral products of the American region. Copper is mined in a number of localities in the Intermontane Province of the United States, with the Bingham Canyon open-pit mine in Utah the most famous. Petroleum and natural gas deposits are being worked in west Texas, New Mexico, Utah, Wyoming, and in northwestern North Dakota. Fields are also being developed in Alberta, Canada (as noted in the discussion of minerals in the Short Summer Humid Continental region). Salts are reclaimed from the deposits of Searles Lake, a remnant of an inland sea in the Mohave Desert of California, as well as from the Great Salt Lake of Utah.

Southern hemisphere regions, except for

FIG. 15.27. A washing and crushing plant in the Witwatersrand District. (The Republic of South Africa, Government Information Office.)

Africa, are comparatively unimportant in the world picture of mineral production. There is a small output of petroleum in Patagonia, which is Argentina's chief source. Gold, diamonds, coal, and iron are mined in the South African region; Witwatersrand, in southern Transvaal is the outstanding gold deposit in the world; the Rand runs about 115 miles east-west and up to 30 miles in a north-south direction. Johannesburg is located in the center of the area. A large coal field lies about eighty miles northeast of this city and iron ore deposits are located about a hundred miles north (Figure 15.27).

MANUFACTURING

Manufacturing has its greatest emphasis in the Eurasian region, with the second major industrial district of the USSR centered in the eastern Ukraine. This district is well situated with respect to resources and markets. The large coal deposits of the Donetz Basin, iron ore of Krivoi Rog, the Kerch Peninsula, and the supplies of manganese from Nikopol have stimulated an iron and steel industry. The products of the steel mills are manufactured into a variety of commodities, ranging from farm machinery to locomotives. Other activities are associated with chemicals, cement, aluminum, sugar refining, wheat milling, leather goods, textiles, and shipbuilding. Hydroelectric power generated on the Dnieper and Volga Rivers and petroleum piped and tanked from the Caucasus and Caspian districts are added industrial attractions. Farther east, Volgograd (formerly Stalingrad), on the Volga, specializes in iron and steel, agricultural machinery (especially tractors), oil refining, and food processing. Southeast of the Aral Sea, modern industrialization has come to many of the ancient oases towns long noted for handicrafts, such as Samarkand, Tashkent, and Fergana, in the form of cotton and silk textile mills, food-processing plants, and a modest steel industry.

In comparison to the great American Manufacturing Belt, the North American region is not a major industrial area, but manufacturing does play a role in the overall economy. Industries rely chiefly on the products of agriculture and include meat packing, the processing of fruits and vegetables, flour milling, and refining of beet sugar. Others have been established to take advantage of local minerals, hydroelectric power, or strategic advantages of space and relative isolation. Many smelters operate near mining areas; copper is smelted at Garfield, Utah. Pueblo, Colorado, possesses a small iron and steel industry, and Provo, Utah, produces almost two million tons of steel annually. Aluminum plants are located at Spokane and Wenatchee, Washington, and The Dalles, Oregon, in response to hydroelectric power. Atomic energy projects are found at Hanford, Washington, Amarillo, Texas, Los Alamos, New Mexico, and near Denver, Colorado. Several industrial centers have developed in Mexico. Steel is produced in significant quantities in Monterrey and Monclova. Cotton textiles manufacturing is increasing in a number of centers.

Industry in the southern hemisphere is represented by the African region. The major area of development extends from Vereeniging northward through Johannesburg to Pretoria. Johannesburg has steelworks, and the other two centers have blastfurnaces as well; their combined output has made South Africa nearly self-sufficient in steel. Vereeniging also processes gold and has developed subsidiary industries including coal-based chemicals, engineering works, and plants for the processing of agricultural products.

FIG. 15.28. The Boulder Canyon Project. Hoover Dam and Lake Mead are major tourist attractions. The principal functions of this multipurpose installation are, however, flood control, irrigation, water storage, and hydroelectric generation. (United States Bureau of Reclamation.)

TOURISM

The Intermontane Province of the United States is rich in resources favoring recreation and tourism. Resources include natural wonders, desert landscapes, Indian ruins, and a variety of other attractions. Many of the spectacular wonders of nature have been preserved for the public in National Parks and monuments such as the Grand Canyon of the Colorado, Bryce Canyon, Zion National Park, the Petrified Forest, Painted Desert, and Carlsbad Caverns. Although the region is distant from large population centers, excellent highways and rail facilities bring thousands of visitors during the vacation season. It also derives advantages from being a "passing-through-land" for east- and west-bound traffic; many of the small cities depend in part on the itinerant trade and some are especially geared to draw tourists. Many Americans are attracted by the experience of crossing a foreign border and by the old-world flavor of Mexico. As a result there is a notable flow of tourists into the Mexican portion of the region.

URBAN CENTERS

Through the bulk of Dry Continental regions, city-forming factors are absent. In Inner Asia cities are small and usually related to favorable positions on trade routes or in oases. In Sinkiang only two are noteworthy: Tihua (Urumchi), the capital on the railroad connecting with Lanchou, the principal industrial center and supply base in northwestern China proper, and Soch'e (Yarkand), an ancient oasis center. Ulan Bator is the capital of the Mongolian People's Republic and center of Lamaism and trade. The Loesslands of China support many towns and at least six cities of 100,000 or more, led by Sian, Lanchow, and Taiyuan.

It is only in the Soviet Union that a cluster of cities is found. There are at least a dozen with populations over 100,000 in the steppe portion of the Ukraine; Donetsk is the chief city within the Donetz Basin. Heavy industry, mining, and processing of agricultural crops have stimulated this growth. Oil exploitation and refining are responsible for the development of Baku into a city of the million class, and industry has caused Tashkent to become the metropolis of Soviet Middle Asia, with a population approaching one million.

Ankara, on the Anatolian Plateau, was established as the capital of Turkey in 1922, supplanting Istanbul. Since that date it has become a modern city of about one million people. It is centrally located but otherwise has little to offer as a capital. Madrid, capital of Spain, occupies a similar position on the Meseta. Despite isolation and lack of resources, it has grown to about 3.3 million people; it functions as the center of government, banking, commerce, and tourism.

Despite the large size of the North American region, only five cities have above 100,000 people. Denver, Colorado, at the eastern base of the Rockies, is the largest and is the principal financial and commercial city between the Missouri River and the Pacific Coast. It is a leading livestock and packing center, a market for a large irrigated area, as well as mining headquarters, chief resort focus for the Rocky Mountain area, and center for federal agencies. El Paso, Texas, has prospered as a result of its location on the middle Rio Grande where the river breaks through the mountains from New Mexico. Albuquerque, the metropolis of New Mexico in the upper Rio Grande Valley at a cross-roads location in the center of the state, is rapidly approaching a population of 400,000, and Lubbock and Amarillo in western Texas are both experiencing accelerated growth. The two largest cities of the Intermontane Province are Salt Lake City, Utah, and Spokane, Washington. Their advantages for assembly and distribution are somewhat better than at other centers of the Province. Spokane has developed a significant amount of manufacturing; flour milling, forest products, and aluminum are the leading industries. Salt Lake City is one of the best-planned cities in the nation and is the center of the Mormon religion. It should also be noted that although Calgary, Alberta, falls within the limits prescribed for the Short Summer Humid Continental region, it is closely associated with the Dry Continental region, serving as the urban center for the Canadian portion. Monterrey, Mexico, with a population of one million, serves as the principal manufacturing and commercial center of northeastern Mexico.

Few cities are found in the southern hemisphere regions. Tucumán and Mendoza, centers of major irrigation districts of the same name, are the only major cities in Argentia. Johannesburg, a gold mining, industrial, transportation center and commercial metropolis, and Pretoria, an iron and steel center, are the important cities of the South African region; both are growing in response to recent economic development. Johannesburg now exceeds 2.5 million in population.

sixteen

MIDDLE LATITUDE HIGHLANDS

Nowhere are the studies of nature and man more involved than in the mountains. Mountains are the most complex, spectacular, and majestic of the landforms of the earth. Here nature manifests her tremendous forces in earthquakes, volcanoes, landslides, and avalanches. Marked climatic changes occur with elevation, as well as from slope to slope. Remarkable contrasts of vegetation are found in a remarkably short space. The environment for the most part is inhospitable to man, but the mountain resources have attracted him to some places. Valuable minerals and forest-clad slopes have stimulated extractive industries. Many attractions for recreation and tourism are present. In favored areas, valley bottoms and smooth slopes may be cultivated, and herds pastured on the alpine meadows. In some places man has modified the mountain environment, but the mountain environment has greatly conditioned his activities. His works are usually molded and dwarfed by the most imposing of earth's features.

Mountain Influences

Climate and weather are universally influenced by mountains. Some highlands are so effective in blocking out precipitation that deserts are formed on lee sides, whereas the heaviest precipitation in the world falls on windward slopes. Mountains act as barriers to man, animals, and plants. Mountains are hindrances to transportation. Road and railroad construction is costly and requires great engineering skill (see Figures 16.1 and 16.2). Isolation is a mountain trait often resulting in the preservation of old customs, languages, and habits. Mountain people are often conservative, industrious, individual-

FIG. 16.1. (top) Saint Gotthard Pass, the most famous and most important of the passes through the Swiss Alps. The difficulties of road building are clearly shown in this scene. (Swiss National Tourist Office.)

FIG. 16.2. (bottom) Aerial cableway in the Swiss Alps of Canton Fribourg. To the left is Bulle and the right is the small town of Gruyères. (Swiss National Tourist Office.)

istic, and independent. The several small independent states that exist among their larger neighbors, such as Andorra, Nepal, San Marino, and Liechtenstein, are mountain countries. Mountains are used for political boundaries: The Pyrenees have been the boundary between France and Spain for centuries; the Himalayas separate India from Tibet. The influences of mountains upon nature and man are so extremely varied that generalizations fitting one area cannot always be applied to another.

Distribution

Highlands cover a significant portion of the earth's surface and are found on every continent (see Figure 16.3). Study of a physical map of the world reveals that the great mountains tend to be located on the peripheries of the landmasses. A rugged ring encircles the Pacific Ocean, and high mountains border the Mediterranean Sea and extend eastward through Eurasia. Four major cordilleran regions contain the principal highlands of the world: (1) the North American Cordillera, which includes the Alaska-British Columbia ranges, the Coast ranges, the Cascade-Sierra Nevada systems, the Basin ranges, and the Rocky Mountain and Sierra Madre systems; (2) the South American Andean Cordillera; (3) the Southern European Cordillera composed of the Atlas Mountains of North Africa, the mountains of Spain, the Pyrenees, Alps, and Carpathians; and (4) the Asian Cordillera, which fans eastward from the Pamir Knot

FIG. 16.3. The Middle Latitude Highlands.

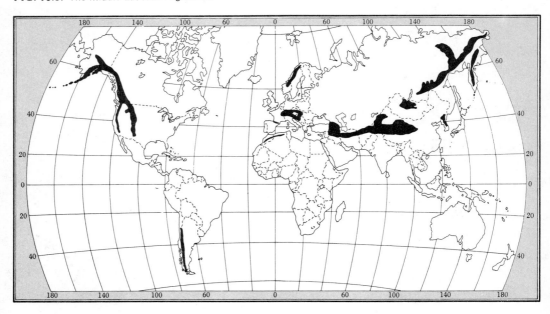

to include the Himalaya and Kunlun Mountains, the Tien Shan, and the intermediate plateaus and ranges, and extends westward in the Hindu Kush, Kopet Dag, Elbruz, and Caucasus Mountains.

Lower uplands frequently occur in isolation, detached from the great cordilleras. Some are ancient and undoubtedly represent the roots of other cordilleras of earlier periods. They now often assume the form of hills or plateaus. Such uplands occur in eastern North and South America, northern and central Europe, eastern and southeastern Asia, and eastern Africa and Australia. In many cases the factor of isolation, usually associated with mountains, is not a great deterrent to man, and the human occupation of these lower uplands is sometimes closely associated with that of neighboring lowlands. For this reason the lower uplands have been discussed almost exclusively in the regions in which they occur.

Physical Environment

CLIMATE

The climatic pattern of the Middle Latitude Highlands, in very general terms, is similar to that of the surrounding lowlands. In reality, however, highlands have an endless variety of climates resulting from differences in latitude, altitude, continentality, mass, and exposure.

Temperature. Temperature decreases with altitude at the average rate of 3.3 degrees per 1,000 feet of elevation. Other major controls operating in varying combinations and intensities make unconformity the general rule. For example, a 32° isotherm, or line of equal temperature, may be used as a reference. The higher the latitude of a mountain, the lower is the location of the 32° isotherm on its slope. A mountain in the interior of a continent has a much lower 32° isotherm than a coast or island mountain in the same latitude. The 32° isotherm is much higher on a large mountain mass than on small, isolated elevations. Contrasts occur between sunny and shady slopes. South exposures in the northern hemisphere receive more direct sun rays and consequently more heat and sunshine than northern slopes. North and south slopes are easily discernible in the spring when snow is melting, since a cover will remain much longer on the shady north side, particularly if forested. The contrasts between the two slopes have considerable effect on human occupance—the sunny slopes are the favorite sites for settlement while the shady sides are often avoided.

Energy actually received at the land surface from the sun increases with elevation. Dust and water vapor, which intercept much of the incoming solar radiation, are concentrated at low elevations, and above 3,000 feet they have little influence. Insolation, therefore, reaches surfaces above this level with little loss. The air is only slightly warmed, but heat-absorbing objects warm rapidly. Rock and soil surfaces in the sun will be as much as 40 to 50 degrees higher in temperature than similar surfaces in the shade. Cooling at night is about as rapid as daytime heating, therefore diurnal ranges are high. Evaporation rates are particularly high on slopes exposed to both sun and wind. Mountain climbers in the high elevations often experience great thirst.

Growing seasons, affected by the same controls as temperature, decrease rapidly with altitude. In some cases, a difference of 2,000 feet in elevation means fifteen days difference in harvest dates.

Precipitation. Precipitation in the moun-

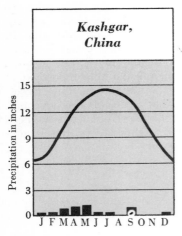
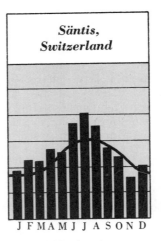
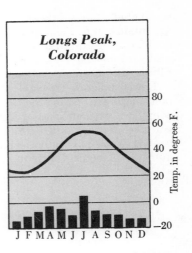

FIG. 16.4. Climatic graphs of Middle Latitude Highland stations.

tains is generally greater than on the neighboring lowlands. Thunderstorms are more frequent and are especially terrifying due to the reverberations. Snow is the characteristic winter precipitation and in high mountains may begin in early autumn and continue intermittently until late spring. High summits have a continuous cover—the lower boundary of this permanent ice and snow is known as the snowline. Snow covers on windward sides are often deep. The western slopes of the Sierra Nevadas in the United States average more than 400 inches a year. In one year a station there recorded a depth of 884 inches.

Snow is of special significance to mountain dwellers. Winter transportation is facilitated by the use of sleds or skis. In some European countries, snow makes it possible to transport hay from small isolated pastures to barns in the main valleys. Loggers cutting on slopes utilize snow for log movement. Some winter sports depend on snow. Furthermore, it acts as a reservoir for water to be used in the lowlands for domestic, irrigation, and hydroelectric purposes. Conversely, snow has adverse effects. An early cover hazards the harvest of a crop and a slow-melting cover in spring causes a delay in crop planting. It blocks roads and rail lines. Masses of heavy snow sometimes avalanche down mountain slopes with great speed and force, destroying forests and occasionally obliterating villages. Thousands of avalanches occur each year in Switzerland. Only a slight vibration—a far-off train whistle or the clanging of a cow bell—may be the start of an avalanche.

Winds. Mountain and valley breezes are common diurnal phenomena in the highlands. A valley breeze is the movement of warmed air up the valley or up mountain sides by day, caused by warm air rising from valley floors and slopes. Mountain breezes occur at night when air near mountain sides cools and the chilled air flows down hillsides and valleys. Winds resulting from a flow of cold air down slope under the pull

FIG. 16.5. View of Andermatt and Hospental from the Oberalp Alpine Road. The picture graphically illustrates the sketch of the U-shaped valley in Figure 16.9. (Swiss National Tourist Office.)

FIG. 16.6. A snow survey team in the Sierra Nevada. Data gathered by these crews will provide information useful in flood control and water-use planning. (United States Weather Bureau and Soil Conservation Service.)

of gravity are known as *katabatic* winds. Foehn winds are also characteristic. These warm, dry winds quickly evaporate snow on mountain sides, often allowing early spring planting and grazing. In the European Alps, certain crops such as grapes are restricted to slopes favored by foehn winds.

SURFACE FEATURES

Mountains display an endless variety of forms; some have jagged crests and are sharp and narrow, some are broad and rounded, others are plateau-like, surmounted by conical peaks. The distinguishing feature of all is a preponderance of area in slope.

Although steepness is a characteristic, slopes seldom exceed 35 degrees, and most average 20 to 25 degrees. Age is reflected in mountain appearances—youthful mountains are sharp and steep, whereas mature and old highlands, long exposed to the forces of weathering and erosion, appear more subdued.

The nature of highlands, as well as other landforms, is related to: (1) the kinds of earth materials or rocks that compose them; (2) the constructional forces that originally produced them; (3) the destructional forces at work to grade them down; and (4) their age, which determines the length of time the gradational processes have been at work. The influences of these factors are best demonstrated in the highlands.

Influences of rocks. Rocks differ in their resistance to deformation, weathering, and erosion. Soft rocks such as shales most often give rise to gentle slopes, plains, and subdued relief features; however, when inclined steeply, deep dissection may occur. Those such as granite, being resistant to deformation, weathering, and erosion, tend to produce more enduring and bold landscapes. Thus the nature of the rocks, their thickness, and lateral extent have direct bearing upon the nature of the landforms of any given area.

Constructional forces. Four types of mountains—folded, faultblock, intrusive igneous, and volcanic—result from processes that derive their energy from within the

FIG. 16.7. Simple folding. The upfolds are called anticlines and the downfolds synclines.

374

earth. *Folded* mountains are produced by stresses and strains that cause the earth's crust to wrinkle and develop a series of great anticlines and synclines (see Figure 16.7). *Faultblock* mountains occur when strain beyond the elastic limits of the rocks causes actual fracturing and blocks are raised or lowered in reference to one another. *Intrusive igneous* mountains are formed when large masses of magma are forced upward into the earth's crust and cause the upper strata to be bowed-up, often in more or less circular pattern. The *volcanic* mountains result from the accumulations of molten materials that have poured out upon the surface. Eruptions from central vents produce individual cones or shield-like shapes; flows from fissures or numerous vents develop plateau-like uplands.

Destructional forces. Opposing the forces that derive their energy from within the earth are destructional forces that derive their energy from without and are at work to bring the land to a uniform low level. These processes, as noted, include the work of atmospheric elements, gravity, water, ice, wind, and plant and animal life. By the process of weathering, which includes mechanical disintegration and chemical decomposition, the coherent rocks are prepared for removal by the processes of erosion that pick up the rock fragments and transport them for deposition in the lower areas.

Running water is the most potent agent of erosion. On the newly elevated surface, runoff is in the form of a sheet of water. Soon it concentrates into rivulets that grow into valley-carving streams. Gradually the valleys grow longer, deeper, and wider until base level is reached and the strong interstream relief features are removed. The streams, too, acquire new characteristics. Their courses become less direct as the water searches continually for the lowest elevations; in the resulting meandering channels the current is lessening its velocity.

Mountain or alpine glaciers, which more extensively developed during the Pleistocene Period, remain destructional forces in some of the high mountains today. Glaciers are bodies of ice formed by great accumulations of snow (see Figure 16.8). They advance as tongues down valleys to points where melting equals forward progression. Moving slowly forward, a glacier carries along loose rock materials, which are dropped when an overload is obtained, deposited as a terminal moraine, or carried down the valley by the meltwaters. These mountain glaciers have a tendency to produce U-shaped valleys, often leaving tributary valleys detrunked and hanging, with waterfalls cascading down the steep slopes (see Figure 16.9). Frequently the terminal moraine blocks the flow of water and causes the development of lakes, which can occur in chains.

Influences of age. Although mountains

FIG. 16.8. A mountain glacier in the Austrian Alps. (Austrian State Tourist Department.)

FIG. 16.9. A mountain valley before and after glaciation. (top), a youthful, V-shaped valley (bottom), the glacier has scoured out the valley, producing a U-shape and sharpening the peaks.

distinctive qualities that influence their habitability and use by man. Young mountains are commonly sharp in their detail, their surfaces highly irregular, and their peaks often snow-covered the year around. Valleys are V-shaped with little room for settlement. Streams are swift and often adaptable to hydroelectric power development. Minerals, however, are frequently deeply buried and out of reach of economic mining techniques.

Mature mountains, longer exposed to the action of the destruction forces, have lost their ruggedness. Elevations have been lowered, and often even the highest peaks are below the timberline. Valleys are broader and slopes more rounded. Further lowering may continue until the mountains have been reduced to a series of low, rolling hills, or even to a near plain-like surface—they have then reached the stage of old age. The mature and old-age mountains have greater possibilities for settlement, because the valleys are broader and slopes less steep. Frequently they are the location of important mineral deposits formed during mountain building and exposed by weathering and erosion.

NATURAL VEGETATION

All mountains display an altitudinal zonation of vegetation. A zone's width, elevation, variety of species, and character of cover depend chiefly on the same controls that affect climate. The vegetation of low altitudes is similar to the surrounding lowlands and changes slowly with elevation. If broadleafs constitute the lowest zone, they are gradually replaced by conifers. Grasslands or desert shrub at the bases are also replaced, depending on amounts of precipitation and heights of mountains, by conifers. Where conifers form the base, such as in the Cascade Mountains of North America, only the species change with higher elevation. Tree growth ceases with lower temperatures and shorter summers. The boundary

appear to be permanent features from the point of view of the lifetime of man, in a broader view of the earth's history no landforms are static. Forces are constantly at work building up and tearing down the earth's surface features. Mountains may be young, mature, or old. In each stage they have

MIDDLE LATITUDE HIGHLANDS 377

marking the termination of tree growth is known as the timberline. Its location is affected by amounts of moisture, exposure, wind, evaporation, and depth of snow. Some mountains, such as the Appalachians, do not extend high enough to have a treeless zone. Above the timberline is the alpine zone, which is subdivided into a shrub zone of stunted and gnarled trees bordering the timberline, followed by the meadow zone known as alps. The alps grade through tundra-like vegetation until bare rock and the snowline are reached. The Rocky Mountains of the United States exemplify altitudinal zonation in the middle latitudes. Six zones of vegetation may be recognized in northeastern Colorado: (1) steppe vegetation up to 6,100 feet, (2) a narrow, interrupted belt of

FIG. 16.10a. Dense lodgepole pine stand in Beaverhead National Forest, Montana. (United States Forest Service.)

FIG. 16.10b. Stand of Ponderosa pine in Boise National Forest, Idaho. (United States Forest Service.)

chaparral or brushland, (3) yellow pine and Douglas fir zone from about 6,100 to 8,130 feet, (4) lodgepole pine zone, 8,130 to 10,170 feet (see Figure 16.10a), (5) Englemann spruce and balsam fir zone, 10,170 to 11,670 feet, and (6) the alpine zone above timberline.

NATIVE ANIMAL LIFE

Animals especially characteristic of Middle Latitude Highlands live in the alpine zone; animal life in the forests of the lower elevations in general is similar to species found in the contiguous lower areas. Alpine animals however, frequently migrate to the forest zone during the winter or when food becomes scarce in the alpine pastures. Most alpine mammals are herbivorous, depending upon the forage of the high mountain pasture. Coarse hair or thick fur protects them from winter cold. Many are able to survive at extremely high elevations; in Central Asia wild sheep and ibex live at about 19,000 feet and yaks even higher. Many are sure-footed, climbing animals with strong, spreading hoofs. Species include the yak, musk deer, ibex, and wild sheep of Central Asia.

Chamois and ibex are found in the Alps, Pyrenees, and Caucasus Mountains of Europe. Rocky Mountain big-horn sheep are the characteristic animals of the Rocky Mountains of North America (Figure 16.12). Several small, rock-inhabiting species, such as the marmot and other rodents, are common in all regions.

SOILS

Mountain soils defy classification, owing to the scattered distribution of mature groups. Slope operates as a major factor in soil accumulation, and only in the small pockets of flatland is development a response to climate and vegetation conditions; thus only patches can be classed as podzols, gray-brown soils, and the like. More commonly, the soils are immature, consisting of a shallow layer overlaying the rock formations and ranging in depth from one inch to one or two feet. Frequent exposures of the bedrock are common, and the soil in general is composed of inorganic materials. They are, therefore, known as lithosols, from the Greek word *lithos,* meaning rock.

Occupance in the Middle Latitude Highlands

Mountains tend to restrict the development of large and widespread population centers. "Man—like air and water—feels always the pull of gravity,"[1] He prefers the plains, and it is only with increased effort and often deprivations that settlements in the highlands have been possible. In spite of adversities, some have chosen the more rigorous habitat for a variety of reasons. There are those who have sought freedom, refuge, or seclusion; others, particularly in

FIG. 16.11. (top) Sheep grazing on a high mountain pass in Wyoming. (United States Forest Service.)

FIG. 16.12. (bottom) Mountain sheep in the Tarryall Mountains of Colorado. (Fish and Wildlife Service: E. P. Haddon.)

[1] Ellen Churchill Semple, *Influences of Geographic Environment* (New York: Holt, Rinehart & Winston, Inc., 1911), p. 521.

the Orient and Mediterranean Basin, have been forced up the slopes by overcrowding on the lowlands. Moreover, mountains are not always completely undesirable. There may be valuable resources of minerals, forests, and pastures. Recreation possibilities are often present. In the tropics, the highland environment frequently presents superior living conditions.

In general, nature places rigid limits on the supporting capacity of the Middle Latitude Highlands. Population is thinly spread and often absent. Noteworthy concentrations are found only in central and southern Europe and in eastern Asia. Although average density is low, there are the isolated areas in many mountain regions, usually in valleys or on plateaus, where, like the oases of the deserts, population numbers are large.

Everywhere numbers decrease sharply with elevation. For example, in Switzerland, the most famous highland country, the majority of the people live on the plateau between the Jura Mountains on the north and the Alps on the south. The Alps have brought the nation much of its fame, but without the plateau, along with the great resourcefulness of the people and central position in Europe the country could not support its nearly 6.5 million people. In contrast, Tibet, 470,000 square miles of plateaus and mountains with three-fourths of its area above 10,000 feet, supports fewer than 1.5 million people—who live under great handicaps.

The range of economic opportunity is seldom as wide in the highland environment as on the plains. Man's adjustments frequently are not as complex or as advanced. In fact, due to difficulties of travel and intercourse with the outside, mountains tend to promote backward cultures. (The name "mountaineer" has long been synonymous with social and economic retardation.) Only in a few areas have modern transportation and communication systems brought any upsurge in development.

AGRICULTURE

The Middle Latitude Highlands are hostile to the widespread development of agriculture. Irregular surface with a sparsity of flatland limits farm sizes, shapes, and numbers, and leads to serious difficulties in the use of machinery as well as in the transport of products to markets. Soils are shallow, often stony, low in fertility, and susceptible to erosion. Climate, changing with altitude, slope and exposure, is found in variety, but a short growing season is a common handicap. The crop possibilities are distinctly limited, with coarse grains, hardy vegetables, tree fruits, and hay most adaptable (Figure 16.14). All of these normally cannot be grown in the same locality. However, on the subtropical margins of the Orient, climate allows the production of rice and other grains, vegetables, mulberries, tea, and a variety of fruits. Quality locations for tree crops and vineyards are found on many of

FIG. 16.13. A Swiss chalet. Notice the steep roof, storm shutters, and wide projecting eaves—all for protection from the heavy snows. The barn commonly is part of the structure. (Swiss National Tourist Office.)

the sunny south slopes facing the Mediterranean Sea.

Pressure of population on the plains of the Old World has caused terracing to be practiced as a means of increasing cultivation in the highlands. It is common in China and Japan; it is practiced in the Himalayas and the Hindu Kush, as well as on the slopes of the Vosges and the Black Forest facing the Rhine River, and in the Alps. Terracing is little practiced, however, by the mountaineers of the newer lands. This system requires and develops an industrious and cooperative peasantry; the cultivators by necessity become high-quality gardeners.

Agriculture is elsewhere restricted primarily to the valleys where lesser slopes, milder climates, and better soils are conducive to production. Stream patterns usually are indicative of the population pattern. Narrow tentacles of population extending up the shoestring valleys of mountains bordering the lowlands, such as in the Willamette Valley of Oregon, denote the diminishing farmlands and increasing physical limitations to agriculture as well as most other activities. In most Western countries there has been a relative decline in highland agriculture in response to the restricted opportunity for it to respond to improving production techniques. In Europe, for example, many of the young people have left the mountain farm to seek employment in the city. Today less than ten percent of the labor forces of Switzerland and Austria are engaged in agriculture.

GRAZING

The pasturing of livestock on slopes, alpine meadows, and open-forest lands of the high elevations is a major feature of the limited agricultural activity in nearly all middle-latitude mountain lands. Each year the animals are driven or transported into the high pastures as soon as the snow cover disappears and forage is available. The move-

FIG. 16.14. Hay harvest in the Tyrol. The mountain farmers must provide for their livestock, which are kept in stables during the long winter season. The difficulties of farming in the highlands are clearly indicated. (Austrian State Tourist Department.)

ment of bands of sheep from the hot, dry plains of the Rhone delta to pastures in the French Alps is a picturesque sight. Like a slowly moving gray cloud, the sheep begin their long journey. Trek-wise rams lead the procession, herders on horseback and scurrying sheep dogs guard the flanks, and the pack asses bring up the rear. A tinkling symphony accompanies the band since recognition bells with distinctive tones are worn by each type of animal. Movement of the band begins during the cool of the evening, and the approximately 200-mile trip is accomplished in easy stages of 15 miles a night to accustom the sheep to the colder temperature and higher altitudes.

FIG. 16.15. Farmland is scarce in Norway and must be devoted to food and winter feed crops; therefore, farmers take their livestock to the nearby mountain pastures for summer grazing. The shelter is a *saeter*, summer headquarters for Norwegian herdsmen. (Norwegian Information Service.)

The dairy cow, the most important animal in the small mountain villages of central Europe, provides milk, butter, and cheese for both home use and market. Goats are of some significance as utilizers of the more abundant forage of the higher elevations and steeper slopes. Many villagers have a ceremony when the cows are released from the long winter captivity in barns to begin their journey to the alps. The animals, carefully groomed and belled, present a picturesque sight as they wend their way up the slopes. Herders chosen by the villagers accompany the cattle to the high pasture and remain with them until late summer. The herders' main chores are milking and cheese-making. When enough cheese has been accumulated, it is brought to the village and prorated among the cattle-owners. Village youngsters in some communities make trips to the alps for the cheese and at the same time take food, tobacco, and other supplies to the herders. In the meantime the villagers below are busy with crops and hay.

The Norwegian farmers who cultivate the narrow bench lands along the fjords send their livestock to *saeters,* or summer "dairies," in the high valleys above their farms (see Figure 16.15.) There they often are tended by the children of the farm families. The saeter system is usually a farm family enterprise, in contrast to the method used by central European mountain villages. This system of transhumance is a planned adjustment whereby the farmer uses his limited farmland in the valley bottoms or on the benches for growing food crops and winter forage instead of pasture. Late snow covers, steep slopes, thin soil, and short growing seasons preclude the use of the high pasture for plow agriculture; but the high pastures ideally fit this special farming scheme and relieve the land-use pressure on the small holdings. In North America, the mountain pastures are the summer grazing grounds for some of the livestock of the Great Plains and intermontane basins and valleys. Livestock, especially sheep, are often shipped by truck or rail to these summer pastures.

FOREST INDUSTRIES

Forests and mountains are almost synonymous, in part because there has been less competition for other uses of the land than on the plains. Several mountain sections of Germany are called forests, such as the Schwarzwald (Black Forest) and Odenwald (Oden Forest). Forestry is associated with all Middle Latitude Highlands in Europe. Conifers are usually the chief commercial species. Many of the forest operations are small-scale, and logging is often practiced by farmers during the winter. The most outstanding forest activity of the mountain regions is concentrated in the western United States. Mature, high-quality timber covers slopes of the Coast Ranges, Cascade-Sierra Nevada Mountains, and the Rocky Mountains. Douglas fir, most important of the commercial trees, prefers slopes that receive heavy to moderate amounts of precipitation; particularly heavy stands are west of the summit of the Cascade Mountains. Western white pine is the most important species cut in northeastern Washington, northern Idaho, and western Montana. Large stands of Ponderosa pine are found on the leeward and drier sides of the Cascades and the interior mountains (Figure 16.16). Several varieties of pine are associated with the Sierra Nevada Mountains. The large mechanized operations in the western United States contrast sharply with methods in Europe, where hand labor and horses furnish much of the power. Winter snows, often restricting work in the woods of the western United States, are welcomed by European loggers, who cut a much smaller tree and utilize the snow for moving logs.

FIG. 16.16. Loading Ponderosa pine on the east slopes of the Cascades. Tractor logging is particularly adaptable in these more open stands of the pine forest. (Weyerhaeuser Company.)

FIG. 16.17. An open-pit iron mine in Styria, Austria. This mine has been operated for hundreds of years and today is still the chief source of ore for Austria's small steel industry. (Austrian State Tourist Department.)

MINING

Metallic minerals are commonly found in the highlands. It is in these regions that weakness in the earth's crust has allowed magmas, formed in the depths, to intrude near the surface, later to be exposed by weathering and erosion. These magmas have often brought minerals that are sufficiently concentrated to be classed as ores and profitably mined. Mining is often the major reason for the existence of population in some of the highland areas. Owing to the importance of minerals in the modern industrial world, rich ore deposits attract men to isolated areas, and despite difficulties of transportation, food supply, and adverse living conditions, a few relatively large centers have developed.

The Rocky Mountains supply a major share of copper, silver, lead, zinc, and molybdenum in the United States. The Ural Mountains of the Soviet Union are highly mineralized and provide a variety of metals, including large quantities of iron ore. Gold and other minerals are taken in considerable amounts from the highlands of southern Siberia.

Mining is based upon a nonrenewable resource. Mineral deposits in time are exhausted, eliminating the base for the mining community; abandonment follows. Black Hawk, Georgetown, and Central City are famous Colorado mining towns of the past that went through a short period of rapid and lively exploitation. The latter is now a noteworthy tourist attraction.

WATER RESOURCE DEVELOPMENT

Mountains perform important roles in water resource distribution and development. They catch and store water and thus nourish and give direction and velocity to many of the world's streams. Artifical storage is facilitated by numerous deep valleys and canyons in which dams and large reservoirs can be constructed economically without damaging settlements or other developments. Thus water supplies for cities, irrigation, and other uses can be implemented. Damsites, regularity of stream flow, and favorable gradient also give the mountainous area large hydroelectrical potentials. These are being developed in the more accessible sections where this form of generation competes favorably with other energy sources. More and more attention is being devoted to mountain watershed management by the progressive nations in the interest of obtaining long-range and maximum benefits from their water resources.

MANUFACTURING

Raw materials having bulk, waste, and excessive weight are often processed or semiprocessed near their sources to facilitate shipping and reduce freight costs. This principle is one of the major stimuli for manufacturing in mountainous areas and is exemplified by sawmills and mineral concentration plants and smelters. Further fabrication is usually outside the mountains. Bases for wide-scale industrial development and diversification are weak. Major handicaps include poor transportation, distance from markets, and shortages of skilled labor.

Handicrafts or home industries have been typical of mountainous areas where the long, cold winters curtail outdoor activities. The mountain dwellers in Switzerland, Germany, and Austria are noted for fine wood carvings, cuckoo clocks, toys, laces, and embroidered goods. Handicrafts not only provide winter hobbies but supplement incomes.

FIG. 16.18. Ashcroft, a ghost town of Colorado. This former small mining community is typical of many centers that flourished in the Rockies during the early mining boom. (State of Colorado Advertising Publicity Department.)

Switzerland is an outstanding example of a small mountainous country that has achieved fame in the industrial world. Over one-third of the Swiss people are engaged in manufacturing activities. Coal and oil are lacking but the country has developed much of its great potential for hydroelectric power. Most of the raw materials have to be imported. The significant Swiss assets are the skill and resourcefulness of the people. The manufacturing emphasis is on high-quality and high value per unit of weight commodities. Industry is concentrated on the Swiss Plateau, where silk, cotton and wool textiles, agricultural and textile machinery, electric engines and turbines, and chemicals are manufactured. The Jura Mountains of Switzerland are the center of the world's clock and watch industries. Individual parts are often produced in home workshops. Swiss chocolate and cheese are also famous products. The varied manufactures are sold widely on world markets.

TOURISM

A country which possesses mountains has a valuable tourist asset. Mountains offer spectacular scenery with forested slopes, colorful alpine meadows, streams, lakes, and glaciers and snowfields, as well as a cool, bracing summer climate. The sheer beauty of their winter landscape is unsurpassed. Forest and alpine trails appeal to the hiker; rugged peaks and slopes challenge the more adventurous climber. Fishing, hunting, and skiing attract the outdoor sportsman.

Resort centers are scattered throughout many of the mountain regions of the United States and Canada. Many in the West were first developed and promoted by railroad companies who realized the tourist potential. Canada and the United States have established large national parks that have stimulated the tourist industry. Many of these noted national parks are located in the western mountains. Two well-known western parks of Canada are Banff, containing beautiful Lake Louise, and Jasper National Park. In the United States, Yellowstone heads an imposing list of national parks that includes Mount Rainier, Crater Lake, Yosemite (Figure 16.19), Glacier, and the Grand Tetons. Although far from population concentrations, the parks are visited by thousands of travelers annually.

The Andes of middle Chile are gaining some significance in tourism. A location marginal to the populated region, the country's Dry Summer Subtropics, has resulted in domestic use, and the area is beginning to attract visitors from many parts of the world.

The greatest economic value of many of the mountains in Europe, especially the Alps, lies in their tourist attractions. The livelihood of many small villages has been radically changed from a meager agricultural existence to a profitable tourist activity based on the feeding, housing, and guiding of visitors. Several small, obscure villages have become world famous, such as St. Moritz, Grindelwald, and Zermatt in Switzerland (Figure 16.21) and the winter resort twin cities of Garmisch-Partenkirchen in Germany. Chamonix, at the base of Mount Blanc in France, and Cortina, Italy, are popular resorts.

Switzerland is one of the major European tourist nations. The Swiss, through careful planning, have systematically developed their tourist activity, and in so doing have turned normal mountain handicaps into assets. Railroads and good highways tap the mountain areas and scenic cable railroads have been constructed especially for tourists. Lodging accommodations fit all pocketbooks, and the food and services are excellent. Success has been the outcome, and the well-

FIG. 16.19. Half dome and Merced River in Yosemite National Park. Thousands of tourists are attracted to this beautiful park in the Sierra Nevada Mountains. (National Park Service.)

FIG. 16.20. One of the picturesque mountain villages in the Austrian Alps. Frequently, as in this view, the ancient church steeple stands as a prominent landmark. These villages have facilities for handling both winter and summer tourist business. (Austrian State Tourist Department.)

executed policies have made tourism a significant segment in the Swiss national economy.

URBAN CENTERS

As a general rule, the Middle Latitude Highland environment is not conducive to city growth. The few cities that do occur have developed as the result of mineral and forest exploitation, or commercial advantage derived from location at the junctions of mountain passes, or at the entrance to mountain passes where lowland routes come to focus. Butte, Montana, is an example of a mining city. Innsbruck, in the Inn Valley

MIDDLE LATITUDE HIGHLANDS   389

of Austria, illustrates the value of the crossing of routes through mountains; here the famous Brenner Pass from Italy to Bavaria meets the valley route from Switzerland to Vienna. The girdle of European cities both north and south of the Alps indicates the importance of the gateway location; none, however, is actually in the mountains.

Most of the cities of Switzerland are on the plateau. Geneva is the gateway city in the west and Basel in the northeast. Zurich, the largest city of the nation, is within the protecting walls of the mountains at the convergent point of north-south and east-west transportation. Bern, centrally located, is the capital. Manufacturing has contributed to the importance of all.

FIG. 16.21. Zermatt. This little Swiss village is almost synonymous with the majestic peak, the Matterhorn, which crowns its backdrop. The village and this picturesque little valley were "discovered" by the English during the Golden Age of mountaineering in the middle 1860s. Today it is a popular resort with accents on skiing in winter and mountaineering in summer. Access to the isolated Valais district, Zermatt's location, is gained by the use of a cog rail line. The buildings are peculiar to the district—round slabs of slate are set under foundation poles of barns and storage sheds to keep out rodents. The weathered coffee-colored buildings are characteristic in the Valais district. (Oliver H. Heintzelman.)

seventeen

SUBARCTICS

The Subarctic regions are the lands of the northern coniferous forests. Dark evergreens appear interminable, sharing the landscape with swamps and bogs, thousands of lakes, winding rivers, and rushing streams, The Subarctics are winter strongholds; summer is brief; spring and fall are fleeting. When winter grips the land, nature seems at rest. The ground is blanketed with snow, streams are stilled, lakes are paved with ice, frost hardens the soft, moist lowlands, birds have migrated, and most animals remain in hiding unless in search of food. The crash and thunder of breaking ice heralds the end of winter. The melting snow soon saturates the low, open lands, vegetation springs into life, countless insects fill the air, and birds return from the south.

Winter is the prime season for many of man's activities in the Subarctics. Ice and snow become highways for travel, and the air is clear of troublesome insects. Loggers, hunters, and trappers invade the forest, but numbers are small—man is always scarce and commonly transient. Isolation keynotes the region as a whole, while forests keynote its economy.

Location

Only the northern hemisphere continents have sufficient landmasses in the high latitudes to allow development of Subarctic climates. The regions straddle the sixtieth parallel with extreme margins extending to latitudes 50° or 55° on the south and 65° or 70° on the north (see Figure 17.2). The southern margin is the zone of contact with the Short Summer Humid Continental regions, and the poleward boundary is the 50° isotherm for the warmest month, which approximately marks the poleward limit of tree growth.

There are two large areas. In North America the Subarctic region, 10 to 15 degrees wide, stretches from the Pacific Ocean through Alaska and Canada to the St. Lawrence River Estuary and Newfoundland on

FIG. 17.1. A scene in eastern Finland. Here is depicted the physical and economic geography of the Subarctics—water, forest, milling, and only patches of agriculture. (Finnish National Travel Office.)

FIG. 17.2. The Subarctics.

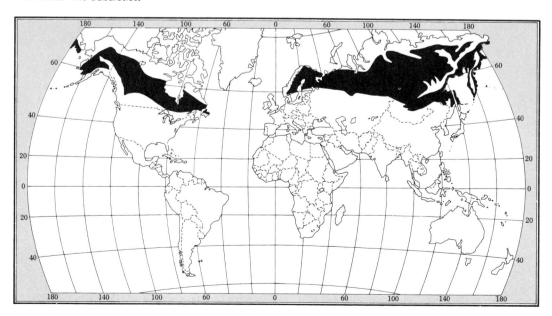

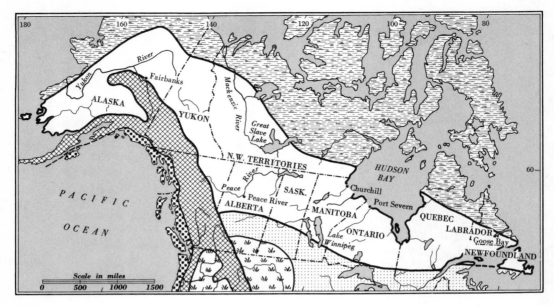

FIG. 17.3. The Subarctics of North America.

FIG. 17.4. The Subarctics of Eurasia.

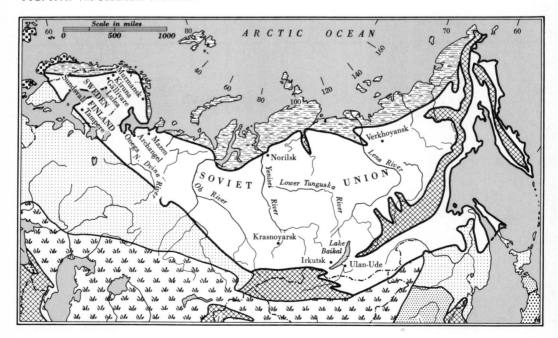

the Atlantic Ocean (Figure 17.3). The Eurasian region extends from the Scandinavian Highlands across Sweden, Finland, and the Soviet Union, terminating on the Pacific Ocean; eastward this region progressively broadens as continentality increases and includes the bulk of Siberia (Figure 17.4).

Physical Environment

CLIMATE

Extreme continentality marks the Subarctic climate, and its seasonal temperature contrasts are the greatest on earth. "King Winter" rules for much of the year.

Temperature. Winters are long and severe; six months or more of the year are below freezing. January temperatures for North American inland stations average below zero, and the extreme minimum of −80° has been recorded in the Yukon. Summers have about three months above 50°. July averages are about 60°, but daytime maxima sometimes exceed 90°. Coast areas modified by open water bodies have warmer winters and cooler summers. Annual ranges are high and in North America average 80 degrees. Growing seasons vary between 90 and 100 days on the southern boundaries and become increasingly shorter with poleward progression. Some favored areas in the north, such as the Mackenzie Valley, may have growing seasons of 50 to 75 days; however, frosts may occur in any month. In the interior of Asia, winters are colder and summers warmer than in North America, due to greater continentality. Of the known temperature recording stations, Verkhoyansk in Siberia holds the record for the high average annual temperature range with a −58° in January and 59° in July—a difference of 117 degrees. Verkhoyansk has recorded a low of −93°, and Oymyakon to the southeast has reported even colder temperatures. This Siberian area, the coldest on record in the

FIG. 17.5. Typical climatic graphs of Subarctic stations.

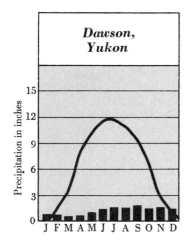
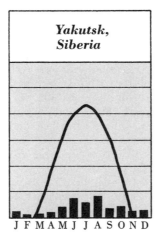
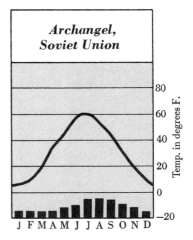

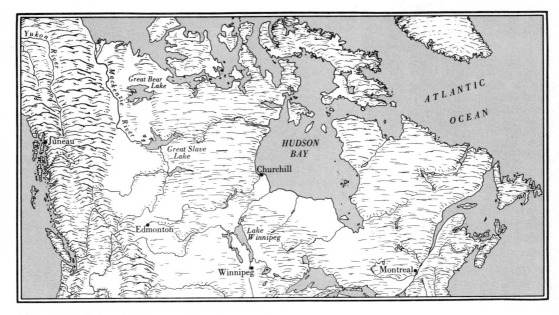

FIG. 17.6. Physiographic diagram of Canada.

world, is often called the "cold pole"—it is seasonably even colder than the Arctic region.

Many hours of sunshine partly compensate for the shortness of summer in these regions, but the plant growth potentials are still low compared to regions farther south, owing to the lower intensity of heat and light. Summer days are long, and on June 21 at 60° north latitude there are $18\frac{1}{2}$ hours of continuous sunshine; however, this is reversed on December 22 when the sun shines six hours.

Precipitation. Precipitation, resulting chiefly from cyclonic activity, ranges from 10 to 20 inches a year for most parts of the regions; however, northeastern Siberia, due to greater continentality, averages only 5 inches. The greater part comes during the summer months, since winter air is too cold to contain much moisture. Snow, dry and granular, falls during the cold season; and, although seldom of any great depth, it remains on the ground from five to seven months, receiving protection from the forest.

Although the low precipitation totals are reminiscent of the Dry Continental climate, the effectiveness is much greater here owing to low evapotranspiration rates. During the short summer standing water is common in topographic depressions.

SURFACE FEATURES

The landforms of the Subarctic regions almost everywhere display evidence of long subjection to the forces of weathering and erosion. Continental glaciation usually has had a major role in shaping the present surface, and hard, crystalline rocks, representing the roots of older highlands, are present in large areas.

The North American region (Figure 17.6) crosses five physiographic provinces: (1) the eastern half is included mainly in the Laurentian Uplands, an ancient highland of metamorphosed rock long exposed to both constructional and destructional forces. Today it appears as a rolling surface with numerous lakes, large stretches of marshland, and scattered low mountains of resistant material. The entire region was glaciated, and in many sectors bedrock is exposed. (2) The southern margin of Hudson Bay is bordered by a lowland covered with sedimentary materials. (3) The Great Plains, the northern part of which is drained by the Mackenzie River, separate the Laurentian Upland from the Rocky Mountains. (4) The Rocky Mountains together with the Brooks Range form a continuous highland to the Arctic Ocean. (5) The westernmost province is the Yukon Plateau, which fills central Alaska. This broad upland surface is characterized by gently rolling topography in which the Yukon River system has cut valleys.

The Eurasian Subarctic, also extending across the entire continent, is composed of several physiographic units (Figure 17.7). The surface of the region is relatively uniform, differing principally in elevation; the major heights are found in the Urals and Far East. The Fenno-Scandinavian shield in the west is an area of ancient crystalline rock, lowered by erosion to an uneven plain-like surface. Severe glaciation has in many places removed the soil covering, and unequal scouring and deposition have produced thousands of lakes. Eastward to the Urals is the northern portion of the Central Russian Lowland, divided into two parts by the drainage

FIG. 17.7. Physiographic diagram of the Soviet Union.

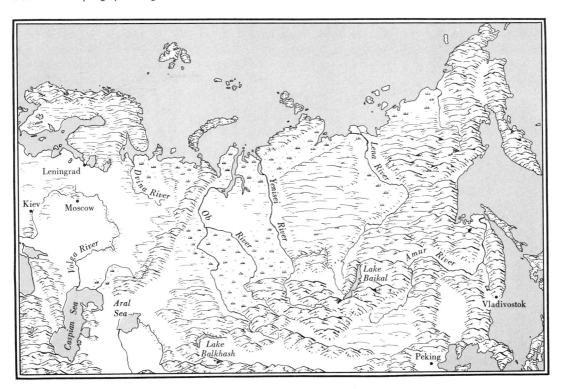

FIG. 17.8. The midnight sun in northern Finland. On the northern margin of the Subarctics the sun does not set during June and most of July. (Finnish National Travel Office.)

basins of the Dvina and Pechora Rivers. The rounded Ural Mountains separate the Central Russian Lowland from the West Siberian Lowland, a large, flat, and poorly drained region through which the Ob River flows. The Central Siberian Plateau, between the Yenisey and Lena Rivers, appears as a flat, but elevated, plain into which the streams have incised deep valleys. Eastern Siberia is extremely mountainous.

The Ob, Yenisey, and Lena Rivers are three of the great streams of the world—in volume, length, and area of drainage basin. These rivers are ice-bound for the long winter season; at Yakutsk the Lena is frozen an average of 210 days per year. Their utility is further reduced by ice choking the lower courses; vast areas are inundated when the warm-season thaw in the southern margins of their drainage basins sends large volumes of water poleward.

NATURAL VEGETATION

Stretching from coast to coast across the broad expanse of the Subarctic regions is the *taiga,* or northern coniferous forest, one of the greatest forest belts in the world. The forest cover is dominated by coniferous trees; spruce, larch, fir, and pine are the major species. Intermingled with the conifers, paralleling stream banks or forming solitary clumps, are willows, alder, aspen, and birch. The forest floors are quite clear, with coverings of moss, lichens, and low bushes. The taiga has its largest trees and greatest forest densities on the southern margins and along river banks. Poleward, with diminishing sunlight intensity and growing season, the forest thins and the trees decrease in size until they are virtually dwarf bushes at the margin of the tundra. Tree growth everywhere is slow, since the frostless season is short and water at the roots is only available from three to five months a year. Trees fifty years old with diameters of only a few inches are common. The average size of conifers in the main portion of the Canadian taiga ranges from heights of 30 to 70 feet and from diameters of six inches to two feet.

The taiga is not a continuous forest region. Where it borders the steppes on the southern margin, the forest is occasionally broken by open prairies. Scattered throughout the taiga are thousands of square miles of sphagnum-moss swamps and muskeg, the accumulated remains of decayed vegetation. During the thaw of the summer months, the swamp and muskeg country is virtually a morass, infested with biting and stinging insects, and travel is practically inhibited.

NATIVE ANIMAL LIFE

Where little molested by man, relatively abundant animal life is a characteristic feature of the Subarctic regions. A variety of species ranges through the stretches of open

FIG. 17.9. The Arctic Circle in Sweden. This photograph was taken from a passenger coach on the railroad between Baden and Gallivare. Notice the small size of the trees. The center structure depicts a Lapp shelter. (Oliver H. Heintzelman.)

FIG. 17.10. Near the tree line. This is the vegetation type in the Kiruna area. Birch is replacing conifers. (Oliver H. Heintzelman.)

FIG. 17.11. A bull moose in a natural meadow in the Subarctic forest. (United States Forest Service.)

forest, which provides both food and protection. Caribou or reindeer, moose (Figure 17.11), and elk comprise the forest-browsers. The fur trade is based on beaver, fisher, fox, marten, mink, muskrat, otter, weasel or ermine, and wolverine. The prime winter pelts of these northern forest animals make the Subarctics the most important fur-producing regions in the world. Bears, wolves, lynx, squirrels, and hares are common. The forest is the habitat of numerous birds such as the grouse, woodpecker, and grosbeak. Streams and lakes are well stocked with game fish. Insect life is superabundant. In summer the muskegs, swamps, ponds, and the general moist condition of the lowlands provide numerous breeding places. Clouds of mosquitoes, stinging gnats, and black flies make life miserable for man and beast.

FIG. 17.12. A Canada goose family. The honk and V-shaped flight pattern of these migratory birds announce the approach of spring to the middle latitudes of North America. (Malak, Ottawa.)

Insects are one of the great drawbacks to living in the taiga.

The forest is the scene of two seasonal migrations. Great flocks of birds arrive from the south in the spring and depart with the approach of winter (Figure 17.12). Caribou, hares, and Arctic birds, summer inhabitants of the tundra, invade the taiga in winter for the forest's protection and food.

SOILS

The greatest development of podzol soils occurs in the Subarctic regions in response to the long subzero winters, cool, moist summers, and coniferous vegetation. Glaciation and imperfect drainage, however, have hindered the development of a uniform pattern. The best soils have formed in the sedimentary-rock areas where breakdown is more rapid, in basins containing glaciofluvial deposits of clays, on lacustrine plains, and along stream floodplains. Soil is often completely lacking where the glaciers scoured the surface and stony in other areas where deposition occurred. In the marshes conditions for soil development are unfavorable, and in some cases peat, several feet thick, has accumulated.

Permafrost—a condition of permanently frozen subsoil—exists in parts of these regions, with probably the widest distribution in northeast Siberia. It causes many problems in construction of houses and roads, or in any digging or excavating. Frost-heaving ejects telephone poles from the ground and forms cracks in brick or concrete walls. With the advent of spring, thawing of the surface layer while the subsoil remains frozen causes a mushy mass of mud to form.

Occupance in the Subarctics

The Subarctic regions are frontiers of settlement. The harsh climate, which is marginal for plant growth and unfavorable to easy living, together with the great distances to major centers of development, produce an environment that is unattractive to man. Thousands of square miles are entirely without people, and nowhere are densities great. Much territory is sparsely occupied by relatively small numbers of indigenous peoples who depend upon hunting, fishing, and fur trading for their living. The greatest white populations are found in the southern margin of the Eurasian region in Sweden,

Finland, and European Russia. In the past forty years, resource development has resulted in some settlement in the area between the Yenisey River and Lake Baykal. Canadians in the Subarctic are confined mainly to the southern fringe in eastern Canada. The average density for much of the realm does not exceed two persons per square mile.

AGRICULTURE

These northlands offer little to attract the farmer. Long, bitterly cold winters, short, cool growing seasons, and soils generally thin and low in fertility restrict farming opportunities. Lack of transportation and distances to market place further limitations in the way of commercial agriculture. Long midsummer days in part compensate for short seasons, so that hardy species of vegetables, grains, hay, and pasture crops can be grown where soils are favorable. Potatoes, turnips, radishes, peas, lettuce, and cabbage, along with rye, oats, and small fruits, are usually the most significant enterprises, but agriculture is by no means widespread (Figure 17.14).

FIG. 17.13. A rural Finnish steam bath or *sauna*. Water is thrown on the heated rocks to produce steam for the bathers, who sit on the raised platform. The cleansing effect of the sauna results from the sweating caused by the heat and the steam; this is often stimulated by beating the body with birch twigs. There are about 350,000 rural saunas in Finland. (Finnish National Travel Office.)

FIG. 17.14. A farm in Finland. Holdings average about 40 acres. Farmers attempt to attain self-sufficiency in food production. Rye, potatoes, hay, and livestock are the major enterprises. The crop shown is flax. (Finnish National Travel Office.)

Canada has not yet exerted a major effort toward agricultural development in this region because of her relatively small population and the possession of other comparatively large productive areas. Three areas in her Subarctic region are showing considerable progress in the production of hardy grains and in animal industries—these are the Vermilion Area of north Alberta, the district north of Prince Albert in Saskatchewan, and the Clay Belt in the vicinity of Abitibi Lake near the Ontario-Quebec border.

Sweden, Finland, and the USSR have given more attention to the development of farming. A major share of the area of each of these countries is in the Subarctics, and of necessity the better districts have had to be opened. Even so, agriculture is of small proportions, and notable progress is being made only along the southern borders. In the Soviet Union, as the result of the development of mineral and forest resources, some advances have been made in establishing small food production bases to provide vegetables, meat, and milk to the workers.

The rigors of climate probably destine the Subarctics to a continued role of a pioneer fringe. Agriculture is severely limited by physical impediments; the farmer who braves these lands operates under as much risk as those who attempt to farm in the marginal drylands. Only the most favorable areas, largely those presently containing some settlement, have immediate future prospects. There is much to be learned about cultivation in this restrictive environment before agriculture becomes economically feasible.

FOREST INDUSTRIES

The coniferous forests of the Subarctic regions are one of the world's great sources of timber and pulpwood. The USSR taiga alone contains 1.5 billion acres. Forests have a major role in the entire economy of each

FIG. 17.15. Timber from wintry forests is the major natural resource of Finland and more than three-fourths of Finnish exports consist of wood products: timber, pulp, paper, newsprint, prefab houses, veneer, and plywood. (Consulate General of Finland.)

FIG. 17.16. The cry of "mush" and the yelps of the Huskies are being replaced by the hysterical chainsaw sound of the snowmobiles. (Standard Oil Company of New Jersey.)

country that shares a part of the taiga. Canada, relying on stands of balsam fir and spruce, has become a world leader in newsprint production. Sweden, Finland, and the USSR account for the bulk of Eurasian surplus forest products.

Logging is concentrated in the accessible southern margins where trees are largest and stands are most dense; the greatest drain has been in the river valleys. Logging operations in both Eurasia and North America are similar. The trees are harvested in the winter and are assembled along frozen stream banks by sled and tractor. The logs are floated downstream to the mills with the thawing of the river ice in the spring. At this time of year and in early summer, the surfaces of the lakes and rivers are covered by logs.

Forest management is advanced in Sweden and Finland, but has been somewhat slower in Canada and the Soviet Union owing to their large forested areas. It is beginning on a systematic basis in both countries. Inventory, improvement of transportation, and concern for fire protection and regeneration are included in the programs. In the Soviet Union there has been growing government interest in the rational distribution of forest-based industries.

FUR INDUSTRY

Furs, not trees, were the earliest attractions of the northern forests. Long before man looked to the taiga for timber, the trapping of fur-bearing animals and the trading of pelts were the dominant and most lucrative activities. Despite centuries of exploitation, the furs of the taiga continue to be a revenue source in some locales.

Winter is the season for trapping, since furs are then in their prime—longer, glossier, and thicker. The trapper must take all his supplies, which include food, blankets, traps, firearms, and ammunition, into the woods. He establishes a base camp near a stream or lake and sets his trap line. The entire winter is spent in checking the string of traps,

FIG. 17.17. An otter—one of the fur-bearers of the Subarctics. (Fish and Wildlife Service: B. Scheffer.)

hunting, and preparing the pelts for market. With the spring thaw he returns with his catch to a collecting center. In Canada the Hudson's Bay Company is still the major fur-trading organization.

Fur farms are becoming a significant source of furs. To be assured of a dependable supply of pelts, trappers began to keep foxes, captured during the warm season, in cages and to hold them until their pelts were prime. This method developed and stimulated the breeding of fur-bearing animals so that it has become a well-established and profitable activity. Animals most usually raised in captivity are fox and mink.

MINING

As easily accessible supplies in other regions are depleted and industrial needs multiply, these northern forest lands are becoming increasingly important as sources of minerals. As yet there is no full information on the possibilities; detailed geologic surveys have covered only small areas. Work to date, however, has revealed widespread occurrences of metals in the old hard-rock areas. The Laurentian Upland of Canada is known to be rich in minerals; the list includes iron, uranium, nickel, copper, cobalt, gold, and silver. Mining camps are being established to exploit these ores. Development of the iron deposits astride the border of Quebec and Labrador is of particular significance. A railroad extending 358 miles southward to the Port of Seven Islands makes ore mined from this district available to United States steel centers via the St. Lawrence Seaway.

Gold in the Yukon Valley was the first major attraction of Alaska and northwestern Canada. Although the more accessible and richer gravels were worked by the turn of this century, the activity continues as a profitable enterprise for a few large companies that use huge dredges.

One of the highest quality iron deposits in the world is in Sweden, north of the Arctic Circle. Sweden exports most of the iron ore from this district, since she does not possess coking coal—her small steel industry cannot consume the total output from her mines. The mining activity here makes Sweden one of the major suppliers of iron ore on the international market. She supplies much of the ore imported by the Western European steel industry, and some is shipped to the United States. Kiruna is the leading center (Figure 17.18); from there ore moves by rail to the ports of Narvik on the coast of Norway and to Lulea on the Gulf of Bothnia. Narvik, always ice-free, is the more important, allowing activity in winter as well as in summer; it has the additional advantage of being close to the mines as well as to the European markets (Figure 17.19).

In the past three decades, the Russians have conducted mineral explorations that have revealed valuable deposits, and several mining centers have developed. Particularly noteworthy is the center in the heart of the Kola Peninsula, where two significant minerals are mined: apatite, a source of phosphate fertilizer, and nepheline, important for its potash content and its use in aluminum and ceramic industries. In the same area, some

FIG. 17.18. (top) The famous Kiruna Mines. The town, with a population of about 11,000, is in the background. A transition from open-pit to underground mining is being made. This is reportedly the world's largest underground mine. (Oliver H. Heintzelman.)

FIG. 17.19. (bottom) Narvik, Norway, port for Swedish iron ore. This ice-free harbor permits ore shipment throughout the entire year. The ore is transported from mines by rail to the ship-loading facilities pictured in the center of this view. The city has a population of about 10,000. (Oliver H. Heintzelman.)

FIG. 17.20. The power house for the Aluminum Company, Ltd. of Canada's Arvida plant. Large quantities of cheap hydroelectric power and access to both tidewater and freshwater navigation have made the Saquenay district the ideal center for the Canadian aluminum industry. (Malak, Ottawa.)

iron and molybdenum are produced, and nickel is mined at Petsamo. Coal and petroleum are being exploited to some extent in the Pechora Basin. Undeveloped reserves of coal are present in the Yenisey and Lena Valleys, and iron is found along the Angara River. Large quantities of gold are mined in Yakutia, mainly along the Aldan and Kolyma Rivers. The settlement at Norilsk in the northern edge of the region near the Yenisey River is based upon the mining of nickel and related minerals.

WATER RESOURCE DEVELOPMENT

Despite the long winter freeze up, the numerous streams and lakes provide these northlands with vast water resources that serve the regions in several ways. From the historic point of view, the transportation function has been most significant, furnishing the avenues of penetration for the trappers who have taken many millions of dollars worth of fur from the forest and marshes. The waterways are still used for this purpose and in addition provide the access routes and shipment arteries for the forest-based industries, as well as others. With a minimum of excavation to connect lakes and streams, the Russians have constructed a Baltic-to-White Sea waterway, which serves the heavy-duty needs of the forest and mines of the northwestern area. This canal system is also connected to the Volga system. Today Moscow, by a series of canals, rivers, and lakes, is connected to the Baltic, White, Caspian, and Black Seas.

Lack of coal and petroleum has encouraged some development of hydroelectric power. Important generation facilities have been established in Ontario and Quebec, Canada, near the populated zone, and in Sweden, Finland, and the adjacent part of the Soviet Union. The potentials are high because of the many natural storages, numerous rapids and falls, and the presence of hard-rock surfaces that minimize sediments and provide firm foundations for dam construction. Distance to major consuming centers is the principal deterrent of more widespread development at the present time.

Use for industrial water supply is of great local importance. Pulp and paper plants frequently are located on streams that supply the transportation facility for bringing in the logs, produce the needed power, and furnish the large quantities of water needed in the process.

MANUFACTURING

Industry in the Subarctics is based mainly on forest products—timber and pulpwood—and an abundance of hydroelectric power. These two factors have made the regions leaders in pulp and paper production.

Canada tops the world in the manufacture of pulp; paper and pulp production is the major Canadian industry. Numerous pulp and paper mills, as well as sawmills, are located along the south edge of the forest and along tributary streams of the St. Lawrence River in Quebec and Ontario. Two of Canada's largest paper mills are at Corner Brook and Grand Falls, Newfoundland. Cheap hydroelectric power has stimulated a second industry—the production of aluminum from imported bauxite. The largest aluminum plant in the world is located at Arvida, Quebec, on the Saguenay River (Figure 17.20).

Forest-based industries in Sweden are located at river mouths along the Gulf of Bothnia, with the greatest concentration around Sundsvall and Harnosand. Norway has a plant at Kirkenes that produces iron concentrates from locally mined low grade ore. Most of this country's forest-product plants are located in the Marine West Coast Region. Sawmills, pulp and paper plants, and other woodworking factories are scattered along the west and south coasts of Finland. The industrial center of Finland is at Tampere, which has some diversification —manufacturing in addition to wood products, textiles, shoes, and metal goods. In the northwestern USSR, rafts of logs move downstream to sawmills, woodworking plants, and pulp mills at Archangel, Onega, and Mezen. Petrozavodsk has diversified forest-based industries, including a ski factory that produces an important share of the nation's skis; there is also a small steel works, and engines and equipment for the forest industry are manufactured. Sawmills are scattered along the rivers traversing the forest in Siberia. There are a number of industrial settlements along the Trans-Siberian Railroad between Krasnoyarsk and Lake Baykal utilizing local coal, iron, other minerals and forest resources for varied manufactures.

URBAN CENTERS

The urban needs of the Subarctic regions are small. As a result there are few compact settlements, and only in Eurasia do any exceed 100,000 in population. Most are resource exploitation centers and result from favorable advantages for mining, milling, assembly, and trans-shipment. Such are the Soviet Union centers along the Baltic-to-White Sea Canal, and along the Trans-Siberian Railroad. Murmansk, the most northern large city in the world with a population of about 300,000, is a major center of the Soviet fishing industry. Its ice-free fjord harbor provides the Soviet Union with a gateway to the open Atlantic. Murmansk also serves as the terminus of the Northern Sea Route, the summer route through the Arctic Ocean to the Pacific. The city is connected to Leningrad by 900 miles of rail line. Archangel has a population of about 320,000. Its situation at the mouth

FIG. 17.21. Fairbanks, Alaska. (Fairbanks Chamber of Commerce.)

of the navigable Dvina River, which provides access to a vast forested area, has resulted in its growth and importance for timber concentration and milling. Krasnoyarsk, 580,000, on the Yenisey, and Irkutsk, with about 425,000 people, are the largest of a number of centers along the Trans-Siberian Railroad. Krasnoyarsk is an important transportation and hydroelectric power center with large electricity-using industries producing cellulose and aluminum. Irkutsk is an old settlement which has recently expanded as the result of industrialization; situated at the head of the Angara Valley near Lake Baykal, it is within easy reach of a variety of minerals, electric power, and forest resources.

Fairbanks, serving the Yukon area of Alaska, is the largest center within the North American region, with about 14,000 inhabitants (Figure 17.21). Several slightly larger centers are located within the margins of the Short Summer Humid Continental region.

PART FOUR

THE POLAR REALM

Man is the world's most adaptable creature—his range of living is from the equator to the poles. He has achieved his greatest development, economic and cultural, in the intermediate lands; here, too, he is present in greatest numbers. Up to now the areas of extremes—those too hot, too wet, too dry, too high, too rugged, too remote, or too cold—rarely have attracted large concentrations on a permanent basis.

Of all the restrictive environments, the Polar Lands offer the least to attract man. Their meager resources and extremely harsh environment indicate scattered and limited possibilities for the future.

eighteen

POLAR LANDS

The barren lands or tundra of the far north and the ice caps of the poles constitute the Polar Lands. Nowhere on earth are winters so long and summers so short. Ice and snow are eternal on the ice caps, and the aspects of summer are almost ephemeral on the tundra. Nowhere on earth does the sun remain continuously above the horizon for as many months and nowhere is its heating less effective. Conversely, no other lands are as long deprived of sunlight. Darkness and light divide the year more than the day. Nature provides two compensations for the bleak and barren landscapes and the long dark winters. In the season of maximum warmth, the tundra displays a mosaic of color when the flowering plants make their hasty appearance, and in winter the skies are brightened by the parading colors of the auroras.

Resemblances can be seen between the Polar Lands and the Tropical Deserts. Both are characterized by scanty precipitation, scanty vegetation, and barren expanses; both are largely uninhabited. One is a desert of drought—the other a desert of cold. But in the Polar Lands man finds no oases, and no major resources have been discovered to attract major numbers of people to these regions.

Location

As their name implies, the Polar Lands are situated in the high latitudes, mainly poleward of 65° (see Figure 18.1). East of Hudson Bay in Canada, however, the tundra bends southward to about latitude 55°. The equatorward margins are marked by the warmest month isotherm of 50°, which approximately denotes the cold boundary of tree growth. It is common to subdivide these regions into the tundra and the ice caps.

Tundra. The tundra is the high latitude cold frontier of vegetation. The warmest month isotherms of 50° and 32° mark, respectively, the equatorward and poleward boundaries. It is found preponderantly in the northern hemisphere fringing the Arctic Ocean. Belts of tundra stretch along the margins of North America and Eurasia and

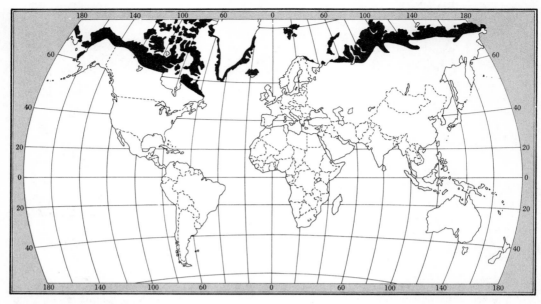

FIG. 18.1. The Polar Lands of the Northern Hemisphere.

FIG. 18.2. The Polar Lands of North America.

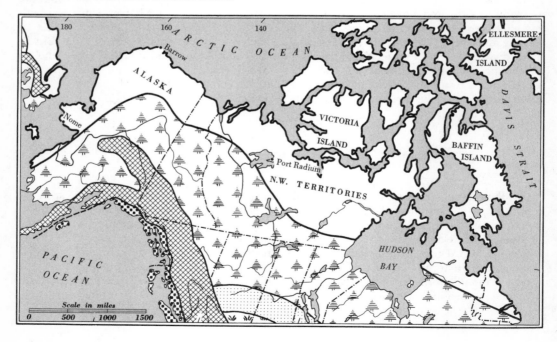

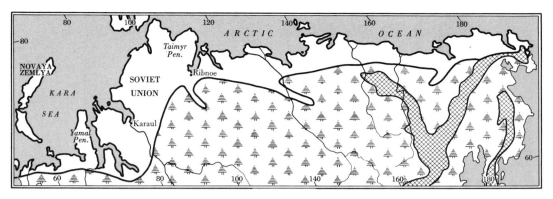

FIG. 18.3. The Polar Lands of Eurasia.

FIG. 18.4. Antarctica.

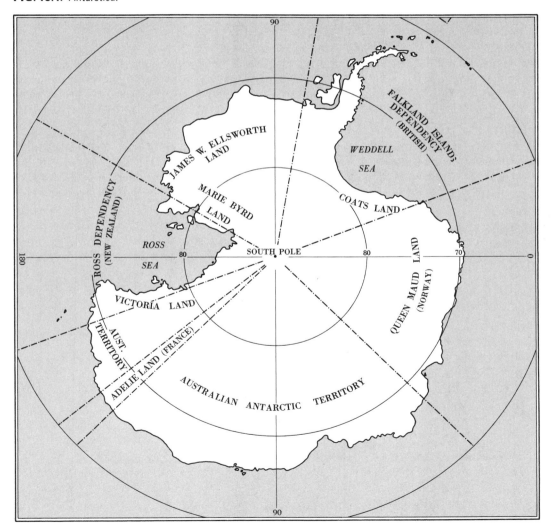

include the coast of Greenland and other Arctic islands (See Figures 18.2 and 18.3). Small and fragmented areas occur along the coast of Antarctica.

Ice caps. The polar lands of perennial ice have reached the greatest development in Antarctica where all the landmass except a few coastal strips is included (Figure 18.4). In the northern hemisphere, the interior of Greenland and the ocean within the vicinity of the North Pole are perpetually frozen.

Physical Environment

CLIMATE

The tundra and the ice caps are divided by the 32° isotherm for the warmest month. The tundra possesses the last vestiges of summer whereas the ice cap is a land of eternal frost.

Tundra. Winters are long and bitterly cold, with about nine months averaging below freezing. January and February have the lowest averages, in the neighborhood of −30° to −40°. Location has considerable effect on winter temperatures, and stations near the coasts are 10 to 20 degrees or more warmer than inland stations and are also much cooler in summer. Summers are cool and short, but days are long. Average monthly temperatures never reach 50° (Figure 18.5); however, daily maxima during the warmest month may be as high as 70° or more. The tundra has many hours of sunshine during the summer, since the sun is above the horizon for two or three months. The sun, however, is low, and the incoming rays are oblique, having little effect as warming agents since much of the energy is consumed in melting snow and ice and evaporating water. Furthermore, much of the insolation

FIG. 18.5. Typical climatic graphs for Polar Land stations.

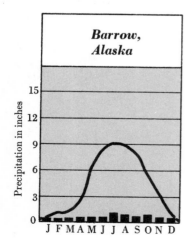
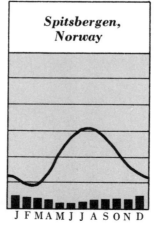
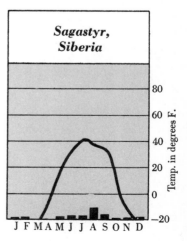

is reflected from the ice and snow surfaces. The region has a period of darkness during the winter when the sun does not make its appearance, but when the sun is below the horizon during these two or three months, the region is not completely dark. There are long periods of twilight following the disappearance of the sun and preceding its return. Light is furnished in this twilight interim by the reflection of celestial bodies on the white ice and snow and by the glowing green or red curtains of the aurora borealis. Precipitation, chiefly of cyclonic origin, is light, averaging about 10 inches throughout the tundra as a whole. The maxima come during the warm season. During this short season, some rain falls; however, snow prevails much of the time. The tundra snow is dry and powdery. The snow does not reach great depths; and, due to strong winds and the lack of protecting vegetation, there are large expanses bare of snow even in winter.

Ice caps. The North and South Poles are the two points on the earth's surface having six months of sunlight and six months of darkness. Equatorward from the poles, the length of continuous light and darkness changes with latitude. March 21 at the North Pole marks the sun's first appearance on the horizon. The barely visible sun follows a circular path around the horizon without setting or rising. It continues to appear higher in the sky until about June 21, when the entire area within the Arctic Circle is bathed in light. Following the summer solstice, the sun gradually sinks in the heavens until, on about September 22, it disappears below the horizon. About December 22, the entire area within the Arctic Circle is covered by darkness. The exact opposite occurs at the South Pole. The winter six months when the sun is below the horizon is not a period of complete darkness. Twilight exists for part of the period, and celestial bodies and the aurora borealis (aurora australis at the South Pole) furnish subdued light.

The Arctic ice cap covers a number of islands, whereas Antarctica is presumably a single landmass of about five million square miles. Climate data for both ice caps are fragmentary. Ice caps have the lowest annual average temperature on earth; the average of the warmest month is below 32°. Temperatures recorded in interior Greenland from November through March averaged −43°. May, June, and July were the warm months with about a 6° average; however, coasts may occasionally have daily maxima above freezing.

The cold air prevailing over ice caps holds little water vapor, consequently precipitation is small in amount and occurs as powdery, dry snow. Little melting allows the snow to accumulate, and in time it becomes loose, granular ice, or *névé*, and solid glacier ice. Both ice caps experience violent blizzards, especially in winter. The wind chill factor lowers significantly the sensible temperatures.

SURFACE FEATURES

The landforms of the Polar regions range from low coastal plains to high ice plateaus, often surmounted by glaciated mountains. In North America the Laurentian Uplands reach the Arctic Coast throughout much of the eastern portion of the continent; here they are low in elevation with relatively slight relief, and often bordered by coastal plains. Fairly wide plains are found along the northern fringe of Alaska and in the Mackenzie River area of Canada. The western islands of the Arctic Archipelago are low, but the eastern islands appear plateaulike and are surmounted by ice-scoured low mountains and hills, many of which are partially covered with glaciers. Greenland, the largest of the world's islands, is a high plateau of 5,000 to 8,000 feet in elevation. Except around the coast, it is covered by a vast ice sheet held within fringing mountain walls.

FIG. 18.6. A summer view of the Alaskan tundra. (United States Geological Survey.)

Flat to rolling plains and low plateaus border the Arctic Coast of Eurasia. The narrow Fenno-Scandinavian portion appears as a dissected plain 500 to 700 feet in elevation, dropping sharply to the Barents Sea. The sea penetrates the land in long, narrow fjords. A low, relatively featureless lowland fringes the shore of the Arctic eastward to the Kolyma River. It is broken only by the Ural Mountains and plateau-like elevations in the Kanin Peninsula, Timan Mountains, and Yamal and Taimyr Peninsulas; rarely do these elevations reach 1,000 feet above sea level.

In the southern hemisphere the continent of Antarctica harbors the world's largest ice sheet. The ice, for the greater part, rises steeply from the sea and levels out to a

FIG. 18.7. The Eurasian tundra. The snow has recently melted exposing the arctic vegetation and expanses of near-barren land, as well as leaving much water on the surface. Soon hordes of insects will make life miserable for man and beast. (Oliver H. Heintzelman.)

rather flat interior with an average elevation of 6,000 feet. Inland from the Pacific Ocean, heights reach 10,000 feet or more.

NATURAL VEGETATION

Conditions in the tundra provide a restrictive environment for plant life. Periods of growth are short, and during the warm season the ground is saturated with water. Despite these handicaps, the tundra has considerable vegetation. It is, however, in extremely fragile balance, and once altered takes many years to return.

The natural cover has three broad divisions: (1) The bush tundra, bordering the taiga, is composed of scrubby alders, birches, and willows only a few feet in height. (2) The grass tundra, beyond the fringe of dwarf trees, occupies the largest area and is most typical. This cover includes mosses, lichens, sedges, flowering plants, and an occasional low bush. (3) The desert tundra on the poleward margin consists of small islands of vegetation growing in the pockets of soil scattered among the bare rock surfaces.

The vegetation pattern of the tundra is a near counterpart of a high mountain alp, the area between the tree line and the permanent snow. During the short summer, plant life is quickened by the warmth. Taking advantage of their brief stay in the sun, plants speed through the various stages of growth. Flowers burst into bloom on the heels of the departing snow, and the bleakness of the landscape is replaced by a colorful carpet of low-flowering plants, poppies, lilies, buttercups, and violets. The true ice caps are barren of vegetation, conditions for plant growth being more hostile than in the driest portions of the Tropical Deserts.

NATIVE ANIMAL LIFE

The majority of animal life in the Polar regions inhabits the tundra lands. Species include the musk-ox, reindeer or caribou (Figure 18.9), wolf, arctic fox hare, and lemming. Millions of insects, especially the mosquito, swarm through the air during the short, warm season. Several species of birds, such as the ptarmigan and the arctic owl, inhabit the tundra throughout the year, and populations are swelled by the great summer migrations from the south. Rookeries along many rocky coasts are crowded with bird life, and water bodies are invaded by waterfowl.

All warm-blooded animals of the Polar

FIG. 18.8. A winter view of the Alaskan tundra. (United States Geological Survey.)

regions have developed means of conserving body heat in order to combat the severe low temperatures. All are equipped with thick fur or feathers. Mammals usually have a silky fur and a thick woolly undercoat. The long hair of the reindeer, which is thicker at the end than at the root, forms an airtight coat. Thick layers of fat provide insulation against loss of heat and act as food reserves for winter. White coloration, typical in winter, radiates less heat and acts as camouflage.

Land animals are almost lacking on the ice caps, but sea life is abundant here as well as along tundra shores. In the Arctic are polar bears, walruses, whales, and seals. The Otary seal of the Pribilof Islands is prized for its fur, and the Arctic seal is hunted for its oil and skin. Sea elephants and whales inhabit the seas of the South Polar region. Whaling is the major activity in the Antarctic waters, which are the world's principal whaling grounds. Breeding colonies of birds are present at both ice caps. The penguin is the typical bird species of Antarctica.

SOILS

Soils occur only in the tundra portions of the Polar regions. Even here the forming processes work very slowly, and barren rock surfaces are common. The tundra soils are characterized by high accumulations of organic matter caused by the extreme retardation of chemical and biological breakdown. As a result, they usually have a peaty surface overlaying a compact subsoil, which in many areas is permanently frozen. Despite low precipitation, the soil is supersaturated with water during the short summer because of low evaporation and poor drainage. Most areas are suitable only as pasturage for reindeer or caribou. The soils of the better-drained, south-facing slopes offer limited possibilities for a few hardy crops that will mature in an extremely short growing period.

Occupance in the Polar Lands

Opportunities for man are indeed small in the Polar Lands. Most areas are as devoid of people as the driest deserts and environments are equally unattractive. Harsh conditions of severe climate, ground that is frozen for most or all of the year, scanty vegetation, and a dependence on animal life compel the few dwellers of these cold lands to exert constant effort to survive. Less than 100,000 people live in these regions, all in the northern hemisphere. None are agriculturalists, and all are more or less migratory. Most of the indigenous occupants of the region have advanced somewhat beyond their earlier closed system of hunting and fishing. They have made modifications as a result of contacts and influences from the outside: For example, an important element in the Eskimo's livelihood was diminished when the whale came under strong exploitation by commercial operators during the nineteenth century. The development of a market for skins, furs, and native artifacts, and the establishment of schools and some alternate employment opportunities in the northlands have induced other changes. Thus, although the Eskimo occupance continues to be largely based upon and geared to hunting and fishing, commercial elements

FIG. 18.9. A herd of caribou in the Alaskan tundra. (Fish and Wildlife Service: Edward F. Chatelain.)

have been introduced into their purpose, and technical elements have affected their practices.

The few nomadic people living in the Eurasian tundra are engaged chiefly in reindeer tending and hunting, especially for sea mammals. Sometimes there is a division of labor with various male members of the family alternately hunting and herding. The reindeer are pastured on the mosses, lichens, and low plants of the tundra during the short, warm seasons but in winter are driven to the edge of the taiga, where the animals are able to expose snow-covered forage plants with their shovel-like hooves. Furthermore, the nomads prefer the forested areas in winter since the trees provide fuel and protection from the Arctic winds. Reindeer furnish meat, milk, and hides, and are used as draft animals. Another domestic animal is the sled dog. In teams of 10 to 14 they are capable of pulling loads up to a ton and can average from 2 to 4 miles an hour, depending upon snow and weather conditions.

The tundra in North America is occupied chiefly by the Eskimos, who live by hunting and fishing. Reindeer were introduced from Eurasia to provide a more dependable food supply, but the experiment has not been too successful. The Eskimo is an excellent hunter and fisherman, and animals are the key to his life—supplying food, heat, light, shelter, and transportation. There is a great dependency upon sea life, especially the seal. The Eskimo's boat, the kayak, is the epitome of seaworthiness. Housing consists of sod igloos, stone huts, or crude shelters made of skin supported by driftwood or whale bones. Temporary snow-block igloos are constructed for shelter during winter hunting trips. Dogs are the only domestic animal and are used in hunting and for transportation; the susceptibility to change, however, is indicated by the increasing number of snowmobiles. The Eskimo has made an excellent adjustment to his harsh and unattractive environment. Eskimos living around the Mackenzie River delta and in Alaska have acquired a veneer of western culture due to their many contacts with Canadian traders and American government workers and servicemen.

Although there may be differences in racial background among Polar peoples, the restricting environment has imposed many similarities in culture. Clothing, food, tools, utensils, and shelter types are quite the same from one tribe to another.

The white man has avoided the Polar Lands. His contacts have been largely in the interests of commerce with the natives, exploitation of aquatic life, military security, and scientific experiments. The airplane has made possible exploration of large areas that were unknown but a few years ago and has given new signficance to these lands in terms of national security. Advancing knowledge of weather and climate has assigned importance to the Arctic region's relationships with the middle latitudes. Weather observation stations have been established within the Arctic Circle.

Since 1930, the Soviet Union has directed some efforts toward development of her Arctic fringe. Several "reindeer farms" have been established and native peoples organized into collectives for systematic and improved use of the tundra pasturages. A number of small ports are operated along the coast for shipment of timber, minerals, and furs from the forest lands to the south over the Northern Sea route. During the brief summer, several scores of ships serve these ports, some making the complete trip from Murmansk to Vladivostok.

The discovery of huge petroleum deposits in the north slope area of Alaska in recent years has resulted in much exploratory activity by American companies. This is not likely to result in significant population growth in the area since the oil will be transported outside the region for refining and use. The movement of the crude oil is a major problem of development. At the present time the most economical means appears to be a pipeline. A proposed 48-inch trans-Alaska line would extend from the locale of Prudhoe Bay 800 miles southward to Valdez. At the present time strong opposition is being voiced by those concerned for the possible damage to the fragile arctic environment, wildlife, and ocean as the result of pipeline, and related road construction, possible breaks in the line, and oil spills in the port.

INDEX

A

Abadan, Iran, 176
Aberdeen, England, 324
Abidjan, Ivory Coast, 133
Abyssinian Mountains, Africa, 184
Acapulco, Mexico, 130
Accra, Ghana, 133
Addis Ababa, Ethiopia, 181, 191
Adelaide, Australia, 204
Aerial photography, 13–14
Aflaj, Saudi Arabia, 171
Africa, 82, 234, 236, 341
  agriculture in, 125, 188–89, 245
  climatic regions of, 111, 178, 222, 341, 371
  forest industries of, 104, 127, 247
  mining in, 129, 249, 363–65
  surface features of, 88–91, 116–17, 184, 230, 347
Ahaggar Highlands, Africa, 169–70
Air pressure, 40–46
Air temperature, 40–45
Alabama, 228, 235, 239–41, 249
Alaska, 91, 307–9, 318, 390, 407
  fisheries in, 324, 331
  oil in, 421–22
  surface features of, 313, 395, 415
Albania, 206
Alberta, 282, 301, 305, 363

Albuquerque, N.M., 366
Alexandria, Egypt, 176
Algeria, 173, 212
Algiers, Algeria, 216
Al Kharj, Saudi Arabia, 171
All American Canal, U.S., 173
Alps (Alpine Mountains, Europe), 261, 370, 374, 379–81, 389
Amarillo, Texas, 365, 366
Amazon Basin, S. America, 85, 89–91, 95, 108–9
Amazon River, S. America, 89–91, 99, 115
American Manufacturing Belt, U.S., 274–77, 279, 282, 302
Amsterdam, Netherlands, 335
Amur Valley, U.S.S.R., 296
Anatolian Plateau, Turkey, 341 352, 366
Andes Mountains, S. America, 85, 156, 158, 178, 183, 185
  agriculture in, 187
  Argentinian, 341, 357
  Chilean, 387
  mining in, 189
  Northern, 116, 182
  Venezuelan, 115
Andorra, 370
Anemometer, 43
Animal life, 61–63
Ankara, Turkey, 366
Annam Coast, Vietnam, 138
Annam Mountains, Indochina, 140, 184

Annapolis-Cornwallis Valley, Nova Scotia, 286, 293
Antarctica, 415, 416, 419
Anticyclones, 49
Antwerp, Belgium, 335
Appalachian coalfields, U.S., 249, 274
Appalachian Highlands, U.S., 228, 261
Appalachian Mountains, U.S., 261, 279, 377
Appalachian Uplands, U.S., 233, 271, 273
Arabian American Oil Company, 161, 171
Arabian Peninsula, 153, 154, 181, 182, 187
Arafura Sea, Australia, 117
Aral Sea, U.S.S.R., 362
Archangel, U.S.S.R., 405–6
Arctic Archipelago, 414, 415
Arctic lands, 411–22
Arctic Ocean, 411–14
Argentina, 110, 222, 341, 365
  agriculture in, 125, 244
  irrigation in, 357–60
  surface features of, 230, 345–46
Argentine Mesopotamia lowland, 230
Arizona, 156, 160, 344
Arkansas, 243
Armenian Highlands, 230–33
Arnhem Aboriginal Reserve, Australia, 121
Aroostook Valley, Maine, 286, 293

Artesian water, 161, 171
Aruba (island), 129
Arvida, Quebec, 405
Assam Province, India, 139, 146
Astoria, Ore., 331
Aswan Dam, U.A.R., 160
Atacama Desert, Chile, 153, 156, 158, 163, 174
Athens, Greece, 216
Atlantic City, N.J., 279
Atlantic Coastal Plain, U.S., 57–59, 228
Atlas Mountains, N. Africa, 169, 370
Atmospheric moisture, 46–50
Auckland, New Zealand, 336
Australia, 163, 327
   agriculture in, 209–11, 245, 352
   climate of, 136, 227, 341
   climatic regions of, 110, 198, 222–25, 307, 341, 371
   forest industries of, 247, 327
   grazing in, 127, 362
   industry in, 220, 251, 332
   mining in, 129, 214, 249
   population of, 121, 236, 320
   surface features of, 117, 184, 204, 230, 347
   urban centers in, 133, 216, 255
   vegetation in, 234, 318, 347
Australian Desert, 156, 160
Austria, 385, 388–89
Azizia, Sahara, 157
Azov Sea, U.S.S.R., 362

B

Bagdad, Iraq, 176
Bahia Coast, Brazil, 102
Bahrein (sheikdom), 173
Bahr-el-Ghazal River, Sudan, 117
Baku, U.S.S.R., 363, 366
Bali, Indonesia, 184
Baltic Sea, 299, 335, 404
Baltic-to-White Sea Canal, U.S.S.R., 404, 405
Baltimore, Md., 276, 279
Bandung, Java, 109
Banff National Park, Canada, 387
Bangka, Indonesia, 106
Bangkok, Thailand, 147, 149
Bangla Desh, 134, 149

Bangweulu Swamps, Zambia, 121
Banks (fishing), 297–98, 324
Barbary Coast states, Africa, 196
Barcelona, Spain, 199, 216
Barometer, 43
Barranquilla, Colombia, 130
Basel, Switzerland, 389
Basra, Iraq, 176
Bay of Bengal, 139, 140
Bay of Biscay, 318, 323
Beirut, Lebanon, 216
Belem, Brazil, 104, 108
Belfast, N. Ireland, 329
Belgium, 319, 320, 328, 330, 335
Belgrade, Yugoslavia, 279, 281
Belitung, Indonesia, 106
Bellingham, Wash., 332
Belo Horizonte district, Mexico, 189, 190
Belts (pressure systems), 43–46
Benares, India, 149
Benguela Current, Africa, 153
Bergen, Norway, 324
Berlin, 303, 305
Bern, Switzerland, 389
Bilboa, Spain, 328
Bingham Canyon, Utah, 363
Birmingham, Ala., 249, 251, 255
Birmingham, England, 330, 334
Black Belt, Ala., 235, 239–41
Black Forest, Germany, 381, 383
Black Hills, U.S., 343
Black Sea, 249, 252, 272, 362, 404
Blue Mountains, Pacific Northwest, 344
Blue Nile, Africa, 117, 123
Blue Ridge Mountains, U.S., 261
Bogor, Java, 88
Bogotá, Colombia, 191
Bolivar Savanna, Colombia, 111
Bolivia, 104, 110, 147, 187, 189
Bombay, India, 148, 149
Borneo, 106, 184
Boston, Mass., 279
Brahmaputra River, 139, 141, 146
Brasilia, Brazil, 132–33
Brazil:
   agriculture in, 125, 188
   climatic regions of, 85, 110, 178, 222
   forest industries of, 104, 247
   industries of, 108 190–91
   mining in, 189, 249
   population of, 95, 121, 185
   surface features of, 183, 230
   urban centers in, 132–33, 191

Brazilian Highlands, 115, 183–84, 230
Breda, Netherlands, 330
Bremerton, Wash., 332
Brenner Pass, Alps, 389
Brisbane, Australia, **129, 225,** 245, 251, 252
Bristol, England, 330
British Columbia, 307, 313, 318, 320, 332
British East Africa, 190–91
British Isles, 306, 319, 320
Bruges, Belgium, 330
Brussels, Belgium, 330, 335
Bucharest, Romania, 281
Budapest, Hungary, 277, 279, 281
Buenos Aires, Argentina, 244, 247, 255
Buffalo, N.Y., 302, 304
Bulgaria, 257, 262, 279
Burma, 134, 140, 142, 144–49
Bushmen, 164–65
Butte, Mont., 388

C

Cairo, Egypt, 176
Calcutta, India, 148, 149
Calgary, Alberta, 305, 367
California, 200, 205, 214–19, 307
   agriculture in, 71, 209, 214
   climatic regions of, 156, 198
   fishing in, 212, 324
   vegetation of, 204, 318, 319
Callao, Peru, 176
Cambodia, 134, 142, 146
Campos (Brazilian grassland), 110, 115
Canada, 292, 299, 390, 400, 402
   forest industries of, 300, 401, 405
   industry of, 274, 302
   Polar lands of, 411, 415, 420
   tourism in, 303–4, 387
   urban centers of, 303–4, 367
Cannes, France, 215
Canton, China, 149
Capetown District, S. Africa, 198, 204, 209–11, 216
Cape York Peninsula, Australia, 129
Caracas, Venezuela, 191
Caribbean America, 85, 104, 130

Caribbean Sea, 115
Carolinas, 241, 243, 269
Carpathian Mountains, Europe, 370
Carrara, Italy, 214
Cartagena, Colombia, 130
Casablanca, Morocco, 216
Cascade Mountains, N. America, 182, 336, 343, 370, 376, 383
Caspian Sea, 337, 362, 363, 404
Caucasus Mountains, U.S.S.R., 230, 371, 379
Central Pacific Islands, 111, 117, 133, 178
Central Russian Uplands, 288
Central Valley, Calif., 202, 209, 214
Central Valley, Chile, 203, 209–10
Ceylon, 85, 142, 146
Chaco (S. American grassland), 110, 129
Chamonix, France, 387
Chaparral woodlands, 61, 204, 379
Chemnitz, Germany, 303
Chengtu Plain, China, 228
Cherrapunji, India, 138
Chesapeake Bay, U.S., 246
Chicago, Ill., 87, 275, 279
Chico, Calif., 200
Chile, 219, 318, 327, 332, 387
    agriculture in, 71, 209, 323
    climate of, 309, 311
    climatic regions of, 156, 198, 307
    mining in, 174, 189, 214
    population of, 205, 320
    surface features of, 313, 314–15
China, 70, 227–28, 235–36, 264, 266
    agriculture in, 237–39, 271–72, 352–53, 381
    climate of, 225–27, 260
    climatic regions of, 134, 222, 257, 337
    fishing in, 245
    industry in, 250, 273
    mining in, 247
    urban centers of, 149–50, 216, 252–53, 366
Chittagong, Bangla Desh, 149
Christchurch, New Zealand, 336
Chungking, China, 250, 253
Cincinnati, Ohio, 279
Ciudad Guayana, Venezuela, 129–30

Cleveland, Ohio, 276, 279
Climate, 39–55, 76–79
Climate control, 50–55, 60
Clouds, 46–48
Coachella Valley, Calif., 173
Coffee, 181, 188, 189
Cologne, Germany, 335
Colombia, 111, 128, 130–32, 188, 190, 191
Colorado, 377, 385
Colorado Plateau, U.S., 344
Colorado River, U.S., 158, 161, 172, 218, 344, 366
Columbia Basins and Plateaus, U.S., 344–45
Columbia River, U.S., 314, 327, 332, 344–45
Congo Basin, Africa, 85, 91–92, 95, 109, 117
Congo River, Africa, 85, 89, 91, 109
Connecticut Valley, U.S., 277
Coosa Valley, U.S., 261
Copenhagen, Denmark, 335
Cordillera mountains, 91, 182, 370–71
Corn belt, U.S., 261, 264, 266–69
Cornwallis Valley, Nova Scotia, 286, 293
Coromandel Coast, India, 140
Cortina, Italy, 387
Cotton, 123, 125, 169–72, 215, 237, 239–41
Coventry, England, 330
Crimea, 198, 215
Cuba, 111, 116, 121, 129, 130
Cultural systems, 22–36
Cumberland Valley, U.S., 261
Curaçao (island), 129
Currents, ocean, 54–55
Cyclones, 49, 138, 227
Czechoslovakia, 303

D

Dacca, Pakistan, 149
Dairen, China, 281
Dakar, Senegal, 176
Dalmatian Coast, Yugoslavia, 202, 215
Damascus, Syria, 176
Danube Basin, Europe, 257, 260, 262, 264, 271, 281
    industry in, 265, 273, 277

Danube River, Europe, 272
Darwin, Australia, 133
Darwin, Charles, 6
Date palms, 170–71
Dawasir Wadi, Saudi Arabia, 170–71
Death Valley, Calif., 157
Deccan Plateau, India, 140, 141
Delhi, India, 149
Denmark, 307, 321–22
Density (population), 29–31
Denver, Colo., 365, 366
Desalinization, 162
Desert El Djouf, Sahara, 156
Deserts, 152–77, 349
Detroit, Mich., 302, 304
Dew point, 46
Djakarta, Indonesia, 109, 133
Dnieper River, U.S.S.R., 272, 363, 365
Dogger Banks, North Sea, 324
Doldrums, 43
Donetz Coal Basin, U.S.S.R., 363, 365, 366
Don River, U.S.S.R., 354, 362
Dortmund, Germany, 331, 335
Dresden, Germany, 303
Dublin, Ireland, 311
Duisburg, Germany, 331, 335
Duluth, Minn., 301
Dunedin, New Zealand, 336
Durban, S. Africa, 252, 255
Dzungarian Basin, U.S.S.R., 342

E

Earth, man and, 37–71
East African Highlands, 92
Eastern Ghats (mountains, India), 140
Eastern Highlands, Australia, 117, 222–25
East Germany, 305
East Indies, 100, 142
East Pakistan, 138, *see also* Bangla Desh
Ecuador, 102, 111, 133, 180
Edmonton, Alberta, 301, 305
Edwards Plateau, Texas, 361
Eel River Valley, Calif., 318
Egypt, 160, 164, 175, 176
Eindhoven, Netherlands, 331
Elbe River, Europe, 313
Elbruz Mountains, Asia, 371

El Paso, Texas, 366
England, 187, 313, 318, 324, 327–30, 332, 334–35
Environment, man and, 22–36
Equinox, 52
Eskimos, 419–20
Essen, Germany, 331
Ethiopia, 111, 117
Euphrates River, Near East, 158, 160, 171
Everglades, Fla., 234, 243

F

Fairbanks, Alaska, 407
Fergana, U.S.S.R., 365
Fiji Islands, 130
Finger Lakes District, U.S., 271
Finland, 299, 301, 304, 393, 399–401, 404–5, 416
Fjordland, Australia, 317, 334
Florence, Italy, 216
Florida, 225, 234, 241, 243, 249, 252
Fortaleza, Brazil, 132
France, 306, 314, 318, 330, 335, 370
  mining in, 212, 274, 328
Frazer River, B.C., 319–20
French Riviera, 202, 207, 215
Fresno, Calif., 200, 217
Frontal precipitation, 48–50, 55
Fronts, 49–60
Fuji (mountain, Japan), 252
Fur industry, 401–2

G

Galati, Romania, 277
Galeria forests, 118, 127
Ganges Plain, India, 139, 140, 149
Ganges River, India, 139, 141, 146
Ganges Valley, India, 144
Garfield, Utah, 365
Garmisch-Partenkirchen, Germany, 387
Geneva, Switzerland, 389
Genoa, Italy, 199, 216
Geography, 3–21
Geography, regional, 7, 72–74
Georges Bank, Atlantic Ocean, 297
Georgetown, Guyana, 109
Georgia, 241, 243, 249
Germany, 282, 289, 303–5, 307, 314, 383
  industry in, 328–29, 331, 385
Ghana, 133
Ghats (mountains, India), 140
Ghent, Belgium 330
Gibson Desert, Australia, 156
Gila River, U.S., 161, 172, 173
Glaciation, 286–89, 375
Glasgow, Scotland, 329, 334
Globes, 10
Gobi Desert, Mongolia, 342, 343, 351
Goias Plateau, Brazil, 115
Gran Chaco (S. American grassland), 110, 129
Grand Banks, Atlantic Ocean, 297
Grand Canyon, U.S., 344, 366
Grand Coulee Valley, Wash., 345
Graphs, 14, 43
Great Barrier Reef, Australia, 118
Great Basin, U.S., 344
Great Britain, 187, 313, 318, 324, 327–30, 332, 334–35
Great Lakes, N. America, 259, 274–75, 285
Great Lakes region, N. America, 279, 282, 287–88, 291, 293
  forest industries of, 290, 299–300
Great Plains, N. America, 57, 347, 350, 353
  climate of, 341, 342
  features of, 228, 261, 285, 343, 395
  livestock industry of, 361, 382
Great Salt Lake, Utah, 363
Great Smoky Mountains, U.S., 261
Great Valley, U.S., 261
Great Valley of Victoria, Australia, 316–17
Greece, 206, 212
Greenland, 414, 415
Green Revolution, 151
Grimsby, England, 324
Growth rates (population), 25–26
Guano, 174–75
Guayaquil, Ecuador, 133, 180
Guayas Lowland, Ecuador, 116
Guaymas, Mexico, 130
Guiana Highlands, Venezuela, 115, 129
Guianas, 104
Guinea Coast, Africa, 95, 98, 100
Gulf Coast, U.S., 228, 234, 249
Gulf of Bothnia, Sweden, 405
Gulf of California, N. America, 172–73
Gulf of Carpentaria, Australia, 117
Gulf of Finland, 299
Gulf of Guinea, Africa, 85, 111, 116
Gulf of Mexico, 246
Gulf Stream, Atlantic Ocean, 297
Guyana, 106, 109

H

Haarlem, Netherlands, 335
Hagerstown Valley, U.S., 261
Haifa, Israel, 171
Hainan, China, 134
Haiphong, N. Vietnam, 149
Hakodate, Japan, 305
Hamburg, Germany, 335
Hanford, Wash., 365
Hangö, Finland, 299
Hanoi, N. Vietnam, 149
Harbin, Manchuria, 305
Havana, Cuba, 130
Hawaiian Islands, 117, 125, 129, 130, 133
Helsinki, Finland, 299
High Plains, Texas, 355
High pressure belts, 43–46
High Veld, Africa, 347
Himalaya Mountains, Asia, 139, 140, 146, 370, 371, 381
Hindu Kush Mountains, Asia, 371, 381
Hofuf Oasis, Saudi Arabia, 171
Hokkaido, Japan, 284, 289, 290, 292, 299, 302–3
  agriculture of, 272, 296–97
  population of, 292, 305
Hollywood, Calif, 215
Hong Kong, 150
Honolulu, Hawaii, 133
Honshu, Japan, 222, 228, 257, 263, 264, 272, 273
Hooghly River, India, 149
Huachipato, Chile, 332
Hudson Bay, Canada, 288, 395

Hudson River, Canada, 275
Hudson's Bay Company, 402
Hull, England, 324
Humboldt, Alexander von, 6
Humboldt Current, 153, 174–75
Hungary, 257, 271, 274, 277
Hunter Valley, Australia, 230
Hwang Ho River, China, 262
Hyderabad, India, 149
Hydrologic cycle, 66–67

I

Ibadan, Nigeria, 133
Ice caps, 414–15
Iceland, 307
Idaho, 383
Imperial Dam, Calif., 173
Imperial Valley, Calif., 161, 172
India, 81, 134, 137–42, 144–49, 156, 370
Indiana, 276
India-Pakistan Peninsula, 137–42
Indochina, 140, 142, 146–48
Indonesia, 85, 92, 101–2, 106, 142, 184
   climatic regions of, 110, 178
   population of, 95, 142
   urban centers in, 109, 133
Indus River, Asia, 158, 160, 172
Indus Valley, Asia, 156
Innsbruck, Austria, 388–89
Interior Lowlands, U.S., 228, 261
Intermontane Province, N. America, 341, 343–45, 361, 363, 366–67
International Date Line, 12
Iowa, 260
Iran, 154, 162, 173, 176
Iraq, 154, 160, 171, 173
Ireland, 329
Irkutsk, U.S.S.R., 407
Irrawaddy River, Burma, 140, 146
Irrigation, 144, 160–61, 164, 168–73, 209, 354–60, 385
Ishikari Plain, Japan, 289, 302, 303
Isobars, 43
Isohyet, 50, 256
Isotherm, 43
Israel, 154, 177, 196, 204, 206–7, 216
Istanbul, Turkey, 216, 366

Isthmus of Tehuantepec, Mexico, 106
Italian Riviera, 202, 207, 215
Italy, 206, 212, 216, 266, 281
Ivory Coast, 133

J

Jamaica, 121, 129, 130
Jamshedpur, India, 148
Japan, 70, 220, 227–28, 235, 252, 266
   agriculture in, 237, 239, 272, 381
   fishing in, 272
   forest industries of, 246, 273
   industry in, 249–50
   mining in, 247, 302
   See also Hokkaido, Japan
Jasper National Park, Canada, 387
Java, 81, 95, 99–100, 106, 109, 133
   surface features of, 117, 184
Jet stream, 49
Johannesburg, S. Africa, 365, 367
Jordan, 154
Jura Mountains, Europe, 380, 387
Jutland Peninsula, Denmark, 289

K

Kalahari-Namib Desert, 153, 156, 157, 160
Kama River, U.S.S.R., 302
Kansas, 257, 260, 261
Kansas City, Kansas, 277, 279
Kant, Immanuel, 5
Kara-Bogas Gulf, U.S.S.R., 363
Karachi, Pakistan, 176
Karaganda deposits, U.S.S.R., 363
Karst topography, 116, 202
Katabatic wind, 53, 374
Katanga District, Africa, 129
Kauai, Hawaii, 116
Kazakh Uplands, U.S.S.R., 343, 363
Kentucky, 269
Kenya, 187, 188, 191
Kerch Peninsula, U.S.S.R., 365
Kern River, Calif., 204
Khabarovsk, U.S.S.R., 305
King River, Calif., 204

Kingston, Jamaica, 130
Kinshasa, Congo Basin, 109
Kiruna, Sweden, 402
Kitimat project, B.C., 332
Kittatinny Valley, U.S., 261
Kobe, Japan, 250, 253
Kola Peninsula, U.S.S.R., 402
Kopet Dag Mountains, Asia, 371
Köppen system, 76–79
Korea, 222, 257, 263, 264, 284, 302
Kounrad, U.S.S.R., 363
Krasnoyarsk, U.S.S.R., 407
Krivoi Rog, U.S.S.R., 363, 365
Kumasi, Ghana, 133
Kunlun Mountains, Asia, 371
Kuwait, 162, 173
Kuznetz Basin, U.S.S.R., 302, 303, 305
Kwanto Plain, Japan, 228, 235, 250
Kyoto, Japan, 253–54
Kyushu, Japan, 222, 247, 250, 253–54

L

Labrador Current, Atlantic Ocean, 285, 297
Lagos, Nigeria, 133
Lahore, Pakistan, 149
Lake Agassiz, N. America, 288, 291
Lake Albert, Africa, 184
Lake Balkhash, U.S.S.R., 362
Lake Chad, Africa, 92, 116
Lake Edward, Africa, 184
Lake Manitoba, Canada, 299
Lake Maracaibo, Venezuela, 115, 128
Lake Michigan, 275, 279
Lake Nyasa, Africa, 184
Lake Rudolf, Africa, 184
Lake Superior, N. America, 301
Lake Tanganyika, Africa, 184
Lake Victoria, Africa, 184
Lake Winnipeg, Canada, 288, 299
Lake Winnipegosis, Canada, 299
Lanchow, China, 366
Landforms, 57–60
Laos, 134, 142
La Paz, Bolivia, 191
Latitude, 11, 14
Laurentian Uplands, N. America, 285–86, 402

Lebanon, 184, 196
Lebanon Valley, U.S., 261
Leduc, Alberta, 301
Leeds, England, 330, 334
Leipzig, Germany, 303
Lena River, U.S.S.R., 396
Leningrad, 302, 303, 305, 405
Lesser Sunda Islands, Malay, 117
Liao River, China, 263
Libya, 160, 173
Liechtenstein, 370
Lille, France, 330
Lima, Peru, 176
Lithosphere, 55–57
Little Rock, Ark., 249
Liverpool, England, 334
Loess Highlands (Loesslands), China, 337, 343, 350, 351, 353, 363, 366
London, 309, 330, 334
Longitude, 12, 14
Lorraine District, France, 328, 330
Los Alamos, N.M., 365
Los Angeles, Calif., 214, 216
Louisiana, 234, 243, 249
Low Countries, 307, 314, 330–31
Low pressure belts, 43
Lubbock, Texas, 366
Luxembourg, 328
Luzon, Philippines, 150
Lythgow, Australia, 251

M

Maceio, Brazil, 109
Mackenzie River, Canada, 395, 415, 420
Mackenzie Valley, Canada, 393
Madeira River, Brazil, 91
Madras, India, 149
Madrid, Spain, 366
Magdalena River, Colombia, 132
Magdalena Valley, Colombia, 115–16, 128
Maine, 299
Malabar Coast, India, 140
Malagasy, 83, 111
Malawi, 127
Malaya, 93, 100–103, 106, 142
Malay Peninsula, 85, 92, 101–2, 106, 109, 140, 184
Malaysia, 147

Man:
 culture and, 22–36
 earth and, 37–71
Manaus, Brazil, 87, 90, 104, 108
Manchester, England, 330, 334
Manchuria, 274, 279, 281, 292, 296
 climatic regions of, 257, 284
 forests of, 264, 273
Manchurian Plain, 262–63, 266, 272
Mandalay Basin, Burma, 140, 144
Manila, Philippines, 150
Manizales, Colombia, 191
Map projections, 10–19
Maps, 9–12, 14–15
Maquis woodlands, 61, 204
Maracaibo, Venezuela, 130
Maracaibo Basin, Venezuela, 111, 115
Maritime Provinces, Canada, 282
Marrakech, Morocco, 162
Marseilles, France, 216
Marsh, George Perkins, 6
Matadi, Congo Basin, 109
Mato Grosso Plateau, Brazil, 115
Mazatlan, Mexico, 130
Mecca, 176–77
Medellin, Colombia, 190, 191
Mediterranean Basin, 200–202, 204–8, 212–16, 380
Mediterranean Sea, 196, 202, 212
Mekong River Valley, Asia, 140
Melbourne, Australia, 336
Menam River Valley, Thailand, 140
Mendoza, Argentina, 357, 367
Merced, Calif., 200
Merida, Mexico, 130
Meridians, 12, 14–15
Mesabi Range, Minn., 301, 363
Meseta (Spanish plateau), 341, 352, 366
Mexican Plateau, 344
Mexico, 111, 121, 156, 173
 agriculture in, 150, 355
 industry in, 128, 190, 365
 mining in, 106, 189
 tourism in, 130, 366
 urban centers in, 130, 191
Mexico City, 190, 191
Michigan, 296, 301
Middle America, 104, 121, 184
 agriculture in, 100–102, 125, 188
 climatic regions of, 91, 111, 178
 surface features of, 116 182–83

Middle Amur Lowlands, U.S.S.R., 289
Middle Atlantic Coastal Plain, U.S., 261
Middle East, 82, 154
 oil in, 70, 173–75, 177
Migration (population), 29–31
Milan, Italy, 277, 279, 281
Milwaukee, Wis., 275, 304
Min River, China, 228
Mindanao, Philippines, 106
Minneapolis, Minn., 305
Minnesota, 301
Minnesota River Valley, 296
Mississippi, 228, 249
Mississippi River, U.S., 228, 275
Mississippi Valley, U.S., 225, 241, 243, 260, 272, 350
Missouri, 274
Missouri Valley, U.S., 350
Mohave Desert, Calif., 363
Moisture, atmospheric, 46–50
Monaco, 216
Monclova, Mexico, 365
Mongolia, 360–61, 366
Monsoons, 54–55, 136–38
Montana, 383
Monterrey, Mexico, 365, 367
Montevideo, Uruguay, 247, 252, 255
Montreal, Canada, 302, 304
Moomie, Australia, 129
Moscow, 303, 305, 404
Moscow Basin, U.S.S.R., 302
Mount Aconcagua, Argentina, 183
Mountains, 55, 59, 182, 369–89
Mount Blanc, France, 387
Mount Waialeale, Hawaii, 115
Mukden, China, 281
Mulhouse, France, 328
Murmansk, U.S.S.R., 405
Murray Basin, Australia, 211, 216

N

Nagasaki, Japan, 250, 254
Nagoya, Japan, 250, 253–54
Nairobi, Kenya, 191
Namib Desert, 153, 156, 157, 160
Nancy, France, 330
Napa Valley, Calif., 209
Naples, Italy, 216

Narvik, Norway, 402
Nashville Basin, Tennessee, 243
Natal Coast, Africa, 222, 227, 251
National parks, U.S., 366, 387
Natural resources, 35, 68–71
Nebraska, 256, 261, 343
Negro Valley, Argentina, 360
Nelson River, Canada, 288
Nemuro Plain, Japan, 289
Nepal, 370
Netherlands, 314, 319–21, 328, 330–31, 335
New Caledonia, 118, 129
Newcastle, Australia, 214, 245, 249, 251, 255
New England, 276, 282, 300–302
Newfoundland, Canada, 390, 405
New Guinea, 117
New Mexico, 344, 355, 363,
New South Wales, Australia, 220, 341
New York (City), 87, 279
New York metropolitan area, 276
New York (State), 300, 302
New York State Barge Canal, 275
New Zealand, 307, 318, 320, 323, 334, 336
 industry in, 327–28, 332
 surface features of, 313, 317
Nice, France, 215
Nigeria, 95, 98, 121, 129, 133
Niger River, Africa, 116
Nikopol, U.S.S.R., 363, 365
Nile River, Africa, 90, 116–17, 123, 158, 160, 164
Nile Valley, Africa, 81, 82, 168–69, 176
Nitrates, 174, 214
Nomads, 164–66, 360–61
Norilsk, U.S.S.R., 404
North China Plain, 262, 271, 281, 343
North Dakota, 363
Northeastern Uplands, N. America, 286, 290
Northern Ireland, 329
Northern Territory, Australia, 127
North European Plain, 288, 289, 292, 296, 305, 314
North Korea, 266
North Pole, 414, 415
North Sea, 323, 335
North Vietnam, 134, 142, 149
Norway, 307–11, 313, 323–24, 332, 335, 380
 forests in, 317, 327, 405

Nouvelle Anvers, Congo Basin, 87
Nova Scotia, 301

O

Oakland, Calif., 216
Oasis agriculture, 168–73
Ob River, U.S.S.R., 396
Occupance, 75–76
Ocean currents, 54–55
Oceania, 85
Odenwald (Oden Forest), Germany, 383
Ohio, 261, 276
Ohio River, U.S., 276
Oil industry, 70, 106, 128–29, 173–75, 177, 214–15, 274
Oklahoma, 222, 249
Old World Deserts, 166, 168–72
Omaha, Neb., 87, 277
Oman, 154, 182, 187
Omdurman, Sudan, 133
Ontario, 301, 302, 304, 404, 405
Orange River, Africa, 158
Ordos Desert, China, 343
Oregon, 307, 313, 318, 327
Orinoco Llanos (river basin, S. America), 110, 115, 126
Osaka, Japan, 250, 253
Oslo, Norway, 335
Ostrava, Czechoslovakia, 303
Ottawa, Canada, 302, 304
Ouachita Highlands, U.S., 228
Oymyakon, Siberia, 393
Ozark Highlands, U.S., 228, 273
Ozark Plateau, U.S., 261

P

Pacific Northwest, N. America, 306–7, 312, 314, 318–20, 322–28, 331–33, 335–36
Pakistan, 134, 139, 144–45, 148, 156
 population of, 81, 142
 urban centers of, 149, 176
Palouse Hills, Wash., 350
Pampas, Argentina, 222, 227, 230, 233–35, 244
Panama City, Panama, 130

Paraguay, 110, 222, 230, 247
Paraguay River, S. America, 115
Paramaribo, Surinam, 109
Paraná Basin, Argentina, 222, 233
Paraná River, S. America, 115, 230
Paris, 330, 332, 335
Patagonia, 182, 341, 347–49, 357, 361, 365
Pechora Basin, U.S.S.R., 404
Pemba (island), 85
Pennine Mountains, England, 330
Persian Gulf, 173
Perth, Australia, 198, 216
Peru, 156, 160, 172, 174–76, 187
Peruvian Current, 174–75
Petroleum industries, 70, 106, 128–29, 173–75, 177, 214–15, 274
Petsamo, U.S.S.R., 404
Philadelphia, Pa., 276, 279
Philippines, 85, 92, 102, 104, 150, 184
 climatic regions of, 134, 178
 population of, 95, 142
Phoenix, Ariz., 173, 177
Photography, 13–14
Pilsen, Czechoslovakia, 303
Pittsburgh, Pa., 276, 279
Plains, 57–58, 260
Plantation system, 101–4, 188
Plateaus, 59, 182
Poland, 282, 289, 303
Po Plain, Italy, 257, 262–65, 271, 281
Population, 23–31, 59
Port au Prince, Haiti, 130
Port Kembla, Australia, 214, 251
Portland, Ore., 309, 332, 336
Port of Spain, Trinidad, 109
Port Phillip Bay, Australia, 317, 336
Port Royal, Jamaica, 130
Portugal, 206
Po Valley, Italy, 277
Powell, John Wesley, 6
Prairie Provinces, Canada, 301, 342
Pral Mountains, U.S.S.R., 384
Precipitation, 46–50, 55, 66–67
Pressure, 40–46
Pretoria, S. Africa, 365, 367
Pribilof Islands, Alaska, 419
Provo, Utah, 365
Pueblo, Colo., 365

Puerto Montt, Chile, 334
Puerto Vallerte, Mexico, 130
Puget Sound, Wash., 336
Puget-Willamette Lowland, Pacific Northwest, 313, 314, 319, 322, 331
Punjab, India-Pakistan Peninsula, 144, 149
Pyrenees Mountains, Europe, 314, 370, 379

Q

Qatar, 173
Quebec, 300, 301, 402, 404, 405
Queensland, Australia, 121, 125, 127, 133, 247
Quito, Ecuador, 180, 191

R

Rand, S. Africa, 365
Rangoon, Burma, 147, 149
Ratzel, Friedrich, 6
Recife, Brazil, 129, 132
Red Basin, China, 227–28, 235, 247, 253
Red River, Canada, 288
Red River Valley, N. America, 293
Red River Valley, N. Vietnam, 140
Red Sea, Near East, 184
Regional geography, 7, 72–74
Relative humidity, 46
Remote sensing (photography), 13–14
Rhine River, Europe, 313
Rhodesia, 125
Rhone River, Europe, 199, 212, 381
Rice, 144, 146, 150–51, 172, 236–37
Riddle, Ore., 328
Rio de Janeiro, Brazil, 108–9
Rio de la Plata, S. America, 222
Rio Grande River, N. America, 343, 366
Rion River, U.S.S.R., 230, 245, 249
Ritter, Carl, 6

Riviera (France and Italy), 202, 207, 215
Rockhampton, Australia, 133
Rocks, 56–57, 374
Rocky Mountains, N. America, 182, 341, 343, 366, 370, 384–85, 395
 vegetation of, 377–79, 383
Romania, 257, 264, 271, 274, 277, 279
Rome, Italy, 216
Rotterdam, Netherlands, 331, 335
Rubber, 100–102
Ruhr District, Germany, 328, 331, 335
Russia, see U.S.S.R.
Ruston, Ore., 331

S

Saar Basin, France, 330
Sable Island Bank, Atlantic Ocean, 297
Sacramento, Calif., 217
Sacramento River, Calif, 204, 209
Sahara Desert, Africa, 114, 153–54, 158–64, 169, 173, 176
Said, Egypt, 176
Saigon, S. Vietnam, 147, 149
St. Lawrence River, N. America, 285, 390, 405
St. Lawrence Seaway, N. America, 274, 275, 402
St. Lawrence Valley, N. America, 304
St. Louis, Mo., 275, 279
St. Paul, Minn., 305
Salem, Ore., 331
Salt Lake City, Utah, 366–67
Salt River, Ariz., 161, 173
Salton Sea, Calif., 173
Salvador, Brazil, 108–9
Samarkand, U.S.S.R., 365
Sambre Valley, France, 328
San Diego, Calif., 175, 176, 214
San Francisco, Calif., 215, 216
San Joaquin River, Calif., 204, 209
San Jose, Calif., 216
San Juan oasis, Argentina, 357
San Marino, 370

Santa Cruz, Bolivia, 341
Santa Cruz, Calif., 200
Santiago, Chile, 216
Santiago del Estero, Argentina, 341
São Paulo, Brazil, 190, 191
Sapporo, Japan, 305
Saudi Arabia, 161, 170–71, 173, 176–77
Savannas, Africa, 63, 118–20, 123–27, 130, 141
Saxony, Germany, 303
Scandinavia, 289, 303, 416
Schwarzwald (Black Forest), Germany, 381, 383
Scotland, 313, 329
Searles Lake, Calif., 363
Seasons, 50–52
Seattle, Wash., 309–11, 312–13, 327, 332, 335–36
Selva (rain forest), 93, 97, 103–4
Semarang, Java, 109
Semple, Ellen Churchill, 6
Sennar Dam, Sudan, 123
Seoul, S. Korea, 254
Seven Devil Mountains, Pacific Northwest, 344
Shanghai, 250, 253, 272
Shantung Peninsula, China, 263, 274
Shari River, Chad, 92, 116
Shatt al Arab, Iraq, 160, 171
Sheffield, England, 330, 334
Shenandoah Valley, U.S., 261
Shikoku, Japan, 222
Sian, China, 366
Siberia, 385, 393–94, 398, 405
Siberian Plain, 288
Sierra Madre Mountains, N. America, 370
Sierra Nevada Mountains, N. America, 202, 218, 343, 370, 372, 383
Silesian Coal Basin, Poland, 303
Simpson Desert, Australia, 156
Singapore, 87, 109
Singkep, Indonesia, 106
Sinkiang region, China, 343, 366
Si River, China, 140, 227
Snake River, U.S., 344–45
Soch'e, China, 366
Sofia, Bulgaria, 279
Soils, 63–66
Solar energy, 38–39, 50–53
Solstices, 52
Sonoran Desert, N. America, 156, 158, 160, 163, 172

South Africa, 210, 252, 362, 365
  fishing in, 212, 246
  urban centers in, 255, 367
  vegetation of, 235, 347
South America, 222
  agriculture of, 244–45
  forest industries of, 103–4
  industry in, 251
  mining in, 249
  population of, 236
  surface features of, 115–16, 230
  vegetation of, 233–35
Southeast Asia Peninsula, 83, 88, 147–48, 162
  agriculture in, 100–102, 145
  climatic regions of, 92–93, 178
  population of, 95, 141–42
  surface features of, 138–40, 184
  urban centers in, 109, 149
South Korea, 228, 254
South Pole, 415
South Sea Islands, Pacific Ocean, 130
South Vietnam, 134, 142, 149
South Wales, England, 330
Southwest Asian Desert, 154, 158, 161
Soviet Union, *see* U.S.S.R.
Spain, 341, 366, 370
  agriculture of, 206, 352
  mining in, 212, 328
Spanish Meseta, 341, 352, 366
Spice trade, 100, 101
Spokane, Wash., 366–67
Stalingrad, U.S.S.R., 365
Stanley Falls, Congo Basin, 92
Stanley Pool, Congo Basin, 92, 109
Stanleyville, Congo Basin, 109
Stavanger, Norway, 324
Steppe lands, 282, 342, 347–49, 361
Stoke-on-Trent, England, 330
Strassfurt, Germany, 328
Stuttgart, Germany, 303
Sudan (African country), 130, 133
Sudan (African region), 111, 114, 116, 121
Sudbury, Ontario, 301
Suez, Egypt, 176
Suez Canal, Egypt, 177
Sukkar Barrage, Pakistan, 172
Sulu Arc, Southeast Asia, 184
Sumatra, Indonesia, 106, 184

Sun, earth and, 38–39, 50–53, 415
Sunda Arc, Southeast Asia, 184
Sungari Plain, Manchuria, 289
Superior, Wis., 301
Surabaja, Indonesia, 133
Surakarta, Indonesia, 133
Surinam, 106, 109
Sweden, 299, 303–4
  climatic regions of, 282, 393
  forest industries of, 301, 401, 405
  Polar lands of, 398–402, 404
Switzerland, 372, 380, 381, 385–89
Sydney, Australia, 245, 249, 251, 252, 255
Syria, 154, 196

T

Tacoma, Wash., 331, 332, 336
Taiga (forests), 396, 400, 420
Taiwan, 134, 145
Taiyuan, China, 366
Takla Makan Desert, Asia, 342
Tampere, Finland, 405
Tampico, Mexico, 128, 130
Tanzania, 190
Tarim Basin, U.S.S.R., 342, 355
Tarpon Springs, Fla., 246
Tashkent, U.S.S.R., 365, 366
Tasmania, Australia, 307, 317, 327
Tea, 146
Tees River, England, 329
Tel Aviv, Israel, 216
Temperature, 40–45
Tennessee, 243, 249
Tennessee River Valley, U.S., 250–51, 261
Texas, 222, 228, 235, 246, 251, 361
  agriculture in, 241, 243
  mining in, 249, 363
Thailand, 134, 145–49
Thames River, England, 313
Thar Desert, India-Pakistan Peninsula, 134, 154–56, 158, 160, 172
Tibesti Highlands, Africa, 169–70
Tibet, 139, 140, 370, 380
Tien Shan Mountains, 371
Tientsin, China, 279, 281
Tierra del Fuego, S. America, 91

Tigris-Euphrates River, Near East, 158, 160, 171
Tihua, China, 366
Tilburg, Netherlands, 330
Tillamook County, Ore., 322
Timor, Southeast Asia island, 184
Tokyo, 250, 253
Tokyo Plain, Japan, 228
Tom River, U.S.S.R., 302
Tombouctou, Mali, 116, 176
Tools of geography, 9–19
Townsville, Australia, 133
Trans-Siberian Railroad, U.S.S.R., 296, 405, 407
Trinidad (island), 93, 109
Trondheim, Norway, 324, 327
Tsingtao, China, 279, 281
Tucumán, Argentina, 357, 367
Tulare Basin, Calif., 204
Tundra, 411–15, 417–20
Tunis, Tunisia, 216
Turan Lowland, U.S.S.R., 343
Turan oases, U.S.S.R., 355
Turin, Italy, 277, 281
Turkey, 196, 212, 341, 352, 366
Tucson, Ariz., 173
Tyne River, England, 329

U

Ubangai River, Africa, 92
Uganda, 188
Ukraine, 257, 264, 265, 337, 350, 352
  agriculture in, 271, 353
  surface features of, 262, 342
  urban centers of, 305, 366
Ulan Bator, Mongolia, 366
United Arab Republic, 160, 164, 175, 176
United Kingdom, 306, 319, 320
United States, 29–31, 260–61, 357–58
  agriculture in, 266–71, 352–54
  fishing in, 212, 245–46, 272
  forest industries of, 299–300, 383, 405
  industry in, 175, 274–77, 302, 365
  mining in, 273–74, 301–2, 363, 384–85, 402
  national parks of, 366, 387
  tourism in, 252, 303–4, 366, 387

United States (*Contd.*):
  urban centers of, 279–81, 304–5, 366–67
  See also New England; Pacific Northwest; United States South; *and individual states*
United States South, 233, 234, 236, 249
  agriculture in, 239–44
  forest industries of, 246–47
  industry in, 250–51
  urban centers in, 254–55
Ural Mountains, U.S.S.R., 288, 303, 305, 395–96, 416
Ural River, U.S.S.R., 354, 362
Urga, Mongolia, 341
Uruguay, 222, 230, 244–45, 252, 255
Uruguay River, S. America, 230
Urumchi, China, 366
U.S.S.R., 252, 342, 404–5
  agriculture in, 245, 296, 352, 354–55
  climatic regions of, 198, 225, 257, 282, 393, 420
  fishing in, 212, 299, 362, 405
  forest industries of, 301, 400–401, 405
  grazing in, 166, 361
  industry in, 303, 351–52, 365
  mining in, 249, 302, 351, 362–63, 384–85, 402–4
  population of, 265, 399
  surface features of, 230–33, 288–89, 342
  urban centers in, 305, 366, 405–7
  See also Ukraine
Ussuri Valley, U.S.S.R., 289, 296
Utah, 344, 363

V

Valencia, Ireland, 312–13
Valencia, Spain, 216
Valparaiso, Chile, 216
Vancouver, B.C., 311, 331–32, 335
Vancouver Island, B.C., 312
Varenius, Bernhardus, 5
Vegetation, 60–62
Veld (African grassland), 347, 349
Venezuela, 104, 111, 128–30
  urban centers of, 132, 191
Venice, Italy, 279, 281
Veracruz, Mexico, 130
Verkhoyansk, Siberia, 393
Vermont, 300
Verviers, Belgium, 330
Victoria, Australia, 307, 316, 320, 328
Victoria, B.C., 312
Victoria, Hong Kong, 150
Victoria Falls, Africa, 117
Vienna, Austria, 279, 281
Virginia, 243, 269
Vladivostok, U.S.S.R., 303, 305
Volga River, U.S.S.R., 342, 354, 355, 362, 365, 404
Volgograd, U.S.S.R., 365
Volta Redonda, Mexico, 190
Vosges Mountains, France, 381

W

Wadi Dawasir, Saudi Arabia, 161, 170–71
Wadis (desert streams), 158–59, 161, 169
Wallowa Mountains, Pacific Northwest, 344
Warsaw, Poland, 305
Washington, D.C., 279–81
Washington (State), 307, 318, 327, 383
Water, 66–68
Weather, 39–50
Wellington, New Zealand, 336
Wenatchee, Wash., 365
Western Europe, 306–7, 314, 319–22, 332
  agriculture in, 320–22
  climate of, 309–12
  fishing in, 323–24
  industry in, 328–31
  mining in, 327–28
  oil and, 173–74
  vegetation of, 317–18
Western Ghats (mountains, India), 140
West Germany, 305, 328, 335
West Indies, 127, 130, 227
  agriculture in, 125, 188
  climatic regions of, 91, 111, 116, 178
  population of, 81, 121
West Siberian Lowland, U.S.S.R., 282, 350
White Nile, Africa, 117
White Sea, Eurasia, 404
Willapa Bay, Wash., 327
Willamette-Puget Lowland, Pacific Northwest, 313, 314, 319, 322, 331
Willamette River, Ore., 314, 332, 336
Willamette Valley, Ore., 314, 322, 381
Winds, 40–46, 53–55
Winnipeg, Manitoba, 284, 305, 311
Wisconsin, 301
Witwatersrand, S. Africa, 365
World regions, introduction to, 72–79
Wuhan, China, 250
Wyoming, 343, 363

Y

Yakutia, U.S.S.R., 404
Yampi Sound, Australia, 129
Yangtze Plain, China, 227, 235, 237, 252–53, 262
Yangtze River, China, 227, 247, 250
Yellow Plain, China, 262, 264
Yellow River, China, 262, 264
Yellow Sea, China, 264, 272
Yellowstone National Park, U.S., 387
Yemen, 154, 182, 187
Yenisey River, U.S.S.R., 396
Yosemite National Park, U.S., 387
Yucatan Peninsula, Mexico, 111, 116, 125, 129
Yugoslavia, 206, 215, 257, 262, 277–79
  mining in, 212, 274
Yukon Plateau, Alaska, 395
Yukon River, Alaska, 395
Yukon Valley, N. America, 402
Yunnan Province, China, 247

Z

Zambezi River, Africa, 117
Zambia, 121, 127, 129
Zanzibar, 85
Zurich, Switzerland, 389
Zwickau, Germany, 303

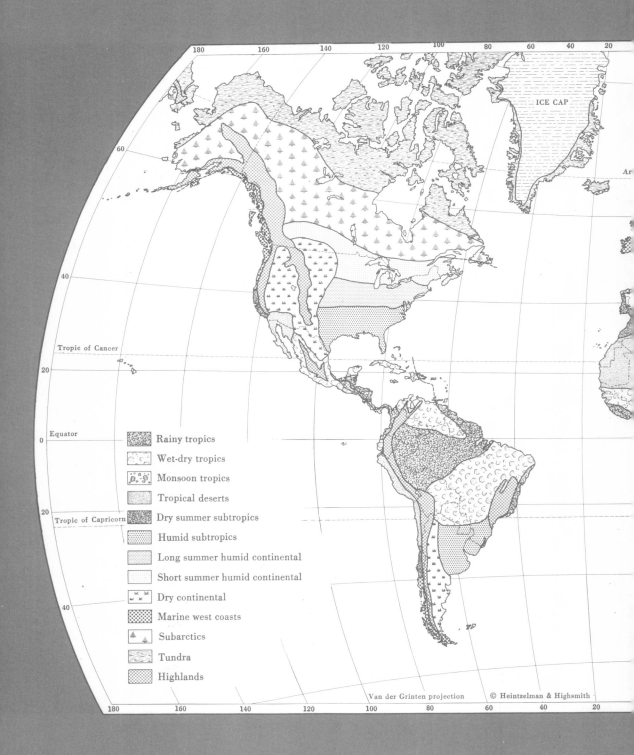

This map, showing the pattern of world climatic regions, is the